Recent Advances in
BIOPHOTON RESEARCH
and its Applications

Recent Advances in
BIOPHOTON RESEARCH
and its Applications

edited by

F. A. Popp, K. H. Li and Q. Gu

Technology Center Kaiserslautern
International Institute of Biophysics
Germany

World Scientific
Singapore • New Jersey • London • Hong Kong

Published by

World Scientific Publishing Co. Pte. Ltd.

5 Toh Tuck Link, Singapore 596224

USA office: 27 Warren Street, Suite 401-402, Hackensack, NJ 07601

UK office: 57 Shelton Street, Covent Garden, London WC2H 9HE

Library of Congress Cataloging-in-Publication Data

Recent advances in biophoton research and its applications / edited by
F.A. Popp, K.H. Li and Q. Gu.
p. cm.
ISBN-13 978-981-02-0855-4 -- ISBN-10 981-02-0855-3
1. Photobiochemistry. 2. Bioluminescence. I. Popp, Fritz
Albert. II. Li, K. H. (Ke-hsueh) III. Gu, Q. (Qiao)
QP517.P45R43 1992
574.19'125--dc20 92-10266
CIP

British Library Cataloguing-in-Publication Data

A catalogue record for this book is available from the British Library.

Preface: Introductory Remarks

Biophoton Emission is now a topical field in contemporary science [1,2,3]. It concerns weak light emission ("dark luminescence", "ultraweak" photon emission) from biological systems. The intensity is of the order of a few up to some hundred photons per cm^2 surface area per second, and the spectrum is almost continous at least within the optical range from 200 to 800 nm.

This universal phenomenon of living systems can now be investigated with case, thanks to the development of photon counting techniques which enable us to measure intensities down to about 10^{-17} W.

The discovery of biophoton emission can be traced back to Alexander Gurvich[4] who also pioneered two ideas on the origin of biophotons that are now set one against the other. These are respectively, the hypothesis that biophotons emanate from free-radical reactions, and the postulate that the emissions come from an organized "morphogenetic" field at the basis of life. The two ideas were not opposed in Gurvich's conception. However, subsequent developments in the life sciences has bifurcated this once unified theory: the more mechanistic mainstream biochemical approach regards biophotons as the sign of random metabolic imperfections, while the more vitalistic approach attempts to understand biophotons in terms of the coherent states of modern quantum optics.

It is likely that the latter, more physical approach, to which most of the papers of this volume relate, will lead back to the unified field theory of Alexander Gurvich, where no contradiction exists between the local aspects of biochemistry and the holistic aspects of modern biophysics.

Recent developments in the life sciences, for instance, Prigogine's dissipative structures[5] and chaos theory[6], demonstrate clear relationship between molecular events and collective phenomena. In particular, Fröhlich's idea of coherent long-range interactions in biological systems[7] and Presman's pioneering work on electromagnetic fields and life[8] have bridged the gap between biochemistry and biophysics in terms of "order parameters" of the system, based on electromagnetic interactions which are most significant in living organization.

In this book we are concerned with the "optical window". Despite their low intensities, biophotons have the advantage of a rather high signal/noise-ratio. In view of their correlations to many, if not all, biological functions they provide a most powerful, non-invasive tool for analyzing biological systems.

Irrespective of how one interprets the results, the very sensitive dependence of biophoton emission on almost all external and internal influences has already opened up many applications, e.g.,

1. the examination and control of food quality,
2. bioindication of pollutants and other environmental factors,
3. research on the efficacy of drugs,
4. diagnostic and therapeutic treatment of different kinds of illness, such as immune diseases and cancer.

In this volume, we also show that a wide range of basic problems in the life sciences may be amenable to investigation by means of biophoton emission. These include molecular interactions[9], immunological and repair processes in aging, growth and differentiation, pattern formation in development, biocommunication and the nature of consciousness[3].

These recent advances in biophoton research may offer new insights into a number of rapidly developing areas in non-equilibrium physical phenomena such as quantum chaos[10], quantum electrodynamics, including the problem of coherence at low level intensities[11], research on the nature of the photon[12], and in particular, cavity quantum electrodynamics[13].

The first five chapters are devoted to the physics of biophoton emission. The treatment is taken from first principles all the way to the details of quantum physics and its relevance to biology. Chapter 6 deals with the methodology of biophoton measurement. Chapters 7-16 report on exciting new findings in a number of biological systems which give evidence of the holistic cooperative character of biophoton emission. Some applications of biophoton research are described in Chapters 17 and 18, while theoretical considerations on coherence and evolution are presented in Chapters 19-21. The final Chapters 22 and 23 review the historical background for biophoton research and the work of Alexander Gurvich in connection with the concept of the "morphogenetic" field.

We are very grateful to the Familie-Ernst-Wendt-Stiftung (Cologne) and the Rut- und Klaus-Bahlsen-Stiftung (Hannover) as well as the Karl-Schweisfurth-Stiftung (Munich) for financial support, and in particular for their engagement in the research on cancer and on food quality, respectively. Financial support was also received from the Ministery of Research (FRG).

Kaiserslautern, February, 1992 — Fritz-Albert Popp

References

1. D. Swinbanks, Body light points to health. *Nature* **324** (1986) p. 203.

2. B. Jezowska-Trzebiatowska, B. Kochel, J. Slawinski and W. Strek (eds.), *Photon Emission from Biological Systems* (World Scientific, Singapore, 1986).

3. F.A. Popp, A.A. Gurvich, H. Inaba, G. Cilento, R. van Wijk, W.B. Chwirot and W. Nagl, Biophoton Emission (Multi-author Review). *Experientia* **44** (1988) p. 543-600.

4. A.G. Gurvich, and L.D. Gurvich, *Die mitogenetische Strahlung* (Fischer Verlag, Jena, 1959).

5. I. Prigogine, Order through fluctuation: Selforganization and social systems, in *Evolution and Consciousness*, eds. E. Jantsch and C.H. Waddington (Addison + Wesley P.C., Reading, 1976).

6. B.L. Hao, *Chaos* (World Scientific, Singapore, 1985).

7. H. Fröhlich, Long-range coherence and energy storage in biological systems. *Int. J. Quantum Chem.* **2** (1968) p. 641-649.

8. A.S. Presman, *Electromagnetic Fields and Life* (Plenum Press, New York, 1970).

9. F. Bistolfi, *Biostructures and Radiation Oder Disorder* (Edizioni Minerva Medica, Torino, 1991).

10. R. Pool, Quantum Chaos: Enigma Wrapped in a Mystery. *Science* **243** (1989) p. 893-895.

11. J.R. Klauder and B.S. Skagerstam, *Coherent States* (World Scientific, Singapore, 1985).

12. R. Kidd, J. Ardini and A. Anton, Evolution of the modern photon. *Am. J. Phys.* **57** (1989) p. 27-35.

13. S. Haroche and D. Kleppner, Cavity Quantum Electrodynamics. *Physics Today* **42** (1989) p. 24-30.

Authors' Addresses

B.W. Chwirot, Department of Plant Cytology and Embryology, Nicholas Copernicus University, ul. Gagarina 11, PL-87100 Torun, Poland.

M. Galle, International Institute of Biophysics, Technology Center, Opelstr. 10, D-6750 Kaiserslautern 25, F.R. Germany.

M. Godlewski, Institute of Physics, Pedagogical University, Cracow, Poland.

F. Grasso, Istituto di Fisica, Università di Catania, Viale A. Doria 6, I-95125 Catania, Italy.

Q. Gu, International Institute of Biophysics, Technology Center, Opelstr. 10, D-6750 Kaiserslautern 25, F.R. Germany; and Department of Physics, Northwest University, Xi'an 710069, P.R. China.

A.A. Gurvich, Institute of Gerneral Pathology and Pathophysiology, Academy of the Medical Sciences, Baltyiskaja Str. 8, Moscow, Russia.

M.W. Ho, Development Dynamics Research Group, Open University, Walton Hall, Milton Keynes, MK76AA, U.K.

K. Lambing, International Institute of Biophysics, Technology Center, Opelstr. 10, D-6750 Kaiserslautern 25, F.R. Germany.

K.H. Li, International Institute of Biophysics, Technology Center, Opelstr. 10, D-6750 Kaiserslautern 25, F.R. Germany.

M. Lipkind, Kimron Veterinary Institute, Beit Dagan, Israel.

W.P. Mei, International Institute of Biophysics, Technology Center, Opelstr. 10, D-6750 Kaiserslautern 25, F.R. Germany; and Department of Physics, Suzhou University, Suzhou 215006, P.R. China.

C. Mieg, International Institute of Biophysics, Technology Center, Opelstr. 10, D-6750 Kaiserslautern 25, F.R. Germany.

F. Musumeci, Istituto di Fisica, Università di Catania, Viale A. Doria 6, I-95125 Catania, Italy.

R. Neurohr International Institute of Biophysics, Technology Center, Opelstr. 10, D-6750 Kaiserslautern 25, F.R. Germany.

H.J. Niggli, Research Center, Cosmital SA, Rte de Chésalles 21, CH-1723 Marly, Switzerland.

F.A. Popp, International Institute of Biophysics, Technology Center, Opelstr. 10, D-6750 Kaiserslautern 25, F.R. Germany.

F. Rohner, Forschungsstelle für Balneologie, Rheumaklinik Zurzach, CH-8437 Zurzach, Switzerland.

S. Ross, Development Dynamics Research Group, Open University, Walton Hall, Milton Keynes, MK76AA, U.K.

P.T. Saunders, Department of Mathematics, King's College, London, Strand, London WC2R, 2LS, U.K.

A. Triglia, Istituto di Fisica, Università di Catania, Viale A. Doria 6, I-95125 Catania, Italy.

H. van Aken, Department of Molecular Cell Biology, State University of Utrecht, Padualaan 8, NL-3584 CH Utrecht, The Netherlands.

R. van Wijk, Department of Molecular Cell Biology, State University of Utrecht, Padualaan 8, NL-3584 CH Utrecht, The Netherlands.

X. Xu, International Institute of Biophysics, Technology Center, Opelstr. 10, D-6750 Kaiserslautern 25, F.R. Germany.

K. Zioutas, Physics Department, University of Thessaloniki, 54006 Thessaloniki, Greece; and CERN-Visitor, European Laboratory for Particle Physics, European Organization for Nuclear Research, Geneve, Switzerland.

Contents

Chapter 1

Some Essential Questions of Biophoton Research and Probable Answers

Fritz-Albert Popp

Abstract

A careful analysis of all the experimental data of biophoton emission (concerning correlations to biological and physiological processes, the temperature dependence, the spectral distribution, the optical transparency, the dependence on external agents, the photocount statistics, and ergodic relaxation behaviour) enforces to reject the hypothesis that this weak radiation originates from spontaneous chemiluminescence. On the other hand, all results agree completely with the alternative hypothesis of a coherent field as the source of biophoton emission. The arguments and objections are discussed from the classical point of view as well as from quantum description. The latter is the proper one due to the obvious fact that single count-emission has to be assigned to quantum theory.

1.1 Imperfection versus Coherence

There are two opposite "theories" about biophoton emission, i.e., the "imperfection theory" [1,2] and the "coherence theory" [3,4] (see also Chapters 3-5).

The imperfection theory explains biophoton emission on the basis of photobiochemistry, i.e, as weak luminescence in terms of rare and random metabolic aberrations which lead to excited compounds in the visible range of the electromagnetic spectrum, e.g. electronic excitations of radicals. The emission of photons can then be assigned to the permanent tendency of excited living matter to return to thermal equilibrium.

The coherence theory, on the other hand, based on the physics of interactions of weak radiation in and with optically dense matter, claims that biophoton emission originates from a delocalized coherent electromagnetic field within living tissue, in particular from its optical modes. In contrast to imperfection theory, this field is claimed to stabilize around a threshold between a "chaotic" and an "ordered" regime far away from thermal equilibrium in the sense of "dissipative structures" [5]. Essentially, coherence theory takes account of the laws of Cavity Quantum Electrodynamics.

Imperfection theory does not take care of non-equilibrium phase transitions in living matter. In contrast, coherence theory characterizes biological systems in terms of the experimentally established $f_\nu \simeq$ constant-rule which means that in the ideal case (where always enough energy is available) all the relevant excited states of the living matter are occupied with about the same probability, independent of the excitation energy. In particular, according to coherence theory, the most essential role in biophoton emission is played by the DNA (with $f_\nu = 1$ over a broad band of frequencies which covers at least the optical range and may even reach down to the extreme low frequence (ELF)-range).

Imperfection theory denies any biological significance of "dark luminescence", while coherence theory assigns to the presumably phase-locked and mode-coupled photons from DNA a permanent regulatory activity within cells (e.g. by triggering metabolic reactions at definite time intervals and in certain positions) and also between cells (e.g. by controlling growth and differentiation, "Gestalt-Bildung").

The imperfection theory appears, at first glance, to be "self evident". It seems not to require any further explanation, since the mechanisms on which it is based are obviously taking place. However, this point of view does not take account of the fact that in an activated matter of high optical density even spontaneously excited molecules do not interact with an unlimited vacuum. Rather, the Einstein coefficients of spontaneous and induced emission may vary over a wide range, giving rise to induced emission as well as to photon trapping.

One of the most essential mechanisms of coherence theory is therefore based on the formation of networks of exciplexes (see Chapters 3-5) or, at least, of exciplex-like coherent excitations in living matter. Actually, there is experimental evidence for exciplex formation in living cells.

The low intensity of biophoton emission can then be explained in terms of the fairly high degree of coherence of the "non-classical" light with very high signal/noise-ratio, rather than becoming understandable as an expression of rare and random chemiluminescence.

However, both, the imperfection theory as well as the hypothesis of coherence, require careful theoretical and experimental analysis.

Before entering into details, let us briefly summarize the experimental results which may support each of the above-mentioned theories of biophoton emission.

The experiments on biophoton emission refer to

- *the source,*
- *correlations to biological and physiological processes,*
- *temperature dependence,*
- *spectral distribution,*
- *optical transparency,*
- *dependence on external agents,*
- *photocount statistics, and*
- *relaxation behaviour after excitation.*

1.2 Some Experimental Indications and Proofs

Roughly speaking, from opposite theories one may expect opposite results. However, as is shown in the following, this is not always true.

One consequence of imperfection theory is that **any** compound can become the target of a rare and random electronic excitation and this possibility cannot be excluded by the theory of a nonlocal electromagnetic field which involves the activation and inactivation of all processes in a cell population. A second point of agreement between the predictions of imperfection theory and coherence theory concerns the spectral distribution and the photocount statistics. As any compound may be involved in spontaneous or coherent excitation, narrow emission bands around definite frequencies are not expected. Rather, some white-noise-like spectral emission from all possible sources, reflecting the variety of electronic excitation energies, should occur.

Also the photocount statistics from a chaotic field (according to the imperfection theory) as well as from a coherent field are expected to be the same, i.e. Poissonian. Although a single mode of a chaotic field displays a geometrical rather than a Poissonian distribution, in the case of a multi-mode field with coherence time τ much smaller than the measurement time interval Δt the geometrical distribution approaches a Poissonian one (see, for instance, Fig.1.1 and [6]).

Consequently, neither the low intensity of biophoton emission nor the spectral distribution (which is significantly different from that of black-body radiation at physiological temperature), nor the photocount statistics (which in quasistationary states is generally Poissonian) may give a definite answer on whether imperfection theory or coherence theory is correct.

However, there are ample indications from other experiments which force us to reject the hypothesis of a rare, random, spontaneous chemiluminescence. Firstly, many – if not all – biological and physiological processes (e.g. growth, differentiation, cell-cycle stages) are correlated to biophoton emission [7]. A rare, random event of the average rate of about 1 per cell and per day should not depend on biological processes. In addition, imperfection theory predicts an increase of "imperfections" with metabolic reactivity, radical activity and stress. However, there are many cases where the increase of metabolic activity does not enhance biophoton emision (see, for instance, Chapter 9) and where strong stress and even radical formation by cosmic radiation incidence do not have significant impact on "dark luminescence" [8].

Secondly, apart from the specific correlations of biophoton emission to physiological processes, the temperature dependence does not follow an Arrhenius law [9], as is expected from a rare random event under quasi-thermal equilibrium conditions. Rather, it follows a hysteresis-like intensity characteristics when the temperature is cyclically increased and decreased [9]. In complete agreement to this nonlinear temperature-response, one observes the same "overshoot"-reactions after an increase in temperature, as is well known from the temperature-dependence of many physiological functions [10]. Consequently, biophoton emission cannot be decoupled from biological processes, but is either a product of physiological processes or/and their actual regulator.

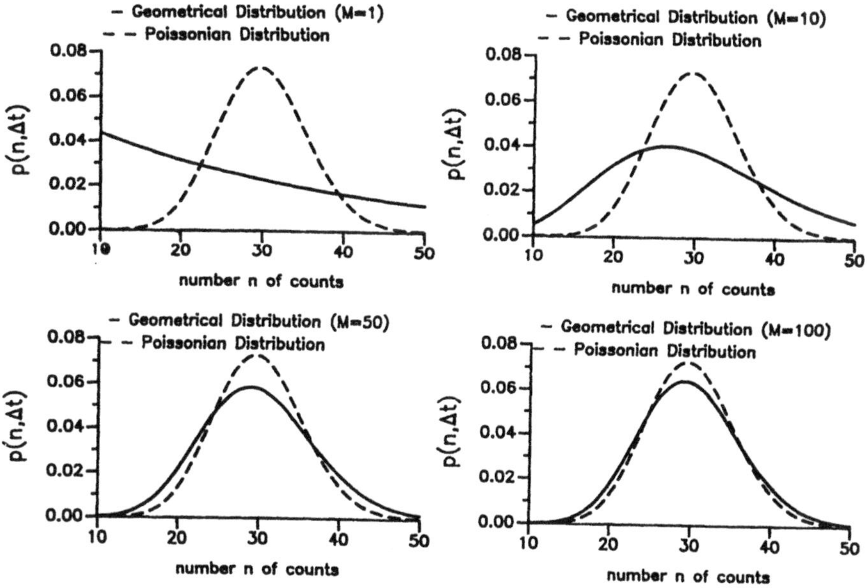

Figure 1.1: *The probability $p(n, \Delta t)$ of registering n photocounts within a preset time-interval Δt is different for a coherent and a chaotic photon field. While a coherent field displays always a Poissonian distribution $p(n, \Delta t)$ (which is always the dashed curve in the four figures), the geometrical distribution of a chaotic field (continous curves) depends on the ratio $M = \frac{\Delta t}{\tau}$, where τ is the coherence time of the chaotic field. For increasing M (decreasing τ), the geometrical distribution approaches more and more the Poissonian one (see figures from the left side, upper part, to the right side, lower part), such that for sufficiently low τ a fully coherent field cannot be distinguished from a chaotic one.*

Apart from the non-linear optical transparency of biophotons through matter[11] (see also Fig.1.2) which indicates at least a partial coherence of this light, evidence of coherence comes from the hyperbolic relaxation kinetics under ergodic conditions after light illumination ("delayed luminescence") (see, for instance, Fig.1.5 and Chapters 2, 3, 10, 13).

Further indication of a (rather) high degree of coherence (which may exceed that of technical lasers by at least 10 orders of magnitude) is displayed by sub-Poissonian photocount-statistics, characterized by significantly lower variances than mean-values, sub-Poissonian photocount statistics and normalized factorial moments $m^{(r)}$ (see (1.63)) which under certain conditions have the tendency to drop to zero with

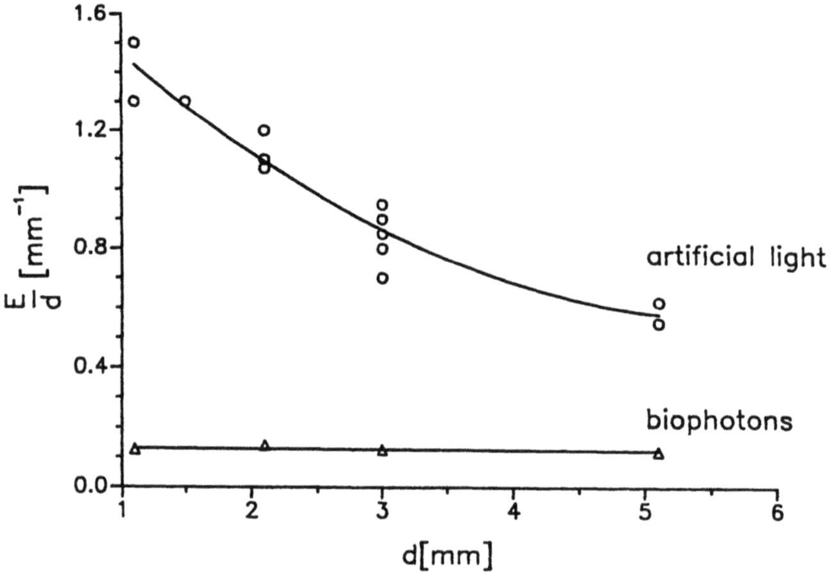

Figure 1.2: *The transparency of a disperse medium depends on the coherence of the transmitted light. The extinction coefficient of the culture of soya cells of various thickness d(mm) is about one order of magnitude lower, if biophotons of cucumber seedling pass through (lower curve), than that of comparable low-intensity light at* $\lambda = 550$ *nm of a Gilford 250 spectrophotometer*[14] *(upper curve).*

increasing order r (Fig.1.3).

The normalized factorial moments $m^{(r)}$ of a fully coherent field are 1 for all orders r. A chaotic field with sufficiently low coherence time τ exhibits the same behaviour. Consequently, no difference can be expected, if one compares the normalized factorial moments of a microlamp ($\tau < 10^{-7}s$, $\Delta t = 100ms$) with those of a biological system in a quasi-stationary state (Fig. 1.3, left side, from $r = 1$ to $r = 8$). With increasing coherence time τ, the $m^{(r)}$ approach the values $r!$. This behaviour can be observed, for instance, in the tail of "delayed luminescence" of polystyrol, indicating collective excitations with rather high coherence time. However, biological systems tend to display "super-Poissonian" statistics, too, as soon as the preset time interval Δt gets sufficiently high ($\geq 1s$). In contrast to non-living matter, the super-Poissonian photocount statistics of biological systems is due to rhythms of the biophoton emission.

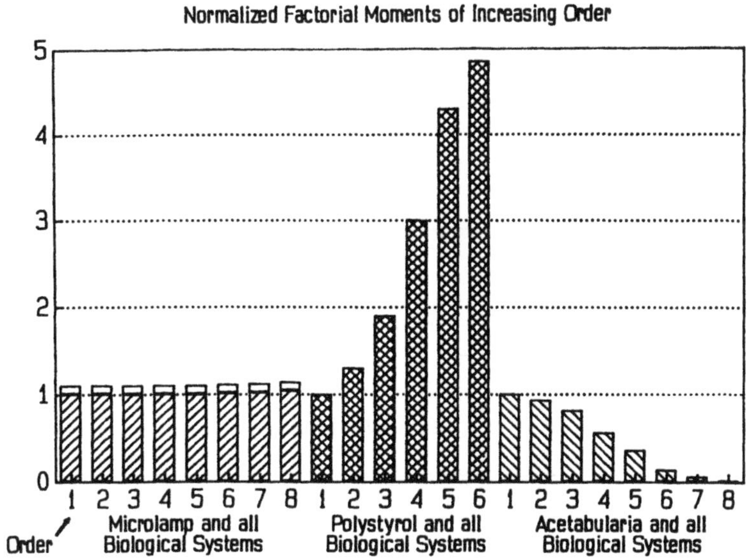

Figure 1.3: *Normalized factorial moments as a function of order from microlamp, polystyrol, acetabularia and all biological systems (more see text).*

A further striking difference between living systems and dead materials concerns the fact that one never observes sub-Poissonian photocount statistics of passive non-living matter ($m^{(r)} < 1$), whereas biological systems show occasionally decreasing normalized factorial moments with increasing r. This happens mainly at rather low intensities of biophoton emission (Fig. 1.3 right side).

From a more biological point of view, the nonlinear increase and decrease of biophoton emission with decreasing distances of the biological units under consideration, for instance cereals, cells or daphnia (see, for instance, Chapters 14, 16) cannot be explained in terms of local events. However, they can truthfully be interpreted as long-range coherent interactions on the basis of Dicke's theory [12].

Table 1.1 displays the differences between expected and actual results with respect to the different approaches.

1.3 The f_ν = constant-rule

Imperfection theory will now be formulated in more detail. It turns out that an Arrhenius-like reaction which accounts for the imperfection the-

Table 1.1: Comparision between expected and actual results with respect to the different approaches.

Subject	Expected Result according to Imperfection Theory	Expected Result according to Coherence Theory	Actual Result	Reference
Source	Radicals and other Oxidation Processes	Essentially DNA	Essentially DNA but also all other Processes	[13 – 16] , see Chapters 7, 8
Correlations to Biological Processes	No Correlation to Biological Functions	Strong Correlations to all Processes	Correlations to all Processes	[7]
Temperature Dependence	Arrhenius-law Dependence	Curie-Weiss-Law Dependence	Curie-Weiss-Law Dependence	[9]
Spectral Distribution	Spectrum of Radicals	f_ν=constant Rule	f_ν=constant Rule	[17, 18]
Optical Transparency	Beer-Lambert-Law Dependence	Sensitive Changes of Transparency	Sensitive Changes of Transparency	[14, 20]
Dependence on External Agents	Weak	Strong	Strong	[19]
Photocount Statistics	Geometrical to Poissonian	Super-Poissonian to Sub-Poissonian	Super-Poissonian to Sub-Poissonian	[21]
Relaxation Behaviour	Exponential	$\frac{1}{t}$-Dependence	$\frac{1}{t}$-Dependence	[3, 21 – 26], see Chapters 10, 13

ory has to be rejected for biophoton emission. One of the contradictions within imperfection theory results from the implicit assumption that the system is governed by thermal equilibrium conditions. Actually, the excitation temperature of biophoton emission is many orders of magnitudes higher than that of a thermal equilibrium system at physiological temperatures, increasing linearly with increasing frequency. In addition, unlike to a heat bath, the efflux of biophoton emission has no corresponding equal influx from the surroundings.

It becomes obvious that the system has to be described by a chemical potential, giving rise to an almost equal probability of occupying electronic states of different energy. This f_ν=constant-rule corresponds to an "ideal" open system, which works with both, the highest possible relative entropy and simultaneously the lowest possible absolute entropy. It reflects an almost noiseless system with the highest possible sensitivity. The situation is discussed in terms of the chemical reaction rate triggered by biophotons, the excitation temperature of the field, the f_ν-values and the entropy of light with and without its interaction with matter. Furthermore we obtain an equation that describes the temperature dependence of biophoton emission.

Imperfection theory can be formulated in terms of the photon intensity i according to Eq. (1.1)

$$i = \sigma_a exp(-\frac{E_a}{kT}), \tag{1.1}$$

where i describes the photocount rate (in s^{-1} per cell), σ_a the cross-section of the metabolic aberration ("imperfection") (in s^{-1} per cell) with an activation energy E_a. kT is the mean thermal energy at temperature T. σ_a is the product of the reaction inhibition factor and the coefficient of impact.

From the Arrhenius plot, where $\ln i$ is displayed versus $(kT)^{-1}$, the average of E_a and σ_a can be found at the same time. Taking into account actual measurements of $i(T)$ which have been carried out on resting spores of certain fungi [27], grain seedlings[28] and cucumber seedlings [21], the estimates of E_a and σ_a, which have been obtained in this way, can be seen in Table 1.2.

The values again indicate that the imperfection theory cannot be correct, because a rare, random and unspecific process is not able to extend over 12 different orders

Table 1.2:

	$E_a(eV)$	$\sigma_a[s^{-1}/cell]$
Resting spores of fungi	$0.04 - 0.16$	$10^{-6} - 10^{-4}$
Barley seeds	0.15	$10^{-6} - 10^{-4}$
Grain seedlings	$0.2 - 0.3$	$10^{-3} - 1$
Cucumber seedlings	$0.4 - 1.2$	$10 - 10^{6}$

of magnitude for σ_a in view of the fairly constant order of magnitude for i itself (of about 10^{-5} s^{-1}/cell).

The fault with (1.1) is due to the assumption that living tissues should be closed systems.

Actually, the spectral distribution i_ν, where ν is the frequency, shows evidence that biological systems are permanently away from thermal equilibrium, and by no means only within negligible time intervals. From a chemical point of view, the activation energy E_a for triggering reactions (e.g. intermediate complex formation) is always available with a much higher probability than that given by the Arrhenius factor $f_T = exp(\frac{-E_a}{kT})$. It originates from the biophoton field itself rather than from a thermal black-body radiation. Consequently, the laws of thermodynamics force us to introduce an excitation temperature $\Theta(\nu)$ instead of the frequency-independent temperature T of a heat bath [29]. The Arrhenius factor $f_T(E_a)$ has to be substituted by the $f_\Theta(E_a)$ value according to

$$f_\Theta(E_a) = exp(-\frac{E_a}{k\Theta(\nu)}), \tag{1.2}$$

where $\Theta(\nu)$ can be calculated from the spectral biophoton intensity i_ν (photons per units of area, time, frequency and solid angle):

$$\Theta(\nu) = \frac{h\nu}{k\,ln\{\frac{2\nu^2}{i_\nu c^2}\}}, \tag{1.3}$$

where c is the velocity of light.

Fig. 1.4 shows the results of a series of measurements on cucumber seedlings, performed with interference filters with a bandwith $\Delta\lambda$ of about 10 nm corresponding to a frequency bandwidth $\Delta\nu = \frac{c}{\lambda^2}\Delta\lambda$.

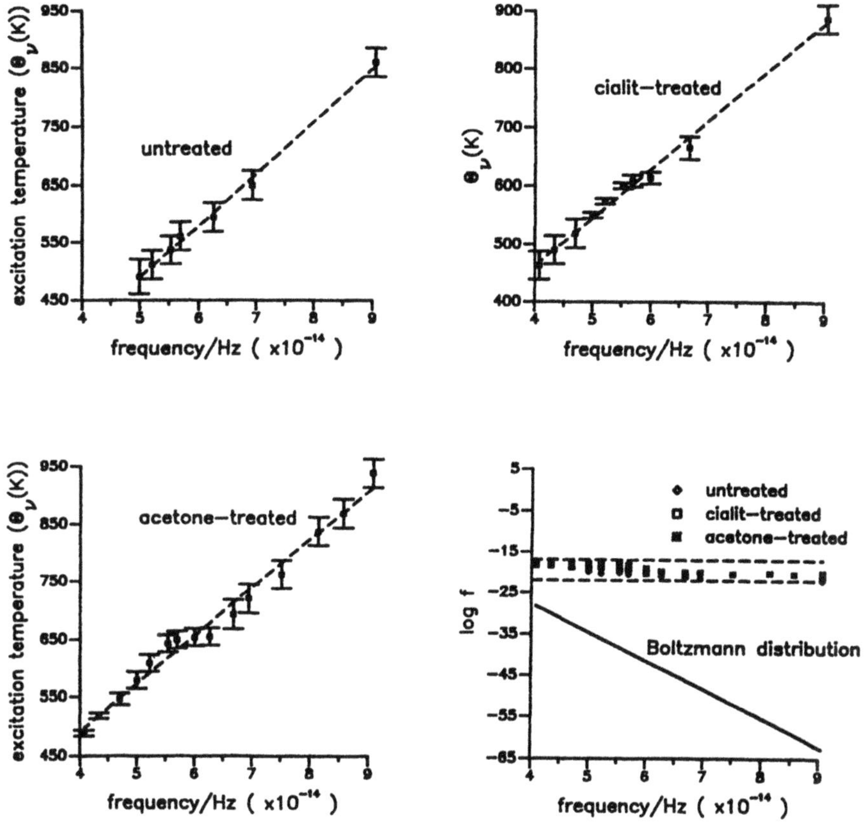

Figure 1.4: *Measurements on cucumber seedlings with the aid of interference filters show that $\Theta(\nu)$ increases proportional to the frequency ν. Left upper side: Untreated outgrown cucumber seedlings. Right upper side: Outgrown cucumber seedlings, treated with cialit. Left lower side: Outgrown cucumber seedlings after treatment with acetone. From the $\Theta(\nu)$-values, one can calculate lower limits of the occupation probabilities $f(\nu)$. Right lower side: compared to the Boltzmann distribution $f(\nu) = exp(-\frac{h\nu}{kT})$ at physiological temperatures T the $f(\nu)$-values of biological systems are (1) many orders of magnitude higher and (2) display no distinct frequency dependence.*

In contrast to black-body radiation for which $i_\nu = \frac{2\nu^2}{c^2} exp(-\frac{h\nu}{kT})$ and $\Theta(\nu) = T$, we get a remarkable result for biological tissue:

$$\Theta(\nu) \propto \nu, \tag{1.4}$$

and in the optical range:

$$\Theta(\nu) \text{ much higher than } T. \tag{1.5}$$

The insertion of (1.4) into (1.2) results in the rule:

$$f_\Theta(\nu) \simeq constant. \tag{1.6}$$

The values of $f_\Theta(\nu) = f_\nu$ for all the biological systems under investigation have so far been

$$10^{-22} \leq f_\nu \leq 10^{-18}, \tag{1.7}$$

where some species-specific "modulations" around $f(\nu) \simeq constant$ can be found. Compared to the Arrhenius-factor f_T, the rule (1.6) is, however, always fairly well fulfilled (Fig.1.4).

In order to understand this f_ν = constant-rule from a very general point of view, let us consider the fact that the living system is a heat engine which by taking up the energy ε from the sun is able to do work w. For the highest efficiency we have always $w = \varepsilon\left(1 - \frac{T_E}{T_s}\right)$, where T_E and T_s are the temperatures of the earth and the sun, respectively. On the other hand, this work can be used for removing heat q from the earth, where at highest efficiency $q = w\left(1 - \frac{T_E}{\Theta}\right)$ with Θ representing the "temperature" of the biological system. Consequently, we have $q = \varepsilon\left(1 - \frac{T_E}{T_s}\right)\left(1 - \frac{T_E}{\Theta}\right)$. It is obvious that the highest possible amount of heat which can be stored is $q = \varepsilon\left(1 - \frac{T_E}{T_s}\right)$. Therefore, we can write $q = \varepsilon\left(1 - \frac{T_E}{T_s}\right) - \chi$, where $\chi > 0$ describes a loss, which does not depend on ε, but only on the properties of the machine under investigation. This means that $\varepsilon\left(1 - \frac{T_E}{T_s}\right)\left(1 - \frac{T_E}{\Theta}\right) = \varepsilon\left(1 - \frac{T_E}{T_s}\right) - \chi$, or, after straightforward calculation

$$\Theta_\nu = \frac{h\nu \cdot T_E\left(1 - \frac{T_E}{T_s}\right)}{\chi},$$

where a spectral analysis ($\varepsilon = h\nu$, $\Theta = \Theta(\nu)$) has been performed. Since the loss $\chi > 0$ is independent of ν, we have immediately the required proportionality between Θ and ν, providing f_ν = constant.

The f_ν = constant-rule can also be obtained by the requirement that biological systems work at an optimum signal/noise-ratio with respect to ν(frequency)[3].

One of the consequences of the rather high f-values is that the biochemical reaction rate can be many orders of magnitude higher in a living tissue than in a thermal equilibrium system at physiological temperatures. This is a well established result [30].

One may ask, however, whether such low intensities of a few photons per second and per cm^2 surface area of a biological system can trigger the high chemical reaction rate of about $10^5/s$ and per cell in a living system.

Actually, the interaction of biological matter with light in the range of biophoton emission cannot be represented by the water-phantom models as in the "radiation-biology" of ionizing radiation. Whereas a cuvette of water does not display any countable selfemission or "delayed luminescence" after light-illumination, living tissue exhibits a characteristic "delayed luminescence" over the whole visible range as well as its permanent "dark glow" of a few photons/$(s \cdot cm^2)$. In addition, the excitation is not a local one, effecting not different molecular reactions for different exciting wavelengths, but always a centralized process of the whole system, where the different modes are strongly coupled together and dropping down continously to the final state of quasi-stationary biophoton emission (Figs. 1.5, 1.6).

For the wavelengths under consideration we get similar changes of the excitability after every treatment (1.7), indicating that the special kind of treatment influences the excitability of acetabularia for <u>all</u> wavelengths in the <u>same</u> way. Consequently, it is improbable that the excitation of the different modes reflects different (photochemical) reactions. Rather, it points to only one process where the different modes are strongly coupled to each other.

As a result, the cross section of the interaction of light with living matter is many orders of magnitude higher than that between water and light. From this "delayed luminescence" it is evident that for a small but measurable part of the light the rescattering process within living systems provides

1. an overall process of the living system under study, and

2. a long-time conservation of the quantum energy over the whole optical range.

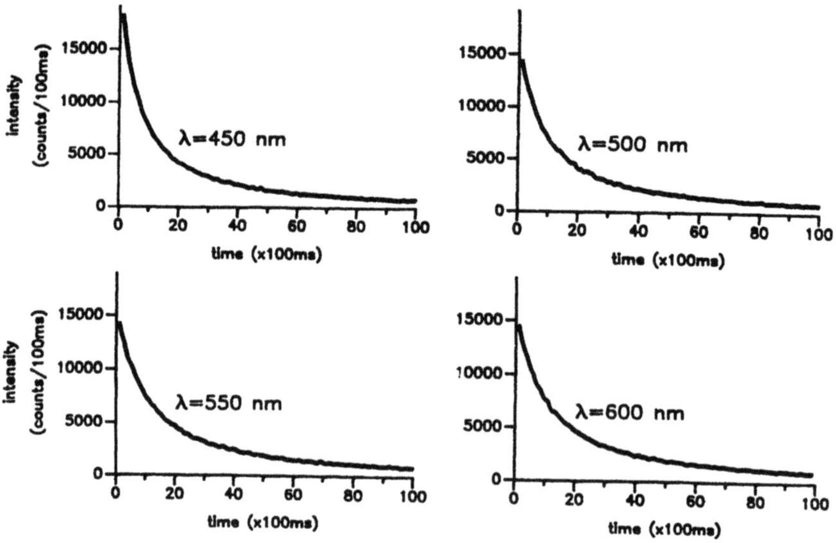

Figure 1.5: *Over the whole visible region acetabularia can be excited with about the same efficiency. The examples display the delayed luminescence of acetabularia which were excited by monochromatic light ($\Delta\lambda = \pm 10$ nm) of different wavelengths. Typical is the long-time relaxation with about the same hyperbolic relaxation kinetics, dropping down to the same final value.*

Provided then, as a limiting case, that ordinary diffusion takes place, an originally localized photon will be distributed after a time t over a range d according to

$$d \simeq \bar{x}\sqrt{\frac{2t}{\tau_R}}, \tag{1.8}$$

where $\bar{x}$ is the mean distance between two consecutive absorbers and τ_R is the time of excitation and reemission. That those migration processes are actually taking place in cells has been demonstrated by G. Cilento [31]. He showed that in biological systems excited molecules can transfer their excitation energy to other biomolecules and thereby trigger amplification mechanisms and/or promote photochemical processes in the dark.

This "cold" migration of light may fairly well be quantitatively responsible for triggering all the biochemical processes in a cell, even at the very low biophoton intensity.

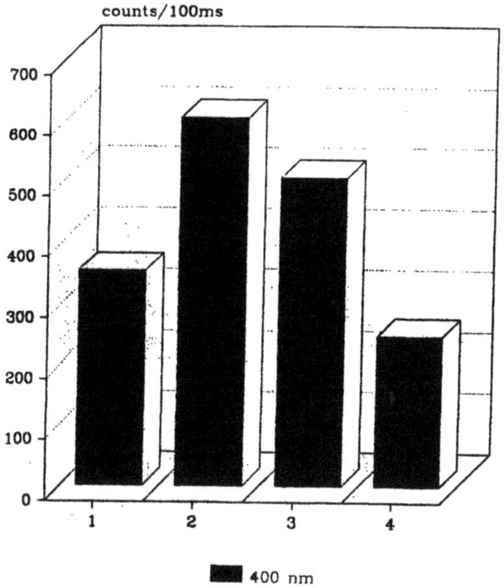

Figure 1.6: *The strong connection between all the modes of delayed luminescence in the visible range is demonstrated here by acetabularia which were differently treated. (1 = Acetabularia in normal medium, 2 = Acetabularia in medium of 90 % vitamine reduction, 3 = Acetabularia in medium of 90 % vitamine reduction plus a remedy which reduces stress at a concentration c_1, 4 = like 3, but instead of c_1 a different concentration c_2, see also Fig. 1.7)*

By taking account of a chemical reaction rate of about 10^5/s per cell, and providing t = 1s, $\tau_R \simeq 10^{-10}s$ and $\bar{x} = 10^{-8}cm$ as the lowest possible distance between two molecules in Eq. (1.8), we obtain $d \simeq 10^{-3}cm$ as the diameter of a cell. This means that even by ordinary diffusion a single photon may be able to trigger about 10^{10} reactions before it is released by a cell in one second.

This consideration invites us to divide the regimes of biophoton interaction with the cell roughly into three regions:

1. Spontaneous luminescence, which will take place for $\bar{x} >> \lambda$, where λ is the wavelength of light under consideration. Additionally the coherence time τ should be much smaller than τ_R: $\tau << \tau_R$.

2. Photon Migration for $\bar{x} \leq \lambda$, $\tau \simeq \tau_R$, and

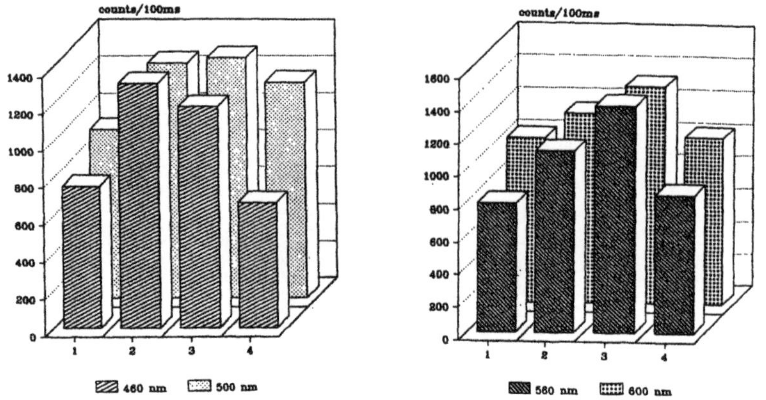

Figure 1.7: *The change of the excitability with light of different wavelengths after three days treatment (first row: before treatment; the other: after treatment). Here the average difference (± 10%) of the first count rate after excitation before and after three days treatment of always five acetabularia is drawn.*

3. Dicke's regime of superradiance and subradiance for $\bar{x} << \lambda$ and $\tau >> \tau_R$, corresponding also to the framework of Cavity Quantum Electrodynamics (see Chapter 5).

In a dense medium such as a cell it is very unlikely that a photon travels over a long distance $\bar{x} >> \lambda$ without hitting an absorbant compound within its coherence time τ. Taking an average reaction time τ_R of about 10^{-10} s as the lowest limit (which is already rather short), the coherence length is then $c\tau_R \approx 3 \cdot 10^{-2} cm$ which is of the order of ten cell diameters. This means that a photon which is created in a cell will certainly transfer information if it is reabsorbed in the same cell. On the other hand, if it travels through many cells without becoming absorbed, it loses the more "information" about its origin the lower the coherence time, e.g. the faster the spontaneous electronic transition of its creation has been. This means that a new absorption can take the role of the origin for transporting new information.

It is useful to consider this problem in terms of the spectral entropy S of a light beam travelling through the cells. According to Planck [32]:

$$S \propto \frac{1}{\tau}[(1 + i\tau)ln(1 + i\tau) - i\tau ln(i\tau)], \tag{1.9}$$

where i is the photon intensity (counts/s) and τ the coherence time [33], and

$$f_\nu = i\,\tau. \tag{1.10}$$

Obviously, S decreases with increasing τ. The f=constant-rule invites then to the interesting result that the optical processes in the cells may become governed by the rule

$$\delta i = -i\frac{\delta\tau}{\tau}, \tag{1.11}$$

which means that a spectral decrease of photon intensity may be due to a prolongation of the coherence time and vice versa. The change in the entropy of the light beam under the general constraint f=constant gives

$$\delta S = -S\frac{d\tau}{\tau} = S\frac{\delta i}{i}. \tag{1.12}$$

This is a strong confirmation of Gurvich's "degradation radiation" (see Chapters 22, 23), since the decay of ordered compounds may lead to an increase of photon intensity. However, the extension of this concept leads to a further result of (1.12): The decrease of relative photon intensity may then indicate an ordering of structures within the living tissue under investigation.

An example confirming the correctness of (1.12) is displayed in Fig. 1.8.

A further consequence of (1.10) is the hyperbolic relaxation after light illumination, as f=constant in (1.10) results in

$$i \propto \frac{1}{t}, \tag{1.13}$$

where we tacitly take account of the fact that τ has to be replaced by the real time t as long as $t < \tau$ (see[22]).

The fact that $\Theta(\nu) \neq T$ forces the introduction of a chemical potential μ according to

$$S = \frac{h\nu - \mu}{T} + const., \tag{1.14}$$

as soon as the biophoton emission is projected onto the state of thermal equilibrium.

Comparison of (1.14) with (1.2) by taking account of $h\nu = E_a$ shows that

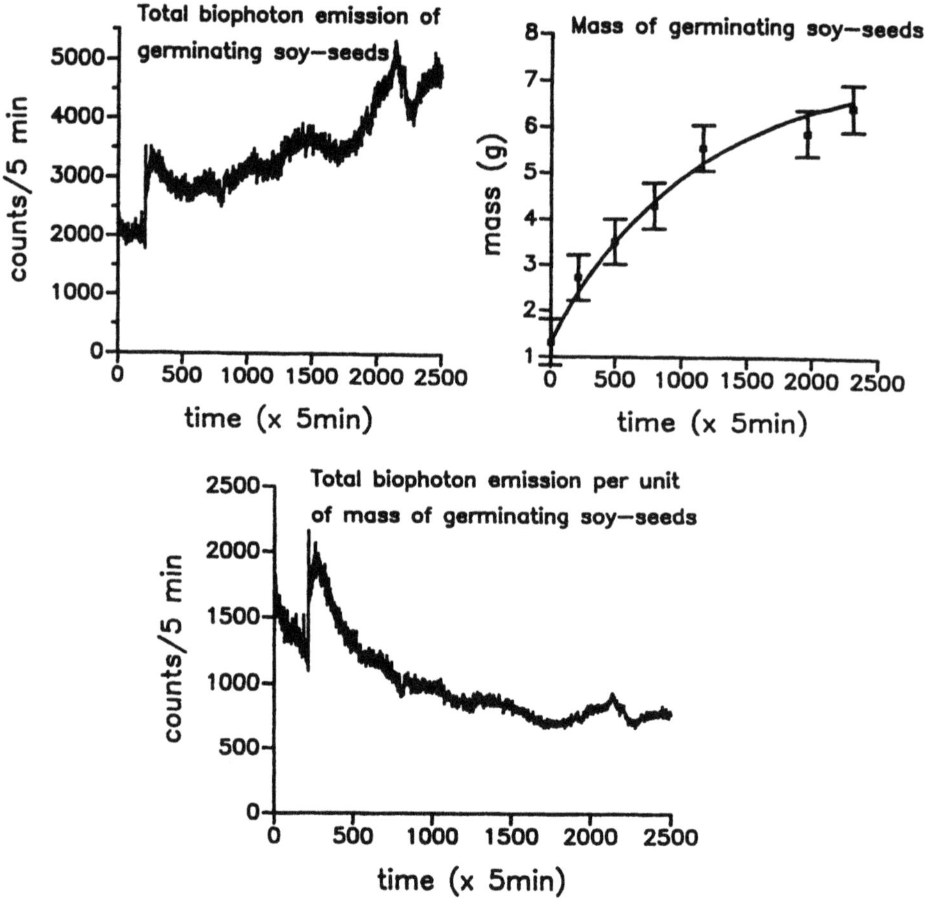

Figure 1.8: *The biophoton intensity from 9 soy-seeds increases during germination in darkness as shown in the left upper figure. The mass* m *during germination increases according to the curve in the right upper figure, where the measured values (points with error bars) have been approximated by the growth curve* $m = \frac{A}{B+(1-B)exp(-Ct)}$, *with* $A = 1.3g$, *corresponding to the mass of 9 soy-seeds,* $B = 0.178$ $C = 2.6 \cdot 10^{-4}$ (min^{-1}), t *representing the time. The lower curve displays the biophoton intensity per unit of mass without taking account of selfabsorption during the germination. It reflects a* <u>*decrease*</u> *of intensity per cell during germination, indicating a corresponding decrease of entropy density (see text).*

$$\Theta(\nu) \simeq T\frac{1}{1-(\frac{\mu}{h\nu})}, \tag{1.15}$$

or, after rearranging

$$\mu(\nu) \simeq h\nu(1-\frac{T}{\Theta(\nu)}). \tag{1.16}$$

One of the most important questions of biophoton emission concerns the nature and mechanism behind this chemical potential, which is, of course, already formally described by $\Theta(\nu)$ or $f(\nu)$. It will be shown that $\mu(\nu)$ is not only a consequence of metabolic "pumping" of the system, but has its deeper reason in the coherence itself. In addition, using (1.14) and (1.9) we get a new interesting insight into the process.

At first, the relation

$$S = \frac{h\nu}{\Theta(\nu)} + const. \tag{1.17}$$

is obtained from (1.14), which shows that an increase in i must not always be correlated to an increase in S, but may have just the opposite meaning. This helps us to understand that biophoton intensity is not an unique parameter of biological processes. The significance of changes in biophoton intensity is very dependent on the boundary conditions.

Secondly, the f = constant-rule (which can also be expressed in terms of $\frac{\partial f(\nu)}{\partial \nu} = 0$) can be translated into

$$\frac{\partial S(\nu)}{\partial \nu} \simeq 0, \tag{1.18}$$

which means that the spectral entropy of the biophoton field does not generally change with the frequency.

We may also find the connection between absolute change of the f-value and the entropy change:

$$f = i\tau = \frac{1}{exp(\frac{\Delta S}{k})-1}, \tag{1.19}$$

which shows that an increase in the f-value corresponds to a decrease in the entropy of the radiation field. In particular, for $\Delta S = k\ln 2$ we create a state $f = 1$. On this basis we get an idea of the power of coherent photons to work towards ordered structures in living tissue.

However, since the coupling between light and matter is essential for understanding the basic mechanisms, let us take into account a simple model based on the exciplex formation which is a very common process in living tissues [34].

If a system consisting of N_0 units of a biopolymer, for instance the basepairs in DNA and of the biophotons, rests in a stationary state, this always leads to N_1 unexcited monomers and N_2 exciplexes (excited dimers) such that [35]:

$$N_1 + 2N_2 = N_0. \tag{1.20}$$

For simplicity we assume that the unexcited monomers are subject to the mean thermal energy kT, whereas the excited molecules have to be assigned to a chemical potential μ. In a stationary state a general energy balance equation holds

$$N_1 kT + N_2(\mu + C_1) = C_0 N_0, \tag{1.21}$$

where C_1 is the energy dissipated over the system per exciplex formation. For $\mu = 0$ we have $C_0 = \frac{C_1}{2} = kT$, representing thermal equilibrium. The coupling between radiation and matter is described by Einstein's famous balance equation,

$$\dot{\rho} = h\nu(AN_2 + (N_2 - N_1)\rho B), \tag{1.22}$$

where ρ is the radiated energy density and A and B are the Einstein coefficients of spontaneous and induced emission, respectively. Thus we have

$$f = \frac{\rho B}{A} = exp\left(\frac{\mu - h\nu}{kT}\right). \tag{1.23}$$

As the system is an open one, it is subject to maximum entropy under the constraint of a stationarity of both $\dot{\rho} = 0$ and exciplex number $N_2 =$ constant. So for $(\frac{\delta N_2}{\delta \rho})_{\rho_0} = 0$ and taking into account

$$N_1 = N_2\left(\frac{A}{\rho B} + 1\right)$$

from Eq. (1.22), and then substituting μ from Eq. (1.23) into Eq. (1.21), we get

$$\frac{N_2}{N_0}\frac{kT}{C_0} = \frac{\frac{\rho B}{A}}{1+\frac{\rho B}{A}(1+\frac{h\nu}{kT}+\frac{C_1}{kT}+ln(\frac{\rho B}{A}))}. \tag{1.24}$$

Differentiating with respect to ρ (or $\frac{\rho B}{A}$) yields

$$\frac{\partial\left(\frac{N_2}{N_0}\frac{kT}{C_0}\right)}{\partial\left(\frac{\rho B}{A}\right)} = \frac{1-\frac{\rho B}{A}}{(1+\frac{\rho B}{A}[1+\frac{h\nu}{kT}+\frac{C_1}{kT}+ln\frac{\rho B}{A}])^2},$$

which takes the value zero only for

$$f = \frac{\rho B}{A} = 1, \tag{1.25}$$

thus confirming the results from the previous considerations.

Consequently, from a thermodynamic point of view, an exciplex system, in a quasi-stationary interaction with radiation adopts for $\mu \neq 0$ a state that can be characterized by $f_\nu = 1$, independent of the frequency.

From (1.24) we also obtain a relation between photon emission and temperature T, i.e.

$$i \propto \dot{N}_2 = \frac{\alpha}{N_0}\dot{T}N_2^2, \tag{1.26}$$

where

$$\frac{\alpha}{N_0} \equiv -\frac{1+\frac{\rho B}{A}\ln(\frac{\rho B}{A})}{(\frac{\rho B}{A})C_0}.$$

This dependence, showing a non-linear relationship between photon intensity and temperature response, has been experimentally confirmed [9]. Fig.1.9 displays an example.

Eqs. (1.20 – 1.22) can be used as a basis for deriving equations for the time-dependence of photon emission in terms of "deterministic chaos". This has already been performed [36] with promising results.

The equations (1.20 – 1.22) can be reformulated as iteration equations for time evolution at time t_i ($i = 1, 2, \cdots$)

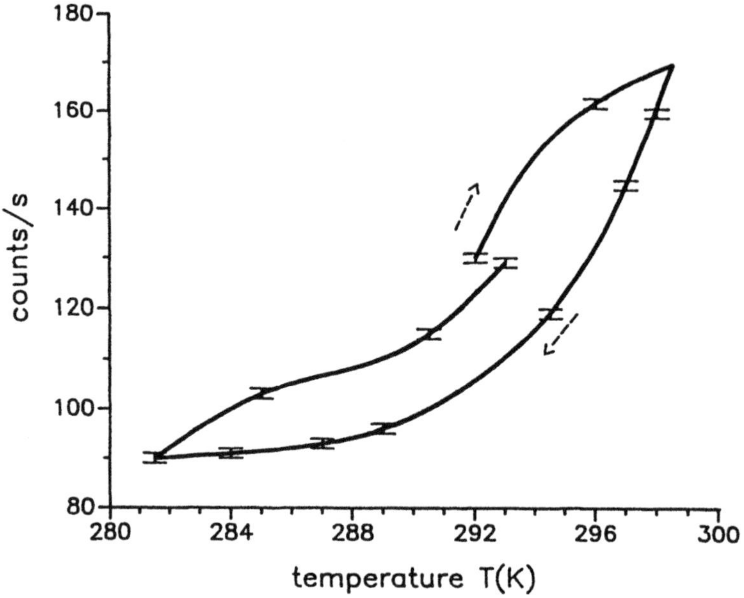

Figure 1.9: *The biophoton intensity of living tissues shows a hysteresis-like dependence on the temperature T, if T is cyclically varied. The example shows etiolated barley, 4 days of germination. The variation of temperature starts at $T = 292K$ with the rate $\frac{\delta T}{\delta t} = 0.5K/min$. At $T = 298.5K$ the rate of temperature change is reversed to $\frac{\delta T}{\delta t} = -0.5K/min$, and again at $T = 281.5K$ with $\frac{\delta T}{\delta t} = 0.5K/min$. This hysteresis-like behaviour of biophoton intensity can truthfully be described as a Curie-Weiss-law-dependence*[9].

$$f_{i+1} = f_i + \frac{\beta}{2 + y_i}(1 + f_i(1 - y_i))$$

and

$$y_{i+1} = y_i + \frac{y_{i+2}}{\left(\frac{\mu + c_1}{kT} - 2\right)} \frac{\frac{f_{i+1}}{f_i} - 1}{1 + f_i},$$

where f_i, according to (1.23), describes the time development of $f(t_i)$, $\beta = h\nu A N_0 \Delta t$ determines the interval Δt of time-evolution and $y_i = \frac{N_1(t_i)}{N_2(t_i)}$ is the ratio of unoccupied monomers to exciplexes at t_i. Fig.1.10 displays an example.

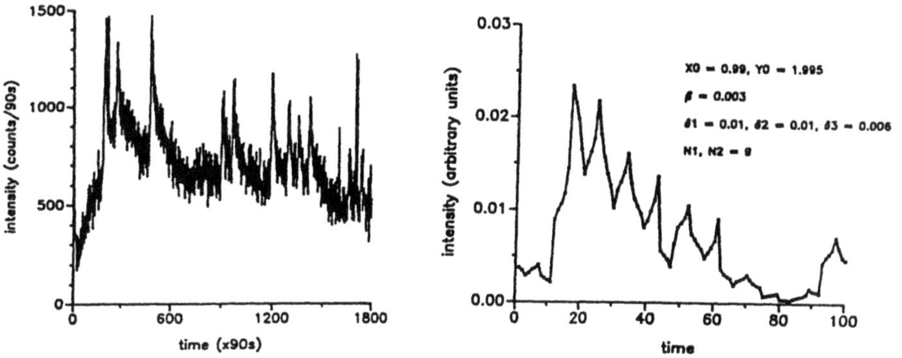

Figure 1.10: *The biophoton emission from* daphnia magna *over a longer time (45h). A similar pattern of photon intensity can be obtained by the exciplex model*[14], *where* $\mu + C_1 = 2kT + \delta_1 f_i^I \cos\omega t + \delta_2 f_i^{II} \cos(\omega t + \varphi) + \delta_3 \cos k(t)t$ *has been used, with* $\omega = \frac{2\pi}{9}$, $\varphi = 0$ *and* $k(t)$ *as numbers of a random number generator between 0 and 10.*

Finally, let us take the simplest derivation of the Boltzmann distribution $f_T = exp(-\frac{h\nu}{kT})$, in order to see what f = constant means on the most basic level of thermodynamics.

The entropy S of a thermodynamic system can be formulated as

$$S = -k \sum_{i=1}^{N} f_i \ln f_i, \tag{1.27}$$

where i accounts for the mode-number and f_i is its excitation probability. N is the number of modes which represent the state under investigation.

The usual derivation of the Boltzmann-distribution is the maximization of the entropy under the constraint of

1. probability conservation

$$\sum_{i=1}^{N} f_i = 1, \tag{1.28}$$

2. energy conservation

$$\sum_{i=1}^{N} f_i \varepsilon_i = E = constant, \tag{1.29}$$

where $\varepsilon_i = h\nu_i$ is the excitation energy of the ith mode and E is the total energy, and, in the case of a chemical potential μ,

3. conservation of the number n of particles

$$\sum_{i=1}^{N} n_i = n = constant\,. \tag{1.30}$$

In order to find the extrema of S under the constraints of (1.28) and (1.29), we apply the usual method of Lagrange parameters β. After multiplying the l.h.s. of (1.28) by $(\beta_N - 1)$ and the l.h.s. of (1.29) by β_E, taking the sum of the results and substracting it from $(\frac{S}{k})$, the variation of this total sum over i has to vanish:

$$\delta\left[\frac{S}{k} - (\beta_N - 1)\sum_{i=1}^{N} f_i - \beta_E\left(\sum_{i=1}^{N} f_i\varepsilon_i\right)\right] = 0.$$

Differentiation results in

$$-\ln(f_i) - 1 - (\beta_N - 1) - \beta_E\varepsilon_i = 0,$$

yielding

$$f_i = exp(-\beta_N - \beta_E\varepsilon_i). \tag{1.31}$$

If, in addition, the constraint (1.30) had been taken into account, instead of (1.31), we would have obtained

$$f_i = exp(-\beta_N - \beta_E\varepsilon_i - \beta_n), \tag{1.32}$$

where β_n is the Lagrange parameter due to the particle number. It is well-known that the Lagrange parameter β_N refers to the partition function, while β_E and β_n have the physical meanings

$$\beta_E = \frac{1}{kT} \tag{1.33}$$

$$\beta_n = -\frac{\mu}{kT}, \tag{1.34}$$

where μ is the chemical potential of the system under study. For $\mu = 0$, we arrive at the wellknown Boltzmann distribution of a black-body system at temperature T.

In order to get f_i = const., the simplest way would be to cancel both the constraints of fixed particle numbers and of fixed energy which holds only for a closed system. This means that we have to turn from a system where input and output intensity (heat flow) are always equal, to an "ideal" open system where always enough energy is available. This is the view of an observer outside the system, who tries to project the results of the measurements onto this experience with a closed system. Of course, the observer discovers a non-thermal biophoton emission which is not accompanied by an equal photon influx.

However, within the system the view of the observer is different. Provided we have stationarity and continuity at the boundaries, there is more than one possibility to achieve f=constant. All the possibilities require the introduction of a chemical potential that cancels the energy dependence of f. This means that the photons become bound in such a way that

$$\mu = \mu(\varepsilon_i) = \varepsilon_i - \chi, \tag{1.35}$$

where χ describes an energy loss which does not depend on the photon energy. It may be dependent on the temperature of the surroundings, reflecting, possibly, some mechanism of homeostasis. In the most general terms, the mechanism behind (1.35) is probably a Bose-condensation of photons, which can also be looked upon, in terms of Dicke's theory, as a consequence of "destructive interference".

Since f = const. always provides an absolute maximum of entropy S of the biophoton field, the ingenious way of nature in optimizing the living state seems to keep an absolute maximum S and at the same time to minimize the value S by means of reducing the number of degrees of freedom. Consequently, for $S = -k\sum_{i=1}^{N} f_i \ln(f_i) = k\ln(N)$ we can even get $S = 0$ for $N = 1$, which always represents a maximum at keeping the lowest absolute value. One of the consequences is a fluctuationless S, providing the highest possible signal/noise ratio. This may account for the high sensitivity of biological systems [3,37].

Finally, in order to enter entropy into a connection with the f = constant-rule and "biological information", (1.27) can be reformulated as

$$S = -k\sum_{i=1}^{N}(f+\delta f_i)\ln(f+\delta f_i), \tag{1.36}$$

where $f = \frac{1}{N}$ is an average excitation probability over the mode-number, δf_i is a fluctuation corresponding to the ith mode and $\delta f_i << f$ is provided. Straightforward calculation under the constraint $\sum_{i=1}^{N}\delta f_i = 0$ results in

$$S \approx k\,\ln(N) - kN\sum_{i=1}^{N}(\delta f_i)^2. \tag{1.37}$$

Apart from the expected result that for a constant number of degrees of freedom any weak modulation of the reference distribution f = constant leads to a reduction of the entropy, i.e., providing thus "information", we learn from (1.37) that adiabatic processes in living systems require a definite interdependence of mode-number N and the spectral distributions, i.e.

$$\delta N = N^2\sum_{i=1}^{N}(\delta f_i)^2. \tag{1.38}$$

In addition, "information" in a biological system is closely linked to a change of the number N of degrees of freedom rather than to the syntax of symbols. The "modulation" of the spectral occupation probabilities f_i has then to be looked upon as the "language" of cells, permanently influencing N. The relation is now subject to further experimental examination.

1.4 On the Coherence of Biophoton Emission

It will be shown now that the classical description of coherence is not sufficient for understanding the coherence of weak light emission.

The quantum description, on the other hand, contains the results of the classical theory but is necessary to understand all the properties of low level luminescence. Of course, as only single quanta are counted, the phenomenon is subject to a real quantum effect.

In view of the low intensity and the broad spectral distribution, measurements by use of interferometry are not possible at present for examining the coherence properties of biophotons, and so the coherence has to be examined in terms of correlation functions like those introduced by Glauber[40].

It will be shown also that for a stationary system, one cannot distinguish between a coherent and a completely chaotic field, as long as the coherence time of the chaotic system is significantly lower than the measurement time interval.

Coherence is usually understood as the ability of electromagnetic waves to interfere. Given two waves, $E_1(t) = e_1 \cos \omega t$ and $E_2(t) = e_2 \cos(\omega t + \phi)$, where e is the field amplitude, ω the frequency, t the time and ϕ the phase difference, superposition results in

$$E(t) = E_1(t) + E_2(t) = e_1 \cos \omega t + e_2 \cos(\omega t + \phi). \tag{1.39}$$

The intensity $i(t)$ at a time t is proportional to the square of the amplitudes

$$i(t) \propto E^2(t) = e_1^2 \cos^2 \omega t + e_2^2 \cos^2(\omega t + \phi) + 2e_1 e_2 \cos \omega t \cos(\omega t + \phi). \tag{1.40}$$

This equation holds during a time interval τ (the "coherence time"), where the phase relation ϕ between the waves is not lost. By contrast, the phase relation $\phi(t)$ is always "forgotten" for $t > \tau$, so the last term, which is responsible for interference, vanishes because positive and negative interferences cancel out on average:

$$\langle \int_{t>\tau} \cos \omega t \cos(\omega t + \phi) dt \rangle = 0. \tag{1.41}$$

The intensity is then simply the sum of the intensities of the single waves.

In the case that $\phi(t)$ is a well-defined function of t, including $\phi(t) =$ constant, the intensity fluctuates in such a way that the two waves alternately cancel out or mutually reinforce to give an intensity which is smaller or greater than the sum of the two separate intensities. These intensity fluctuations around the sum of the intensities of the single waves, are referred to as temporal coherence, while the corresponding spatial interference patterns are described by the equation

$$i(x) = e_1^2 \cos^2 kx + e_2^2 \cos^2(kx + \phi) + 2e_1 e_2 \cos kx \cos(kx + \phi), \tag{1.42}$$

where $k = \frac{\omega}{c}$ with c representing the velocity of light.

By analogy to this classical definition of coherence, Schroedinger [38], and later, Glauber [39] introduced the "coherent state" into quantum optics. It is defined as an eigenstate $|\alpha\rangle$ of the annihilation operator according to

$$a_k \, |\alpha_k\rangle = \alpha_k \, |\alpha_k\rangle, \tag{1.43}$$

where k describes the k-th mode of the field.

This state has its classical analogy just in the wavetrains with infinite coherence time τ. A classical electric field is described by

$$\vec{E}(\vec{r}, t) = \sum_k c_k \vec{U}_k(\vec{r}) exp(-i\omega_k t), \tag{1.44}$$

where $\omega_k > 0$ is the frequency of the k-th modes with the vector mode function $\vec{U}_k(\vec{r})$ at a space-point $\vec{r}$, and c_k is a complex number which describes the field amplitude.

In quantum theory, the electric field operator is introduced according to

$$\vec{E}_+(\vec{r}, t) = i \left(\sum_k \sqrt{\frac{1}{2\hbar\omega_k}} a_k \vec{U}_k(\vec{r}) exp(-i\omega_k t) \right), \tag{1.45}$$

where a_k is the annihilation operator of the k-th mode, $\vec{E}_+$ is determined by a_k and a_k^+ which obey the commutation relations

$$\begin{aligned} [a_k, a_{k'}^+] &= \delta_{kk'}, \\ [a_k, a_{k'}] &= [a_k^+, a_{k'}^+] = 0, \end{aligned} \tag{1.46}$$

and $\vec{E}_{(\pm)}$ are mutually adjoint:

$$\vec{E}_- = (\vec{E}_+)^+. \tag{1.47}$$

In order to elucidate the connection between classical and quantum descriptions, let us confine the discussion to a single mode of two waves, represented by the states $|\beta_1\rangle$

and $|\beta_2\rangle$. The corresponding density operators are $|\beta_1\rangle\langle\beta_1|$ and $|\beta_2\rangle\langle\beta_2|$, respectively. The intensity of the two states is proportional to

$$i(\vec{r},t) = T_r\{\rho E_-(\vec{r},t)E_+(\vec{r},t)\}, \tag{1.48}$$

where $\rho \equiv |\beta\rangle\langle\beta|$ is the density operator of the whole field under consideration. Splitting the electric field operators into the two corresponding parts $E_+^{(1)} + E_+^{(2)}$, we then have

$$\begin{aligned} i(\vec{r},t) &= T_r\{\rho(E_-^{(1)} + E_-^{(2)})(E_+^{(1)} + E_+^{(2)})\} \\ &= \langle\beta|E_-^{(1)}\, E_+^{(1)}|\beta\rangle + \langle\beta|E_-^{(2)}\, E_+^{(2)}|\beta\rangle \\ &\quad +\langle\beta|E_-^{(1)}\, E_+^{(2)}|\beta\rangle + \langle\beta|E_-^{(2)}\, E_+^{(1)}|\beta\rangle. \end{aligned} \tag{1.49}$$

Obviously, the first two terms on the r.h.s. of (1.49) have to be assigned to the separated waves, while the third and fourth term describe the interference between them.

If $|\beta\rangle$ represents a coherent state, (1.43) holds, which means that by taking account of (1.45) we obtain

$$i(\vec{r},t) = |\beta_1|^2 + |\beta_2|^2 + \beta_1^*\beta_2 + \beta_1\beta_2^*. \tag{1.50}$$

Consequently, in the case of Eq. (1.42) the classical result for $t < \tau$ is obtained, indicating that Eq. (1.43) is sufficient for a fully coherent field.

Glauber [40] extended the concept of Eq. (1.48) to nth order correlation functions. Consider n ideal broad-band counters at position $\vec{r}_1, \vec{r}_2, \cdots, \vec{r}_n$. In sufficiently small time intervals Δt around the sequence of times $t_1, t_2, \cdots, t_n$ the probability of always registering precisely one photon at $x_1 \equiv \vec{r}_1, t_1;\ x_2 = \vec{r}, t_2, \cdots;\ x_n = \vec{r}_n, t_n$ is then given by

$$p^{(n)}(x_1, x_2, \cdots, x_n) = T_r\{\rho E_-(x_1)E_-(x_2)\cdots E_-(x_n)E_+(x_n)\cdots E_+(x_2)E_+(x_1)\}. \tag{1.51}$$

More generally, the nth order correlation function of field amplitudes is defined by

$$g^{(n)}(x_1, x_2, \cdots, x_n; y_1, y_2, \cdots, y_n) = T_r\{\rho E_-(x_1)\cdots E_-(x_n)E_+(y_1)\cdots E_+(y_n)\}, \tag{1.52}$$

where by comparison of (1.51) and (1.52) it follows

$$p^{(n)}(x_1, x_2, \cdots, x_n) = g^{(n)}(x_1, x_2, \cdots, x_n; x_n, \cdots, x_2, x_1).$$

From (1.43) it is evident that for a coherent state $g^{(n)}$ can always be factorized in the form

$$g^{(n)}(x_1, x_2, \cdots, x_n; y_1, y_2, \cdots, y_n) = E^*(x_1)E^*(x_2)\cdots E^*(x_n)E(y_1)E(y_2)\cdots E(y_n), \tag{1.53}$$

where $E(x_i)$ is now the classical field amplitude at $\vec{r}_i, t_i$.

For practical purposes it is useful to measure the number of photons within a preset time interval Δt at a position $\vec{r}$. The measurement of the number of photons $\dot{n}(t)\Delta t$ is a powerful tool for investigating the coherence of light. It is possible to get the data of $g^{(n)}(x_1, \cdots, x_n; x_n, \cdots, x_1)$ then directly by investigating the results of the photocount statistics (PCS), where the probability $p(n, \Delta t)$ of measuring $n = 0, 1, 2, \cdots$ photons within the time interval Δt is evaluated from $n(t) = \dot{n}(t)\Delta t$ during time t at any position.

Let us at first confine the discussion to a stationary state of biophoton emission, where the probability of counting n photons within a time interval Δt can be written in the form [41]

$$p(n, \Delta t) = \int_T K(|\beta|^2, \Delta t, n)W(\beta)d^2\beta, \tag{1.54}$$

where $W(\beta)$ is the amplitude probability distribution of the stationary field under investigation and K describes the process of photodetection, e.g. the transformation of the field amplitude β into the count number $n = \dot{n}\Delta t$.

Whereas a chaotic field is described by

$$W(\beta) \propto \frac{exp(-\frac{|\beta|^2}{\langle n\rangle})}{\pi\langle n\rangle}, \tag{1.55}$$

with $\langle n\rangle$ representing the average of photons over Δt during the measurement time T, a fully coherent field follows

$$W(\beta) = \delta^{(2)}(\beta - \beta_0) \tag{1.56}$$

with constantly fixed amplitude β_0.

From (1.43) already we know the result of $p(n, \Delta t)$ in the case of a fully coherent field, registered by an ideal detector.

Of course, if we expand the coherent state $|\alpha\rangle$ in terms of the number states $|n\rangle$ (where $a|n\rangle = \sqrt{n}|n-1\rangle$):

$$|\alpha\rangle = \sum_{n=0}^{\infty} |n\rangle \langle n|\alpha\rangle, \tag{1.57}$$

by use of Eq. (1.43) we obtain

$$\begin{aligned} a|\alpha\rangle &= \sum_{n=0}^{\infty} a|n\rangle \langle n|\alpha\rangle \\ &= \sum_{n=0}^{\infty} \sqrt{n}|n-1\rangle\langle n|\alpha\rangle \\ &= \alpha \sum_{n=0}^{\infty} |n\rangle \langle n|\alpha\rangle. \end{aligned} \tag{1.58}$$

After equating identical coefficients on the r.h.s. of (1.58) and normalization $\langle\alpha|\alpha\rangle = 1$ we then get

$$\begin{aligned} p(n, \Delta t) &= |\langle n|\alpha\rangle|^2 \\ &= \frac{exp(-|\alpha|^2)(|\alpha|^2)^n}{n!} \\ &= \frac{exp(-\langle n\rangle)\langle n\rangle^n}{n!}. \end{aligned} \tag{1.59}$$

Thus we arrive at the important result that a fully coherent field in a stationary state is always subject to a Poissonian photocount distribution.

Consequently, by taking account of (1.56) and (1.59), for an ideal detector:

$$K(|\beta|^2, \Delta t, n) = \frac{exp(-|\beta|^2)(|\beta|^2)^n}{n!}. \tag{1.60}$$

For a chaotic field according to (1.55) the result (1.60) can be inserted into (1.54) and integrated to give:

$$p(n, \Delta t) = \frac{\langle n \rangle^n}{(1 + \langle n \rangle)^{n+1}}. \tag{1.61}$$

One might take the difference of (1.59) and (1.61) as a powerful tool, in order to decide whether a field is coherent or chaotic. However, in calculating (1.54) we implicitely assumed that $\Delta t \leq \tau$, where τ is the coherence time of the field under consideration.

Over longer time intervals, $\Delta t >> \tau$, where τ is the coherence time of the chaotic field, the integration of $d^2\beta$ concentrates more and more around the average value $\langle n \rangle = |\beta|^2$ such that from (1.61) a result approaching (1.59) will be obtained. A detailed calculation yields[42]

$$p(n, \Delta t) = \frac{\Gamma(n + M)}{n! \Gamma(M)} \left(1 + \frac{M}{\langle n \rangle}\right)^{-n} \left(1 + \frac{\langle n \rangle}{M}\right)^{-M}. \tag{1.62}$$

where $M = \frac{\Delta t}{\tau}$ is a measure of the "degree of freedom" of the chaotic field under investigation.

For $M \to 1$ (1.62) reduces to (1.61) again, but for $M \to \infty$, (1.62) approaches a Poissonian distribution according to (1.59) (see Fig. 1.1).

We would like to emphasize here that it is principally impossible to decide whether we are concerned with a fully coherent field or a chaotic field with $\tau << \Delta t$, as soon as a stationary photon emission displays a Poissonian distribution (Fig. 1.1).

In practice, instead of calculating $p(n, \Delta t)$ it is advantageous to evaluate the normalized factorial moments according to

$$m^{(r)} = \frac{\sum_n \frac{n!}{(n-r)!} p(n, \Delta t)}{[\sum_n p(n, \Delta t)]^r}, \tag{1.63}$$

where $r = 1, 2, \cdots$.

The $m^{(r)}$ are independent of the special geometry, the detector efficiency and the beam attenuation, since by normalization the dependence of the kernel function $K(|\beta|^2, \Delta t, n)$ on the properties of the equipment cancels out. The normalized factorial moments of the actual photocount distribution registered by a photomultiplier and that of the ideal photon-number distribution are then identical.

It has been shown [42] that

$$m^{(r)} \propto \int_0^T \cdots \int_0^T g^{(r)}(t_1, t_2, \cdots, t_r; t_r, \cdots, t_1) \prod_{j=1}^{r} dt_j, \tag{1.64}$$

where $g^{(r)}$ according to (1.52) is taken over a sufficiently small area of the detector system $\vec{r}_1 \simeq \vec{r}_2 \simeq \cdots \vec{r}_k$.

For the normalized factorial moments $m^{(r)}$ (1.52), a Poissonian distribution $p(n, \Delta t)$ according to (1.58) yields

$$m^{(r)} = 1 \qquad for\ all\ r, \tag{1.65}$$

whereas a chaotic field, which follows (1.61), results in

$$m^{(r)} = r! \qquad for\ r = 1, 2, \cdots. \tag{1.66}$$

The difference of (1.65) and (1.66) which increases with increasing r, provides a tool for distinguishing between a fully coherent and a chaotic field. However this also holds only as long as $\Delta t \leq \tau$. Instead, for a chaotic field with $\Delta t >> \tau$, from (1.63) we obtain:

$$m^{(r)} = \frac{\Gamma(r + M)}{\Gamma(M)M^r}, \tag{1.67}$$

where $M = \frac{\Delta t}{\tau}$.

For $M \rightarrow \infty$, (1.67) turns into (1.65), as (1.62) did, when it approached (1.59) for $M \rightarrow \infty$.

Fig. 1.3 displays some examples of $m^{(r)}$ for quasi-stationary photon emission.

Since an expansion of (1.67) for $\Delta t >> \tau$ gives the approximate expression

$$m^{(r)} \approx exp\left\{r(r-1)\frac{\tau}{(2\Delta t)}\right\}, \tag{1.68}$$

(1.68) provides a useful tool to estimate the upper limit of the coherence time of a chaotic field if a Poissonian distribution has been found. As long as $m^{(r)}$ remains 1 for increasing r we can provide that

$$\tau << \frac{2\Delta t}{r(r-1)}. \tag{1.69}$$

If, for instance, the normalized factorial moments remain 1 up to the tenth order by choosing $\Delta t = 10$ ms, either a chaotic field with $\tau << 0.2$ ms or a coherent field then becomes the subject of investigation. Theoretically, by choice of Δt and r, any coherence time τ of a chaotic field can be estimated in this way.

Investigations on biophoton emission [21] gave evidence that if the biophotons originated from a chaotic field, τ should be at least lower than 10^{-5} s[21]. Since even extremely short lifetimes of radicals are between 10^{-4} and $10^{-2}s$ [43], it is therefore very unlikely that the photon emission originates "spontaneously" from radicals. Once we agree, however, that biophoton emission has to be assigned to radical formation, the sources should be very fast processes like polymerization or similar chain reactions. The other possibility, which has to be concluded from the experimental results, is that under stationary conditions biophoton emission always displays $m^{(r)} \leq 1$ even at high r, that is a Poissonian or sub-Poissonian field.

However, it is very difficult to decide from experimental results on stationary biophoton emission whether a chaotic field of $\tau << 10^{-5}s$ or a fully coherent field is the real origin.

A puzzling problem arises from experiments, where the photocount statistics (PCS) has been evaluated for both small time intervals Δt (with $\Delta t\, \dot{n} \approx 1$, where $\dot{n}$ is the count rate) and rather high time intervals $\Delta t\, \dot{n} \approx 10000$. For a stationary state of a chaotic light source one expects that with increasing $\Delta t = M\, \tau$ (where τ is the coherence time of the chaotic field) the PCS approaches more and more a Poissonian distribution. But just the opposite happens in biological systems, as, for instance, in the case of *Gonyaulax polyedra* and soybean seedlings (*Glycine max*).

Of course, for instance, Acetabularia acetabulum exhibits even a sub-Poissonian PCS, when only one cell in a quasi-stationary state ($\Delta t\, \dot{n} \approx 1$) has been the subject of investigation (Fig. 1.11).

A similar result can be obtained for *Gonyaulax polyedra* and for soybean seedlings. However, for higher time intervals Δt ($\Delta t\dot{n} >> 1$), the PCS turns more and more into a geometrical distribution rather than a Poissonian one (see Figs. 1.12, 1.13) with a M-value (correponding to the best fit with the experimental data) of the order of $M \simeq \dot{n}\, \Delta t$.

This means that the character of biophoton emission may turn from a coherent into a chaotic state, depending on the "memory"-time within the system under investigation. In line with this interpretation is the result that instead of hyperbolic relaxation we get in case of a sufficiently long time interval Δt an exponential relaxation of Gonyaulax polyedra after an internal excitation-phase (Fig. 1.12). However,

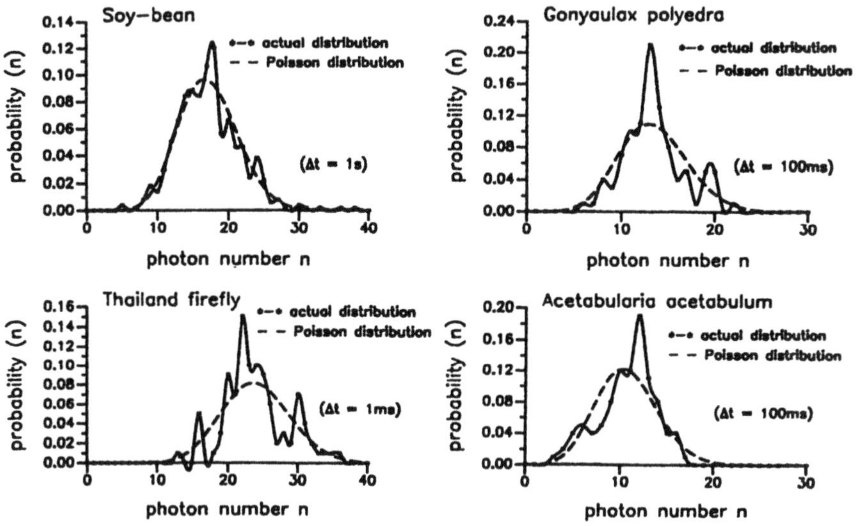

Figure 1.11: *The photocount statistics (= probability $p(n, \Delta t)$ of registering n counts in a preset time interval Δt, where $n = 0, 1, 2, \cdots$) is very similar for all biological systems. If Δt is sufficiently small such that the mean number of photons in the field becomes lower than about 100, $p(n, \Delta t)$ displays a Poissonian (and sometimes even a sub-Poissonian) photocount distribution. There are 4 different examples with different Δt. 100 measurement values have been used always for evaluation, where the biological state was kept quasi-stationary.*

the interpretation becomes even more complicated, if one takes account of long time oscillations of the biological sample under investigation. Fig. 1.13 displays an example.

The result may be interpreted in terms of non-classical light which is originally squeezed and gives rise to oscillations of the sub-Poissonian PCS [44]. With increasing Δt, the field becomes Poissonian and turns partially into a chaotic one, loosing its phase-memory by random rescattering. The sum of the variance of the oscillation plus that of the fluctuations around the oscillation equals the total variance of the photon emission. This means that the multiplier seems to register completely independent processes, e.g. squeezed photons of a "vibrating" source (see Chapter 5) and photons from a chaotic field.

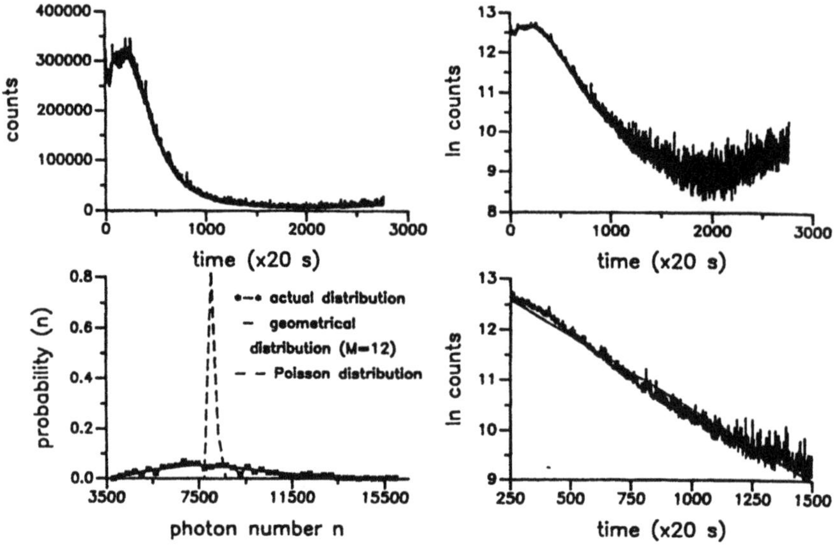

Figure 1.12: *Photon emission from an ensemble of* Gonyaulax polyedra *over longer time (upper, left). The right side of the upper part displays instead of the count rate (as on the left side) the logarithm of the count rate. From the maximum on it relaxes according to an exponential law (lower, right), indicating that the memory of the whole system is always lost during a time-interval of 20s. The system behaves there as an ensemble of single cells. In accordance to that is the photocount distribution* $p(n, \Delta t)$ *(lower, left), which is a geometrical distribution for a coherence time* $\tau = \frac{\Delta t}{M} = 1.67s$*. The evaluation is based there on a quasi-stationary state between 1900 and 2100* ($\times 20s$)*.*

However, since the measurement takes account of all the different time intervals from the initial production of photons up to the "age" Δt of photons within the tissue and even longer times, this PCS-fingerprint is likely due to the transformation of originally squeezed light into rather chaotic one, working in this way for minimizing the entropy of the tissue. This interpretation is in line with the sub-Poissonian statistics for small Δt, the Poissonian one for increasing Δt and finally the geometrical distribution for rather high Δt.

Whether the sub-Poissonian statistics for very small Δt is due to "antibunching" remains at present an open question. However, even if this "non-classical" light corresponds to number-states, there is no doubt whatsoever about its "informational"

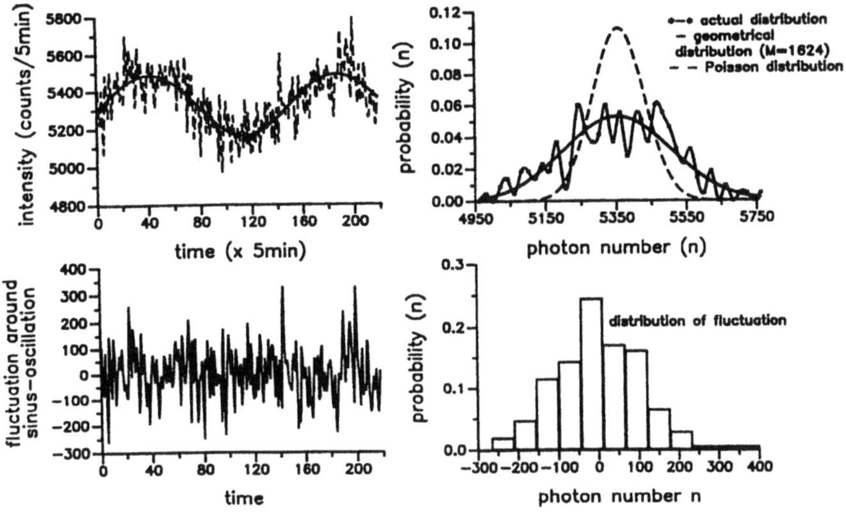

Figure 1.13: The biophoton emission from soybean-seedlings when they arrive after 8 days of germination at a quasi-stationary state. The intensity follows there a long-time sinus oscillation of a period of about 12 hrs. The photocount distribution $p(n, \Delta t)$ (upper part, right side) corresponds to a chaotic field of a coherence time $\tau = \frac{\Delta t}{M} \simeq 0.2$s. However, the biophoton emission consists of a fluctuationless sinus-oscillation ($\delta^2 \approx 12000$) and a fluctuation around the sinus-oscillation with $\delta^2 \approx 11000$ (lower part). They add up to the total variance of about 23000 which is about 4 times higher than the mean value.

character, i.e. its high signal/noise-ratio [45,46].

The informational function of this light-emission can be registered actually by the following arrangement (Fig. 1.14).

Two multipliers collect the photons from almost always one of two samples which can be optically connected and disconnected by a shutter between them. One calculates the linear correlations of photon emission without and with optical contact. In a series of experiments on Thailand fireflies, *Gonyaulax polyedra*, *Acetabularia acetabulum*, chicken embryos and drosophila eggs we confirmed that there is a considerable increase of correlations if we switch from the separated state to the state of optical contact.

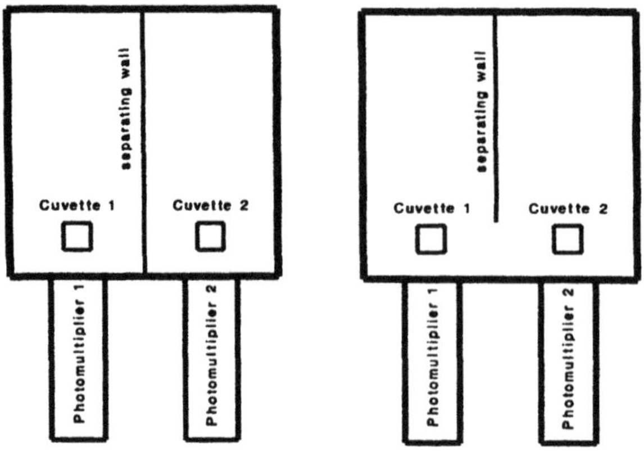

Figure 1.14: *In order to investigate communication between cell populations and organisms, this arrangement is used. Two cuvettes, containing the samples, are either separated (left side) or connected by light contact (ride side). The multipliers register the biophoton intensities of both the samples. The differences in the correlations of the signals without and with contact of the samples are subject of communication experiments.*

The Fourier analysis displays at the same time an increase of the overlap of the "vibration"-modes (Fig. 1.15) of the different samples under investigation.

The Figs. 1.16, 1.17 show examples.

This means that the "informational" function of biophoton emission is due to a trigger effect on possible "biological oscillations" ("biological clocks") in such a way that systems under light contact may synchronize each other in the amplitudes and phases of their slow oscillations.

A very striking feature of biophoton emission is the fractal behaviour of the intensity-pattern in dependence of time t and/or frequency ω. This means that the variation of the sampling time Δt over many orders of magnitude preserves the similarity of the intensity (t, ω)-pattern. Even the Fourier-pattern is self-similar to its original intensity-time-pattern. An example demonstrating again that biological photon emission can never be assigned to ordinary chemiluminescence is displayed in Fig. 1.18. It shows a typical emission from *Gonyaulax polyedra* which (as other living biological systems do either) comes in form of wave-trains which are repeated several times.

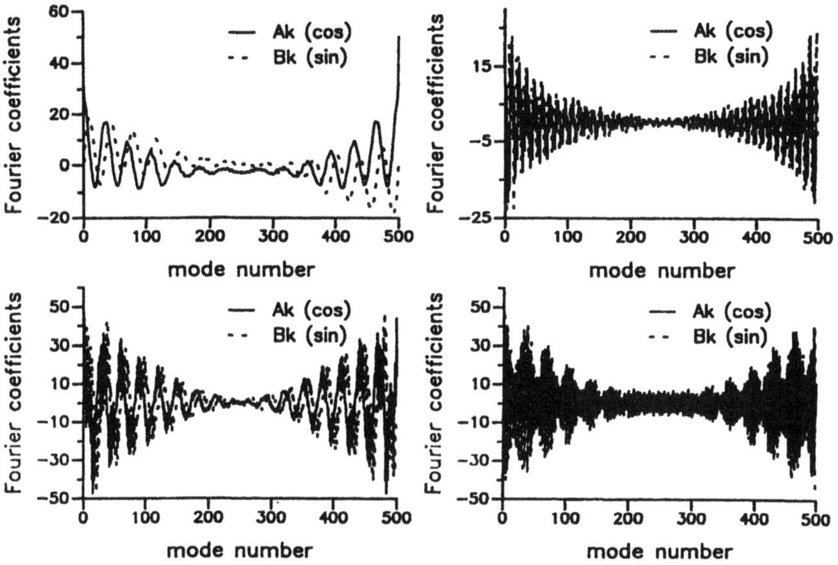

Figure 1.15: *Fourier coefficients (A_k:* cos*-part, B_K:* sin*-part) of photon emission from* Gonyaulax polyedra *for different samples at different times. The Fourier coefficients display a variety of different patterns, indicating varying occupation of "vibrational" modes of the photon source (s). The evaluation is always carried out for 500 values of a preset time interval Δt = 100 ms.*

In order to prove that for moderate Δt biophoton emission originates from a fully coherent field, we have finally to examine the relaxation behaviour of light-emission after exposure of the living system to weak light-illumination (or other weak excitation) under ergodic conditions ("delayed luminescence").

Let us, for simplicity, first discuss the simple classical case of an exponential decay function, which is always obtained under ergodic conditions if the intensity $\dot{n}(t)$ of radiation becomes proportional to the number n of radiators, i.e.

$$\dot{n}(t) = -\lambda n(t)\,. \qquad (1.70)$$

Actually, the solution of (1.70) takes the form

$$n(t) = n(0)exp(-\lambda t). \qquad (1.71)$$

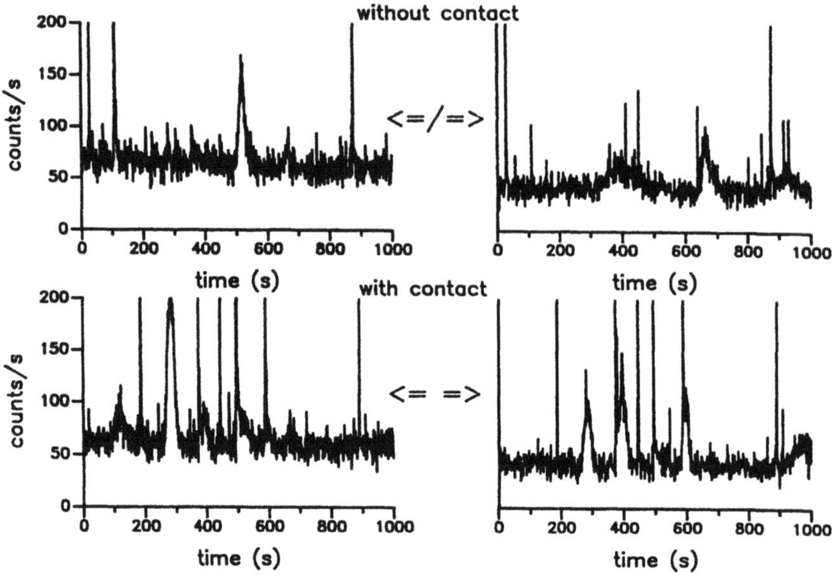

Figure 1.16: *Photon emission from* Gonyaulax polyedra, *where always half of them are kept in a cuvette (see Fig. 3.12). The upper two figures display their emission, when the shutter between them is closed, the lower two curves show the emission after opening the shutter (see Fig. 3.12). It is evident that the light-contact induces an increase of simultaneous emission of the whole system. The photon emission is observed over 1000s, where* $\Delta t = 1s$.

However, as soon as a coherent system, where any radiator is coupled to any other one, relaxes after excitation, the intensity is proportional to the square of the number of radiators. Consequently, instead of (1.70) we have to take account of

$$\dot{n}(t) = -\mu n^2(t), \tag{1.72}$$

where the solution becomes "hyperbolic"

$$n(t) = \frac{1}{\mu}(t + t_0)^{-1}, \tag{1.73}$$

with t_0 as a constant, e.g.

$$t_0 = \frac{1}{n(0)\mu}. \tag{1.74}$$

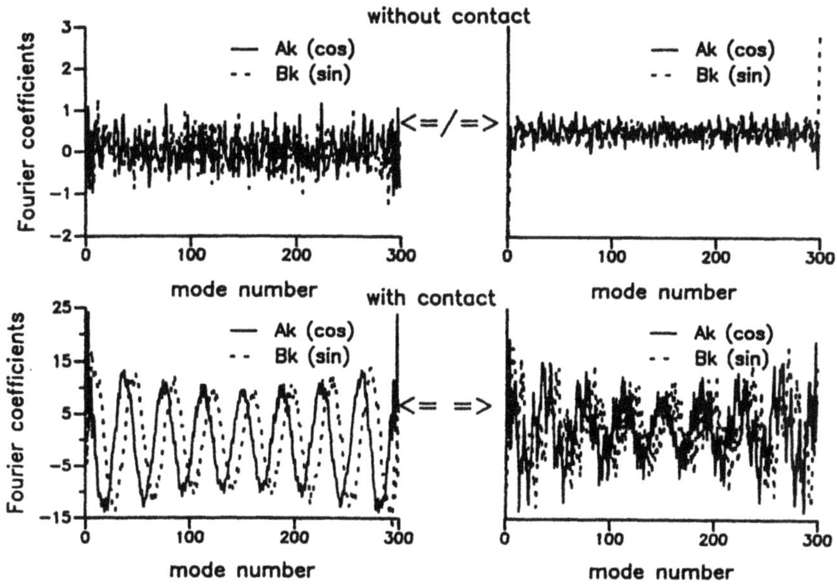

Figure 1.17: *Fourier spectrum of a part of the photon emission of Fig. 1.14 (180-480s). Whereas the samples without contact display not a distinct similarity of their Fourier pattern, the correlations between the Fourier coefficients of the samples with contact become quite obvious. It should be noted, however, that internal rythms of the systems may likely lead to similar Fourier spectra, too. In addition, the results indicate again the validity of the f_ν = constant-rule even down to the ELF-range.*

A more detailed investigation of this problem is discussed in next chapter. It gives evidence that under ergodic conditions a hyperbolic decay is always a sufficient condition for a fully coherent field. This means that as soon as the hyperbolic decay behaviour of a mode of biophoton emission ("delayed luminescence") after exposure of the system to weak light illumination has been demonstrated, we are forced to conclude that the biophoton emission originates from a fully coherent field.

The Fig. 1.5 and Fig. 2.2 of chapter 2 indicate that

1. we actually have hyperbolic instead of exponential relaxation, and

2. the ergodicity condition is satisfied.

A variety of papers [3,21–26], (see also Chapters 10, 13) confirm this hyperbolic relaxation behaviour. Thus, we have to conclude that biophoton emission originates from

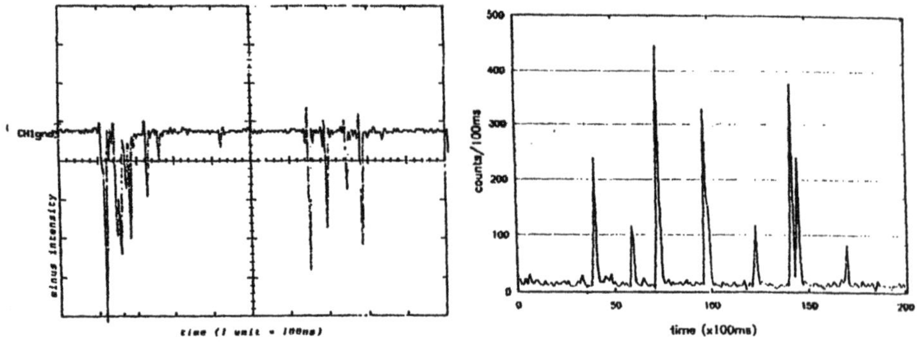

Figure 1.18: *Typical wave-trains of biophoton emission. This example (left side) displays the emission of* Gonyaulax polyedra *on a short time scale (1 unit ≐ 100 ns). The repetitions of wave-trains can occur many times. This measurement has been performed at CERN/LAA. I am grateful to Prof. Zioutas and Prof. Zichini for their courtasy. On the right side, the emission pattern is shown for a time interval* Δt *= 100 ms. The "fractal" behaviour becomes obvious.*

a fully coherent field.

We conclude that apart from the important research on the biochemical aspects of biophoton emission [47,48], in view of the development of modern physics, the time is ripe of investigating biophoton emission carefully in terms of the physics of weak (non-classical) light.

1.5 References

1. A.I. Zhuravlev, Ultraweak luminescence in biology. *Trans. Moscow Soc. Naturalists* **vol. 39** (Nauka, Moscow, 1972) (Russian).

2. H.H. Seliger, Applications of bioluminescence and chemiluminescence, in *Chemiluminescence and Bioluminescence*, eds. M.J. Cormier, D.M. Hercules and J. Lea (Plenum Press, New York, 1973) p. 461-478.

3. F.A. Popp, Photon Storage in Biological Systems, in *Electromagnetic Bio-Information,* eds. F.A. Popp, G. Becker, H.L. Koenig and W. Peschka (Urban & Schwarzenberg, Muenchen, 1979) p. 123-149.

4. K.H. Li, F.A. Popp, W. Nagl and H. Klima, Indications of optical coherence in biological systems and its possible significance, in *Coherent Excitations in Biological Systems,* eds. H. Froehlich and F. Kremer (Springer, Berlin, 1983) p. 117-122.

5. I. Prigogine, G. Nicolis and A. Babloyantz, Thermodynamics of Evolution. *Physics Today* **11** (1972) p. 23-28.

6. J. Perina, *Coherence of Light* (D. Reidel P.C., Dordrecht, 1985).

7. F.A. Popp, A.A. Gurwitsch, H. Inaba, J. Slawinski, G. Cilento, R. van Wijk, W.B. Chwirot and W. Nagl, Biophoton Emission (Multiauthor Review). *Experientia* **44** (1988) p. 543-600.

8. R.N. Tilbury and T.I. Quickenden, The Effect of Cosmic-Ray Shielding on the Ultraweak Bioluminescence Emitted by Cultures of *E. coli. Radiat. Res.* **112** (1987) p. 398-402.

9. J. Slawinski and F.A. Popp, Temperature hysteresis of low level luminescence from plants and its thermodynamical analysis. *J. Pl. Physiol.* **130** (1987) p. 111-123.

10. H. Precht, J. Christophersen, H. Hensel and W. Larcher, *Temperature and Life.* (Springer, Berlin, 1973).

11. F.A. Popp, W. Nagl, K.H. Li, W. Scholz, O. Weingaertner and R. Wolf, Biophoton Emission: New Evidence for Coherence and DNA as Source. *Cell Biophys.* **6** (1984) p. 33-52.

12. R.H. Dicke, Coherence in spontaneous radiation processes. *Phys. Rev.* **93** (1954) p. 99-110.

13. M. Rattemeyer, F.A. Popp and W. Nagl, Evidence of photon emission from DNA in living systems. *Naturwissenschaften* **68** (1981) p. 572-573.

14. F.A. Popp and J. Deny, Biophotonen-Information und Chaostheorie, in *Ganzheitsmedizin*, ed. A. Stacher (Facultas Universitätsverlag, Wien, 1991).

15. W.B. Chwirot, New indications of possible role of DNA in ultraweak photon emission from biological systems. *J. Pl. Physiol.* **122** (1986) p. 81-86.

16. W.B. Chwirot and R.S. Dygdala, Ethidium Bromide-Induced Changes in Intensity of Ultraweak Photon Emission from Microsporocytes of Larch in Selected Stages of Development. *J. Pl. Physiol.* **134** (1989) p. 762-765.

17. B. Ruth and F.A. Popp, Experimentelle Untersuchungen zur ultraschwachen Photonenemission biologischer Systeme. *Z. Naturforsch. C* **31** (1976) p. 741-745.

18. T.E. Quickenden and S.Q. Hee, The Spectral distribution of the luminescence emitted during growth of the yeast *Saccharomyces cerevisiae* and its relationship to mitogenetic radiation. *Photochem. Photobiol.* **23** (1976) p. 201-204.

19. F.A. Popp, *Bericht an Bonn* (Verlag fuer Ganzheitsmedizin, Essen, 1986).

20. W.B. Chwirot and R.S. Dygdala, Light transmission of scales covering male in fluorescence and leaf buds in larch during microsporogenesis. *J. Pl. Physiol.* **125** (1986) p. 79-86.

21. F.A. Popp, B. Ruth, W. Bahr, J. Grass, P. Grolig, M. Rattemeyer, H.G. Schmidt and P. Wulle, Emission of visible and ultraviolet radiation by active biological systems. *Coll. Phenomena* **3** (1981) p. 187-214.

22. F.A. Popp, On the Coherence of Ultraweak Photonenemission from Living Tissues, in *Disequilibrium and Self-Organisation*, ed. C.W. Kilmister (D. Reidel P.C., Dordrecht 1986).

23. F.A. Popp, Principles of quantum biology as demonstrated by ultraweak photon emission from living cells. *Proc. Int. Conf. Lasers'85*, Las Vegas 1985 (STS Press, Lean, Mc., Va, 1985) p. 311-316.

24. W.B. Chwirot, R.S. Dygdala, and S. Chwirot, Optical coherence of white-light-induced photon emission from microsporocytes of *Larix europea*. *Cytobios* **44** (1985) p. 239-249.

25. W.B. Chwirot, R.S. Dygdala and S. Chwirot, Quasi-monochromatic-light-induced photon emission from microsporocytes of larch shows oscillationing decay behaviour predicted by the electromagnetic model of differentiation. *Cytobios* **47** (1987) p. 137-146.

26. R. Mathew and S. Rumar, The non-exponential decay pattern of the weak luminescence from seedlings in *Cicer arietinum L.* stimulated by pulsating electric fields. *Experientia*, in press.

27. J. Slawinski, E. Grabikowski and I. Majchrowicz, *Biophoton Physics I* (1978) p. 77 (ISBN 3-922372-01-5).

28. A. Sh. Agaverdiyev, Ya. Ye. Doskoch and B.N. Tarusov, *Biophysics* **10** (1965) p. 920-924.

29. R.N. Thomas, *Some Aspects of Non-Equilibrium Thermodynamics in the Presence of a Radiation Field* (Boulder, Colorado, 1961).

30. A.L. Lehninger, *Biochemie* (Verlag Chemie, Weinheim, 1975) p. 156.

31. G. Cilento, Electronic Excitation in Dark Biological Processes, in *Chemical and Biological Generation of Excited States*, eds. W. Adam and G. Cilento (Academic Press, New York, 1982).

32. M. Planck, Ueber irreversible Strahlungsvorgaenge. *Ann. d. Phys.* **1** (1900) p. 69-122.

33. H. Gamo, Matrix Elements of Partial Coherence, in *Progress in Optics* **3**, ed. E. Wolf (North-Holland, P.C., 1964) p. 289-293.

34. P. Vigny and M. Duquesne, On the fluorescence properties of nucleotides and polynucleotides at room temperature, in *Excited States of Biological Molecules*, ed. J.B. Birks (J. Wiley, London, 1976).

35. J.B. Birks, Excimers. *Rep. Progr. Phys.* **38** (1975) p. 903-974.

36. F.A. Popp and J. Deny, Biophotonen-Information und Chaostheorie, in *Ganzheitsmedizin. Zweiter Wiener Dialog*, ed. H. Stacher, (Facultas-Universitaetsverlag, Wien 1991) p. 53-66.

37. F.A. Popp, Coherent Photon Storage of Biological Systems, in *Electromagnetic Bio-Information, 2nd ed.*, eds. F.A. Popp, U. Warnke, H.L. Koenig and W. Peschka (Urban & Schwarzenberg, Muenchen, 1989) p. 144-167.

38. E. Schroedinger, Der stetige Uebergang von der Mikro- zur Makromechanik. *Naturwissenschaften* **28** (1926) p. 664-666.

39. R.J. Glauber, Coherent and Incoherent States of the Radiation Field. *Phys. Rev.* **131** (1963) p. 2766-2788.

40. R.J. Glauber, Coherence and Quantum Detection, in *Quantum Optics*, ed. R.J. Glauber (Academic Press, New York, 1969) p. 15-56.

41. F.T. Arecchi, Photocount Distributions and Field Statistics, in *Quantum Optics*, ed. R.J. Glauber (Academic Press, New York, 1969) p. 57-110.

42. J. Perina, *Coherence of Light* (D. Reidel P.C., Dordrecht, 1971).

43. H.R. Christen, *Grundlagen der organischen Chemie* (Diesterweg, Frankfurt, 1970).

44. P. Tombesi and E.R. Pike (eds.), Squeezed and Nonclassical Light. *NATO ASI series. B: Physics* **Vol. 190** (Plenum Press, New York 1989).

45. E.R. Pike and H. Walther (eds.), *Photons and Quantum Fluctuations.* (Adam Hilger, Bristol, 1988).

46. Q. Gu, Novel Features in Photon Statistics from a Single-Mode Laser. *Science in China A* **33** (1990) p. 1460-1468.

47. B. Yoda, R. Abe, Y. Goto, A. Sacki, C. Takyu and H. Inaba, Spontaneous chemiluminescence os smoker's blood, in *Analytical Applications of Bioluminescence and Chemiluminescence*, eds. L.J. Kricka et. al. (Academic Press, London, 1984) p. 587-590.

48. X. Shen, X. Han, J. Tian, F. Zhao, L. Xu and X. Li, Spontaneous luminescence from soybean Rhizobium bacteroids. *FEMS Microbiology Letters* **81** (1991) p. 335-340.

Chapter 2

Hyperbolic Relaxation as a Sufficient Condition of a Fully Coherent Ergodic Field

Fritz-Albert Popp and Ke-hsueh Li

Abstract

In three cases, one originating from a classical model, the second from the time-evolution operator, and the third from photocount statistics, it is shown that an initially excited coherent field which remains coherent in time development relaxes according to a hyperbolic rather than to an exponential law.

This has particular relevance for the analysis of biological systems.

2.1 Introduction

It is well known that relaxing ergodic systems, subject to a linear dependence of the intensities $\dot{n}$ upon the number of radiating atoms or molecules n, display an exponential decay law $n \propto exp(-\lambda t)$, where λ is the decay constant and t the time. If the coupling becomes non-linear, e.g. $\dot{n} \propto -n^2$, it is obvious that a hyperbolic law of the form $n \propto \frac{1}{t}$ then represents the corresponding solution of the problem.

A variety of papers[1–7] has been devoted to the quantum description of exponential decay and its basic origins and problems. However, there is no investigation at present which reveals the basis of the hyperbolic relaxation within the framework of quantum theory.

In addition to a semiclassical approach[7] three striking cases are analysed here, in order to show that a hyperbolic relaxation of an ergodic system is sufficient (but not necessary) for keeping its coherence, while an exponential one is necessary (but not sufficient) for a chaotic field.

2.2 A Classical Model

Let us turn to the most simple case of an oscillator of amplitude $x(t)$. Instead of a constant damping λ, we introduce more generally a time-dependent damping factor $\lambda(t)$ requiring that the eigenfrequency ω_0 of the system shall remain exactly constant under relaxation.

Of course, the exponential relaxation is a consequence of a constant damping λ which changes the eigenfrequency ω_0 of the system to $\omega = \sqrt{\omega_0^2 - \lambda^2}$. The damping energy is dissipated into heat. However, a constant eigenfrequency ω_0 provides at the same time a constant degree of coherence in terms of the visibility of the interference fringes. Systems with constant eigenfrequency ω_0 and combinations of them can communicate by means of frequency modulations independent on their amplitudes. Actually, by use of the equation

$$\ddot{x}(t) + 2\lambda(t)\dot{x}(t) + \omega_0^2 x(t) = 0\,, \tag{2.1}$$

where $2\lambda(t)$ describes the time-dependent damping of the system, $x(t)$ can be separated into an oscillating part $y(t)$ that keeps the frequency ω_0 stable, and a decaying fraction $exp(-\int \lambda(t)\,dt)$:

$$x(t) = y(t)\,exp\left(-\int \lambda(t)\,dt\right). \tag{2.2}$$

After insertion of (2.2) into (2.1) we obtain

$$\ddot{y} + (\omega_0^2 - \lambda^2 - \dot{\lambda})\,y = 0. \tag{2.3}$$

Stability of ω_0 requires

$$\lambda^2 = -\dot{\lambda}. \tag{2.4}$$

The requirement that the frequency ω_0 is independent on the changes of oscillator amplitude y is just in consistency to the classical coherence consideration by Schrödinger (1926)[8].

The solution of Eq. (2.4) reads

$$\lambda(t) = \frac{\lambda_0}{1 + \lambda_0 t}, \tag{2.5}$$

where λ_0 is a constant.

Consequently, the solution of (2.1) takes under the constraint of (2.5) the final form

$$x(t) = \frac{y(t)}{1 + \lambda_0 t}, \tag{2.6}$$

where $y(t)$ is periodic in time t with frequency ω_0.

Eq. (2.6) describes a hyperbolic relaxation behaviour.

2.3 Time-Evolution

As Fonda et al. (1978)[5] have shown, exponential decay is one of the possible consequences of the semigroup law

$$< A(t_1)A(t_2) > = < A(t_1 + t_2) > \tag{2.7}$$

of the time-evolution operator $A(t) = exp(-\frac{i}{\hbar} \int^t H(t')\, dt')$ for chaotic fields. Actually, the solution of (2.7) is exponential: $\langle A(t) \rangle \propto exp(-\lambda t)$, where the real value of λ follows from the unitarity of $A(t)$[5].

Let us now start with the number of photons of a coherent field of amplitude $\kappa\, \alpha$, where κ shall represent a real parameter, for instance a real function of time, while α accounts for a complex function of time.

Obviously we have

$$n = \kappa^2 \, |\alpha|^2. \tag{2.8}$$

In terms of the displacement operator $D(\kappa\alpha)$ [9] Eq. (2.8) has to be rewritten as

$$n = \langle 0|\, D^*(\kappa\alpha)\, a^+ a\, D(\kappa\alpha)\, |0\rangle \,, \tag{2.9}$$

where $|0\rangle$ is the vacuum state, and a^+, a are the creation and annihilation operators, respectively. They are subject to

$$[a, a^+] = 1 \,, \tag{2.10}$$

and a^+a shall not explicitly depend on κ.

From (2.9) we obtain immediately

$$\frac{\partial n}{\partial \kappa} = \langle 0| \left(\frac{\partial}{\partial \kappa} D^*(\kappa\alpha)\right) a^+ a\, D(\kappa\alpha)\, |0\rangle + \langle 0|\, D^*(\kappa\alpha)\, a^+ a \left(\frac{\partial}{\partial \kappa} D(\kappa\alpha)\right) |0\rangle. \tag{2.11}$$

Using Glauber's derivation (1963)[9]

$$\frac{\partial}{\partial \kappa} D(\kappa\alpha) = (\alpha a^+ - \alpha^* a) D(\kappa\alpha) \tag{2.12}$$

and

$$\frac{\partial}{\partial \kappa} D^*(\kappa\alpha) = D^*(\kappa\alpha)(\alpha^* a - \alpha a^+),$$

we then get from (2.11)

$$\frac{\partial n}{\partial \kappa} = \langle 0|\, D^*(\kappa\alpha)((\alpha^* a - \alpha a^+) a^+ a + a^+ a(\alpha a^+ - \alpha^* a))\, D(\kappa\alpha)\, |0\rangle. \tag{2.13}$$

For $(\alpha^* a - \alpha a^+) a^+ a + a^+ a(\alpha a^+ - \alpha^* a)$ we can write $\alpha a^+ + \alpha^* a$ on account of (2.10). Consequently

$$\frac{\partial n}{\partial \kappa} = \langle 0|\, D^*(\kappa\alpha)\, (\alpha a^+ + \alpha^* a)\, D(\kappa\alpha)\, |0\rangle \,. \tag{2.14}$$

Mehta et al. (1967)[10] have shown that a coherent state remains always coherent, when the Hamiltonian $H_c(t)$ has the following form:

$$H_c(t) \;=\; \omega(t)a^+a \;+\; (f(t)a^+ \;+\; f^*(t)a) \;+\; \beta(t)\,, \tag{2.15}$$

where $\omega(t)$ and $\beta(t)$ are arbitrary real functions of t, while $f(t)$ is an arbitrary complex function of t. Any term on the r.h.s. of (2.15) accounts for the conservation of coherence.

Consequently, we are free in choosing $f(t)$ for $\omega \neq 0$ such that

$$f(t)a^+ \;+\; f^*(t)a \;=\; \omega(t)\mu(t)(\alpha a^+ \;+\; \alpha^* a). \tag{2.16}$$

After substitution of (2.15) into (2.14), by taking account of (2.16), we get

$$\frac{\partial n}{\partial \kappa} \;=\; \frac{1}{\omega(t)\mu(t)}\,\langle 0|\; D^*(\kappa\alpha)((H_c(t) \;-\; \beta(t) \;-\omega(t)a^+a)\; D(\kappa\alpha)\;|0\rangle, \tag{2.17}$$

where we have as well on the l.h.s. as on the r.h.s. of (2.17) only real functions.

Since

$$\langle 0|D^*(\kappa\alpha)\,(H_c(t) - \beta(t))\; D(\kappa\alpha)|\,0\rangle \;=\; \frac{\epsilon_c(t) - \beta(t)}{n(t)}\, n(t)\,,$$

where $\epsilon_c(t) = \langle\kappa\alpha|H_c(t)|\kappa\alpha\rangle$, equation (2.17) can be written as

$$\begin{aligned}\frac{\partial n}{\partial \kappa} \;&=\; \frac{1}{\mu(t)\omega(t)}\,(\frac{\epsilon_c(t) - \beta(t)}{n(t)} - \omega(t))\langle 0|\; D^*(\kappa\alpha)\; a^+a\; D(\kappa\alpha)\;|0\rangle \\ &=\; -\frac{1}{\gamma(t)}\, n(t).\end{aligned} \tag{2.18}$$

We are not completely free in the choice of $\gamma(t)$. Of course, the ergodicity condition requires that the ensemble average $\langle n(\bar{t})\rangle$ at any instant $\bar{t}$ equals the time average according to

$$\langle n(\bar{t})\rangle \;=\; \lim_{t_1,t_2\to\infty} \frac{1}{t_1 + t_2}\int_{\bar{t}-t_1}^{\bar{t}+t_2} n(t)\, dt. \tag{2.19}$$

Eq. (2.19) holds obviously for a stationary field $\langle n(\bar{t})\rangle = constant$ for any t_1 and t_2.

Let us provide that (2.19) shall be fulfilled also for the relaxing coherent field after excitation. n shall be a function of t only, which means that we can make the special choice $\kappa = t + t_1$ in (2.18), in order to evaluate the decay kinetics $\frac{\partial n}{\partial \kappa} \equiv \frac{\partial n}{\partial t} = \frac{dn}{dt}$ under the condition (2.19).

After insertion of (2.18) into (2.19) and taking $d\kappa = dt$, we then have

$$\langle n(\bar{t})\rangle = -\lim_{t_1,t_2\to\infty} \frac{1}{t_1+t_2} \int_{\bar{t}-t_1}^{\bar{t}+t_2} [\gamma(t)\frac{dn}{dt}]\, dt. \tag{2.20}$$

After integration by parts we obtain

$$\begin{aligned}\langle n(\bar{t})\rangle &= -\lim_{t_1,t_2\to\infty} \frac{1}{t_1+t_2}[\gamma(t)n(t)]_{\bar{t}-t_1}^{\bar{t}+t_2} \\ &+ \lim_{t_1,t_2\to\infty} \frac{1}{t_1+t_2} \int_{\bar{t}-t_1}^{\bar{t}+t_2} (\frac{d\gamma}{dt})\, n(t)\, dt. \end{aligned}\tag{2.21}$$

Since $\langle n(\bar{t})\rangle$ shall not depend on the special choice of t_1, t_2 even if t_1, t_2 remain finite, the solution of $\gamma(t)$ has to satisfy the condition

$$[\gamma(t)n(t)]_{\bar{t}-t_1}^{\bar{t}+t_2} = \{\gamma(t)n(t)\}_{\bar{t}+t_2} - \{\gamma(t)n(t)\}_{\bar{t}-t_1} = 0. \tag{2.22}$$

The equality of (2.21) and (2.19) at any t_1 and t_2 requires then

$$\frac{d\gamma}{dt} = 1$$

or

$$\gamma(t) = t + t_0 = \kappa + (t_0 - t_1)\,, \tag{2.23}$$

where t_0 and t_1 are arbitrary constant times.

Consequently, for an ergodic field, the relaxation of $n(t)$ after excitation satisfies the condition (2.18) under the constraint of (2.19):

$$\frac{dn(t)}{dt} = -\frac{1}{t+t_0} n(t). \tag{2.24}$$

Eq. (2.24) results in a hyperbolic relaxation

$$n(t) = \frac{n(0)\, t_0}{t+t_0}, \tag{2.25}$$

where $n(0)$ is the photon number at time $t = 0$.

The condition $[\gamma(t)n(t)]_{t-t_1}^{t+t_2} = 0$ according to (2.22) is as well satisfied. Thus, the solution is self-consistent.

Equation (2.25) is sufficient for coherent relaxation under ergodic conditions, since the selection of a special case (defined by (2.19) and the choice $\kappa = t + t_0$) from a collection of possible cases which do not overlap with chaotic (exponential) relaxation involves a sufficient but certainly not a necessary condition. This means that if a hyperbolic relaxation under ergodic conditions has been registered, the decay products can originate only from a fully coherent field, while the measurement of an exponential decay function cannot necessarily be assigned to a chaotic field. On the other hand, a chaotic field which is an eigenstate of a^+a under ergodic conditions decays always according to an exponential function[5].

It should be mentioned that in contrast to the classical oscillator for this solution it is not more necessary to keep the frequency constant.

In (2.24) the real time t can be assigned to the coherence time τ. As long as the relaxation of a fully coherent field takes place, there is no difference between τ and the real time t, while for a chaotic field $t+t_0$ on the r.h.s. of (2.24) has to be replaced by the reciprocal of the decay constant[11].

In order to show that (2.25) reflects a limiting case like that of the equality in case of the Schwarz inequality equation, let us briefly discuss the semigroup law (2.7). It can be reformulated under the constraint of a fully coherent field in such a way that again a hyperbolic law is obtained.

Of course, if Eq. (2.7) is replaced by

$$A(t_1)A(t_2) = A\left(\frac{t_1+t_2}{2}\right)\frac{1}{2}(A(t_1)+A(t_2)), \tag{2.26}$$

the exponential solution turns for $t_1 \neq t_2$ into the hyperbolic $A(t) = \frac{c}{t}$, where c is a constant:

$$\frac{c}{t_1}\frac{c}{t_2} = \left(\frac{2c}{t_1+t_2}\right)\frac{1}{2}\left(\frac{c}{t_1}+\frac{c}{t_2}\right).$$

Because of

$$A(t_1)A(t_2) = A\left(\frac{t_1+t_2}{2}\right)A\left(\frac{t_1+t_2}{2}\right), \tag{2.27}$$

the change from (2.7) to (2.26) corresponds to the substitution of

$$A\left(\frac{t_1+t_2}{2}\right) = \frac{1}{2}(A(t_1)+A(t_2)). \tag{2.28}$$

By use of the identities:

$$A\left(\frac{t_1+t_2}{2}\right) = \sqrt[2]{A(t_1+t_2)} = \sqrt[2]{A(t_1)A(t_2)},$$

we can see from (2.28) that the geometrical mean $\sqrt[2]{A(t_1)A(t_2)}$ has then to be substituted by its arithmetic mean $\frac{1}{2}(A(t_1) + A(t_2))$, in order to turn from (2.7) to (2.26). This reminds again on the ergodicity condition. Of course, the average value $\bar{t} = \frac{t_1+t_2}{2}$ can always be fixed, while t_1 and t_2 may take arbitrary values. Consequently, (2.28) can be interpreted as an adjustment in such a way that the ensemble average at a definite time $\bar{t} = \frac{t_1+t_2}{2}$ equals its time-average.

By the way, this includes also the case of "destructive interference". Of course, since the displacement operator $D(\gamma(t))$ represents $A(t)$ in the coherent-state representation, the replacement of

$$D(0) = \sqrt[2]{D(\gamma)D(-\gamma)}$$

by its arithmetic mean $\frac{1}{2}(D(\gamma) + D(-\gamma))$ provides that the vacuum state will be obtained, if the field amplitudes γ and $-\gamma$ are superimposed within the coherence volume. This perfect coherence indicates again that the hyperbolic relaxation is sufficient for a fully coherent ergodic field.

2.4 Photocount Statistics (PCS)

The following proof of hyperbolic relaxation for a fully coherent ergodic field is based on the well-known fact that a fully coherent stationary field displays always a Poissonian distribution of the probability $p(n, \Delta t)$ of registering n photons in an arbitrarily small, but fixed time interval Δt. A chaotic field, on the other hand, follows a geometrical distribution of $p(n, \Delta t)$. If $p(n, \Delta t)$ remains Poissonian at any instant t, where the probability distribution shall generally account for an ensemble of identical samples, it is evident that this is then sufficient for a fully coherent ergodic field, even if it is not stationary.

Following Loudon (1983)[12], we introduce the probability $p(t)\, dt$ of registering a particle of the decay product (for instance, a photon), originating from a relaxing

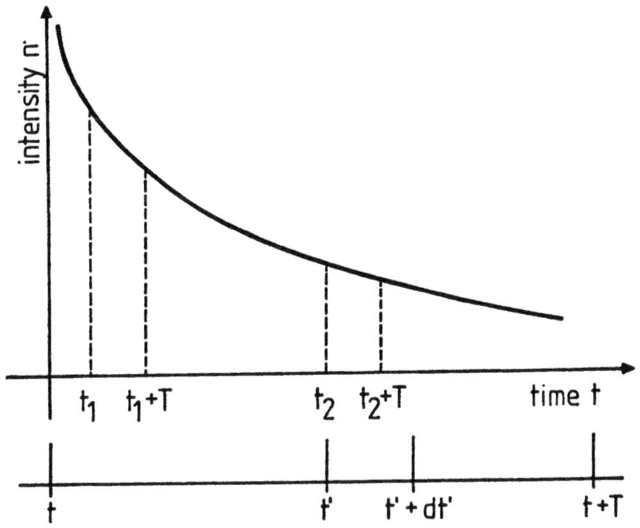

Figure 2.1: *A typical relaxation of an excited field, where the intensity $\dot{n}(t)$ decays. The number of particles of the decay product ($\dot{n}T$) is always registered between time intervals t_i and t_i+T, $i=1,2,\cdots$. These intervals can be divided into smaller ones t'_j and t'_j+dt'; $j=1,2,\cdots N$, where either no particle or only one can be counted.*

field, within a small time interval between t and $t+dt$. The measurement of the number of particles $\langle \dot{n}T\rangle$ is always performed within the fixed time interval t and $t+T$, where $T > dt$ (see Fig. 2.1).

According to Loudon[12], we have generally the relation

$$P_0(t,T) = exp(-\int_t^{t+T} p(t')\,dt'), \tag{2.29}$$

where $P_0(t,T)$ is the probability of registering no photon in the interval t and $t+T$. At any instant t, P_0 is normalized: $P_0(t,0) = 1$. The probability of measuring no photon between t' and $t'+dt'$ is then (see Loudon[12], or expand the r.h.s of (2.29)):

$$P_0(t',dt') = 1 - p(t')\,dt'. \tag{2.30}$$

An ergodic coherent field displays (in contrast to chaotic fields) no bunching. This means that the probability of registering no photon within t and $t+T$ is, according to the probability theory of independent processes, the product of all the N single probabilities $(1-p(t')\,dt')$ of measuring no photon in the consecutive small time intervals dt' between t and $t+T$, where $N\,dt' = T$ (see Fig. 2.1):

$$P_0(t,T) = (1-p(t'_1)\,dt')\,(1-p(t'_2)\,dt')\cdots(1-p(t'_N)\,dt'). \tag{2.31}$$

Since the $(1-p(t'_i)\,dt')$ *for* $i = 1,2,\cdots N$ are positive real numbers, we then have

$$1-p(t'_m)\,dt' \le \sqrt[N]{\prod_i (1-p(t'_i)\,dt')} \le 1-\frac{1}{N}\sum_i p(t'_i)\,dt', \tag{2.32}$$

where $p(t'_m)$ represents the maximum of $p(t')$ within t and $t+T$. (2.32) reflects the fact that the geometrical mean is always lower or at most equal to the arithmetic mean. According to the mean value theorem we then can rewrite (2.31), taking account of (2.32)

$$P_0(t,T) = [1-p(t+\mu T)\,\frac{T}{N}]^N, \tag{2.33}$$

where $dt = \frac{T}{N}$, and μ with $0 \le \mu \le 1$ is a time-independent quantity, since within the interval T the measurements do not allow the time resolution of $\bar{n}(t)$.

Consequently, from Eqs. (2.28) and (2.32), we get

$$(1-p(t+\mu T)\,\frac{T}{N})^N = exp(-\int_t^{t+T} p(t')\,dt'). \tag{2.34}$$

The time-dependent solution for $\mu = 1$ takes the form

$$p(t) = \frac{N}{t+t_0}, \tag{2.35}$$

where t_0 is a constant. The special choice of $\mu = 1$ cannot have any impact on the physical significance, since T is optional. (2.35) represents again a hyperbolic law.

We may conclude as follows: If a relaxing field keeps its Poissonian PCS-distribution at any instant, it decays according to a hyperbolic law. If, on the other hand, an ergodic field relaxes according to a hyperbolic function, one has to conclude that the field is a fully coherent one.

This hyperbolic relaxation seems to be a rather common phenomenon in biological systems, e.g. the white-light induced reemission of photons from cells[13,14].

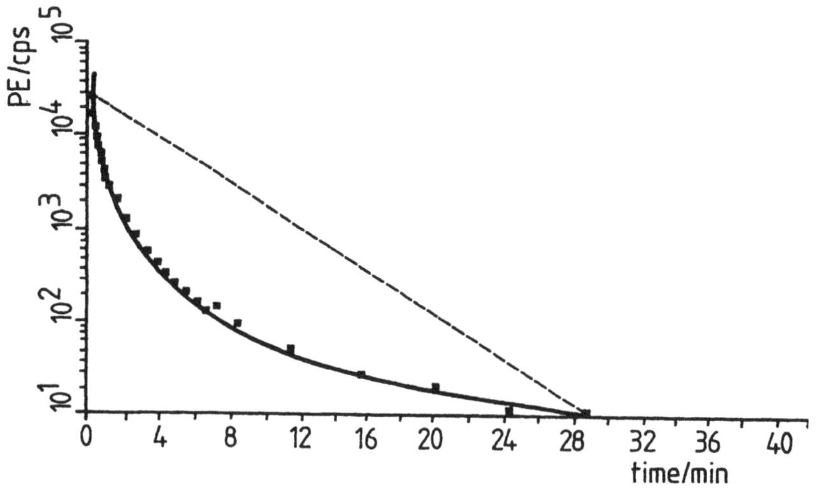

Figure 2.2: *"Delayed luminescence" of a* Bryophyllum daigremontanum *leaf at a wavelength of* 676 ± 10*nm*[13]*. The measured count rate after exposure to a red-light illumination can be approximated by a hyperbolic decay law according to Eq. (2.25) (lower curve), while an exponential decay (dashed line), or even a superposition of three exponential functions with different decay constants cannot be adjusted.*

Fig. 2.2 demonstrates this for the spectral reemission of a leaf of the plant *Bryophyllum daigremontanum* after illumination with red light of a wavelength of 676 ± 10nm. The exponential function (dashed line) evidently cannot account for this decay, but the result clearly indicates coherent rescattering of light within living systems. This has also been indicated by "light-piping in plant tissues"[15], which has been traced back to a high degree of coherence[16].

Hyperbolic relaxation may thus become a powerful tool for analysing the living state in terms of coherence[11]. This has been confirmed also by use of canonical coherent states in order to describe the long-range forces between human blood cells[17] and the long-range phase coherence in the bacteriorhodopsin macromolecules[18].

2.5 References

1. V. Weisskopf and E.Wigner, *Z. Phys.* **63** (1930) p. 54; **65** (1931) p. 18.

2. W. Heitler and S.T. Ma, *Proc. Roy. Irish Acad.* **52** (1949) p. 109.

3. I. Ersak, *J. Sov. Nucl. Phys.* **9** (1969) p. 263.

4. E.B. Davies, *Helv. Phys. Acta* **48** (1975) p. 365.

5. L. Fonda, G.C. Ghirardi and A. Rimini, *Rep. Progr. Phys.* **41** (1978) p. 99.

6. M. Bunge and A.J. Kalnay, *Nuovo Cimento B* **77** (1983) p. 1.

7. K.H. Li and F.A. Popp, *Phys. Lett. A* **93** (1983) p. 626.

8. E. Schrödinger, *Die Naturwissenschaften* **28** (1926) p.665.

9. R.J. Glauber, *Phys. Rev.* **131** (1963) p. 2766.

10. C.L. Mehta, P. Chand, E.C.G. Sudarshan, and R. Vedam, *Phys. Rev.* **157** (1967) p. 1198.

11. F.A. Popp, in *Disequilibrium and Self-Organisation,* ed. C.W. Kilmister (D. Reidel Publishing Company, Dordrecht, 1986).

12. R. Loudon, *The Quantum Theory of Light*, 2nd edition (Clarendon Press, Oxford 1983) p. 229-232.

13. W. Scholz, U. Staszkiewicz, F.A. Popp, and W. Nagl, *Cell Biophysics* **13** (1988) p. 55.

14. F.A. Popp, B. Ruth, W. Bahr, J. Boehm, P. Grass, G. Grolig, M. Rattemeyer, H.G. Schmidt, and P. Wulle, *Collective Phenomena* **3** (1981) p. 187.

15. D.F. Mandoli, and W. Briggs, *Proc. Natl. Acad. Sci. USA* **79** (1982) p. 2902.

16. H. Smith, *Nature* **298** (1982) p. 423.

17. R. Paul, *Phys. Lett. A* **96** (1983) p. 263.

18. L.J. Dunne, A.D. Clark, and E.J. Brandas, *Chem. Phys. Lett.* **97** (1983) p. 573.

Chapter 3

Quantum Theory of Biophoton Emission

Qiao Gu

Abstract

Quantum theory of biophoton emission from a system of N biomolecules according to the Dicke model is presented by taking account of the interaction of collective biomolecules with both "pumping field" and biophoton field. Analytical expressions describing the steady-state behaviour of both the collective biomolecules and the emission field are derived, such as populations of biomolecules in all levels and their fluctuations, emission intensities, intensity correlations, and spectral distributions. Dynamic evolution of the system after excitation is also investigated and a set of expressions for the decay of the collective biomolecules and the corresponding emission rate is obtained.

The analytical results predict, in the limit $N \rightarrow \infty$, a nonequilibrium first-order phase transition of all quantities under consideration, with respect to a parameter x related to all factors of the system. Critical point $x = 1$ of the phase transition marks a transition of the working point of the system from "cooperative region" ($x < 1$) to "individual region" ($x > 1$), or from "ordered state" to "chaotic state". The individual region corresponds to a group of molecules without correlation to each other, whereas the biomolecules in the cooperative region display the long-range and long-time cooperative effects in all their aspects. In general, living systems may be expected to work within a very narrow region of variation of the parameter x around $x = 1$.

Finally, a comparison of the theoretical results with the corresponding experimental observations from several biological samples is shown.

3.1 Introduction: Cooperativity

A lot of research fields are concerned with composite systems consisting of many similar subsystems, e.g., a strong magnet made up of a large number of spins, a laser medium comprising a large number of electric dipoles, a polymer composed of a large number of molecules, a living tissue made up of a large number of cells, a forest of countless trees, or a society of numerous people. In most cases the subsystems interact with one another, so that it is difficult to explain the macroscopic qualities of a composite system by a simple summation of the microscopic qualities of the individual subsystems. In fact, all the subsystems are integrated through a correlation that results in a composite system with a certain macroscopic order. In other words, the composite system may be considered to be constructed in such a way that every subsystem seems to obtain an "information" on the total state of the composite system [1]. This is often termed the cooperative effect [2].

The cooperative effect can also occur in a radiation process from a system of many atoms (or molecules), which is called cooperative emission [3]. Two typical examples are superradiance and superfluorescence [4–27]. Superradiance is a cooperative emission from a system with initial correlation between the dipoles belonging to different atoms. In contrast, superfluorescence originates from a system without initial correlation. It starts with normal fluorescence, and then a correlation develops, this being induced by the "noisy", spontaneously emitted photons of normal fluorescence. Cooperativity of the two phenomena lies essentially in the same fact that a spontaneous phase-locking of the atomic dipoles leads to an emitted intensity proportional to the square of the total number N of atoms. In contrast, independent emission of spontaneous photons would give a linear dependence of emitted intensity on N. The cooperative effect in which atoms behave collectively to produce processes is very different from that exhibited by a group of atoms without correlation to each other. The latter essentially displays a single-atom behaviour [28].

It is evident that in living systems the cooperative effect can play an important role in retaining the activity of life [29]. Szent-Györgyi already pointed out that we can only begin to understand the characteristics of living systems if we take into account the collective properties of molecular aggregates [30]. Fröhlich already predicted that there are long-range coherence and energy storage in biological systems [31]. One of the indications of coherence in biological systems is the coherent biophoton emission [32–41]. Coherence of biophoton emission may arise from the cooperative action of collective biomolecules. Thus, the cooperativity of living systems inspire us to present here a quantum theory of biophoton emission.

In this paper, on the basis of biomolecular exciplex formation, we consider a system of collective biomolecules interacting with "pumping field" and biophoton field, according to the Dicke model[4]. We describe the system using the quantum mechanics master equations. Some analytical expressions for describing the steady-state properties of collective biomolecules and biophoton field are derived, such as the populations of biomolecules in all levels and their fluctuations, emission intensities, intensity correlations, and spectral distributions. We also study the dynamic evolution of the system after excitation and obtain a set of expressions for the biomolecular decay and the corresponding emission rate. Finally, we show the comparison between the theoretical results and the corresponding experimental observations on biophoton emission from several biological samples.

3.2 Exciplex Model and Three-Level System

Firstly, let us consider a biomolecular system with exciplex formation [42,43]. Fig. 3.1(a) gives a representation of its potential energy profiles as a function of the separation of monomers. The potential curve Σ^* represents the bound excited state and Σ represents the repulsive ground state. The radiative behaviour of such a system may be simplified to that of a four-level laser (the exciplex model [33,39–41]), as shown in Fig. 3.1(b). With the help of the exciplex model, the biophoton emission process may be understood as follows.

The molecules from the lowest level $|\alpha_1\rangle$ are pumped into state $|\alpha_e\rangle$, which is an arbitrary excited state confined within potential curve Σ^*. The pumping energy supply is being maintained by the metabolism (glycolysis, ATP, etc.) and even direct pumping with sunlight being an inexhaustible source [39]. One may imagine the energy supply to be from a "pumping field". From state $|\alpha_e\rangle$ the molecules then decay very rapidly through a nonradiative transition to state $|\alpha_2\rangle$ which is a metastable state having a long lifetime. Thus the pump effectively transfers molecules from $|\alpha_1\rangle$ to $|\alpha_2\rangle$ through $|\alpha_e\rangle$. From state $|\alpha_2\rangle$ the molecules decay by photon emission to state $|\alpha_3\rangle$. One can see from the exciplex model that the main biophoton emission follows

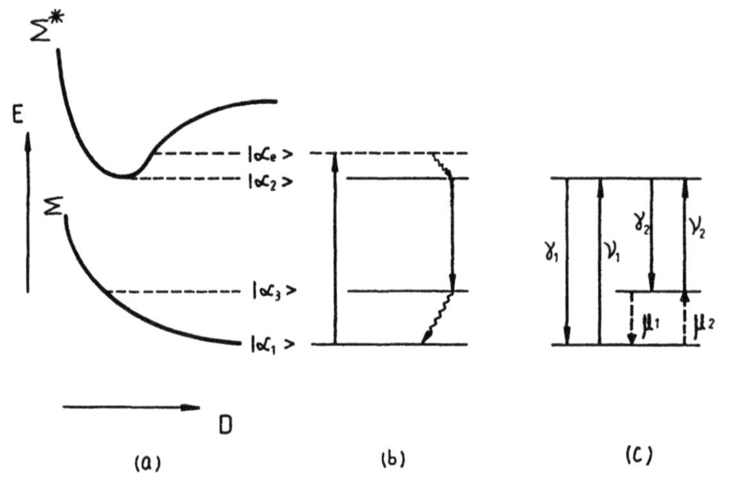

Figure 3.1: *(a) Representation of potential energy profiles of a biomolecular system with exciplex formation, where Σ^* and Σ represent the bound excited state and the repulsive ground state, E: exciplex energy, D: distance of monomers; (b) The exciplex model; (c) Simplified three-level system.*

from the transition from $|\alpha_2\rangle$ to $|\alpha_3\rangle$. However, in general, the radiative transition from $|\alpha_2\rangle$ to $|\alpha_1\rangle$ is also possible. After the molecules arrive in state $|\alpha_3\rangle$, most of them may relax down through a nonradiative transition into state $|\alpha_1\rangle$, ready for being pumped again to $|\alpha_e\rangle$, while the rest may be pumped immediately by another "pumping field" to $|\alpha_e\rangle$. Note that $|\alpha_3\rangle$ should be considered as an arbitrary state on potential curve Σ, which can take all possible values of vibrational energy of the lattice system of biomolecules [39,41].

According to the above-mentioned understanding of biophoton emission, it seems to indicate a four-level system. However, the four-level system can usually be simplified to a three-level system. In fact, the transition $|\alpha_1\rangle \rightarrow |\alpha_e\rangle \rightarrow |\alpha_2\rangle$ under consideration can be effectively simplified to $|\alpha_1\rangle \rightarrow |\alpha_2\rangle$. So, in our approach only three levels ($|\alpha_1\rangle$, $|\alpha_2\rangle$ and $|\alpha_3\rangle$) need to be taken into account, as shown in Fig. 3.1(c).

Now we consider a system of N identical three-level molecules, which occupy a region small compared to the wavelength of all the relevant radiation modes (the Dicke model [4]). The molecules interact with two pumping fields E_1 and E_2 inducing the molecular transitions from $|\alpha_1\rangle$ to $|\alpha_2\rangle$ and from $|\alpha_3\rangle$ to $|\alpha_2\rangle$, respectively. In living systems enough pumping energy is always available, which permits the two pumping fields to be treated classically, so that the Hamiltonian of the system can be written in the rotating-wave approximation in the form ($\hbar = 1$) [44]

$$H = \sum_{i=1}^{3} \omega_i A_{ii} + \left(G_1 A_{12} e^{-i\Omega_1 t} + G_2 A_{32} e^{-i\Omega_2 t} + HC\right), \tag{3.1}$$

where ω_i ($i = 1, 2, 3$) represent the molecular energy levels, Ω_1 and Ω_2 are the resonant frequencies of the two pumping fields, respectively,

$$G_1 = -d_1 E_1 \tag{3.2}$$

and

$$G_2 = -d_2 E_2 \tag{3.3}$$

are the Rabi frequencies of the two pumping fields, respectively,

$$A_{ij} = \sum_{k=1}^{N} |\alpha_i\rangle_k \, {}_k\langle\alpha_j| \quad (i, j = 1, 2, 3) \tag{3.4}$$

are the collective operators of the three-level molecules, which obey the commutation relations

$$[A_{ij}, A_{kl}] = A_{il}\delta_{jk} - \delta_{li}A_{kj} , \tag{3.5}$$

and HC denotes the Hermitian conjugate terms.

In terms of Eq. (3.1), the master equation of the reduced density operator ρ of the molecular system can be expressed by [45]

$$\begin{aligned}
\frac{\partial \rho}{\partial t} = & -i[H_0, \rho] \\
& -\gamma_1(1 + \bar{n}_1)(A_{21}A_{12}\rho + \rho A_{21}A_{12} - 2A_{12}\rho A_{21}) \\
& -\gamma_1\bar{n}_1(A_{12}A_{21}\rho + \rho A_{12}A_{21} - 2A_{21}\rho A_{12}) \\
& -\gamma_2(1 + \bar{n}_2)(A_{23}A_{32}\rho + \rho A_{23}A_{32} - 2A_{32}\rho A_{23}) \\
& -\gamma_2\bar{n}_2(A_{32}A_{23}\rho + \rho A_{32}A_{23} - 2A_{23}\rho A_{32}) \\
& -\nu_1(A_{12}A_{21}\rho + \rho A_{12}A_{21} - 2A_{21}\rho A_{12})
\end{aligned}$$

$$-\nu_2(A_{32}A_{23}\rho + \rho A_{32}A_{23} - 2A_{23}\rho A_{32})$$
$$-\mu_1(A_{31}A_{13}\rho + \rho A_{31}A_{13} - 2A_{13}\rho A_{31})$$
$$-\mu_2(A_{13}A_{31}\rho + \rho A_{13}A_{31} - 2A_{31}\rho A_{13}), \tag{3.6}$$

where γ_1 and γ_2 are the coefficients of single-molecular spontaneous emission from $|\alpha_2\rangle$ to $|\alpha_1\rangle$ and from $|\alpha_2\rangle$ to $|\alpha_3\rangle$, respectively, ν_1 and ν_2 are the corresponding absorption coefficients, μ_1 and μ_2 are the coefficients of single-molecular nonradiative transitions from $|\alpha_3\rangle$ to $|\alpha_1\rangle$ and from $|\alpha_1\rangle$ to $|\alpha_3\rangle$, respectively,

$$\bar{n}_1 = \frac{1}{e^{\frac{\Omega_1}{kT}} - 1} \tag{3.7}$$

and

$$\bar{n}_2 = \frac{1}{e^{\frac{\Omega_2}{kT}} - 1} \tag{3.8}$$

are the mean photon numbers of the thermal fields at temperature T, corresponding to frequencies Ω_1 and Ω_2, respectively, k is the Boltzmann constant, and H_0 is the effective Hamiltonian given by

$$H_0 = G_1(A_{21} + A_{12}) + G_2(A_{23} + A_{32}). \tag{3.9}$$

In general, Eq. (3.6) cannot be solved exactly. In order to obtain its approximate solutions, it is convenient to introduce the dressed states, which are the normalized eigenstates of the single-molecule Hamiltonian $H_0^{(k)}$,

$$H_0^{(k)}|\beta_i\rangle_k = \lambda_i|\beta_i\rangle_k. \tag{3.10}$$

The eigenvalues λ_i are given by the roots of the cubic equation:

$$\lambda_i^3 - G^2\lambda_i = 0, \tag{3.11}$$

where

$$G = (G_1^2 + G_2^2)^{\frac{1}{2}}. \tag{3.12}$$

It is easy to obtain the eigenstates

$$\begin{aligned}
|\beta_1\rangle_k &= \frac{1}{\sqrt{2}}g_1|\alpha_1\rangle_k + \frac{1}{\sqrt{2}}|\alpha_2\rangle_k + \frac{1}{\sqrt{2}}g_2|\alpha_3\rangle_k \; (\lambda_1 = G), \\
|\beta_2\rangle_k &= g_2|\alpha_1\rangle_k - g_1|\alpha_3\rangle_k \; (\lambda_2 = 0), \\
|\beta_3\rangle_k &= \frac{1}{\sqrt{2}}g_1|\alpha_1\rangle_k - \frac{1}{\sqrt{2}}|\alpha_2\rangle_k + \frac{1}{\sqrt{2}}g_2|\alpha_3\rangle_k \; (\lambda_3 = -G),
\end{aligned} \tag{3.13}$$

where $g_1 = \frac{G_1}{G}$, $g_2 = \frac{G_2}{G}$.

Now, we can define the dressed operators $|\beta_i\rangle_k \, {}_k\langle\beta_j|$ for the kth molecule and the corresponding collective operators

$$B_{ij} = \sum_{k=1}^{N} |\beta_i\rangle_k \, {}_k\langle\beta_j|. \tag{3.14}$$

The new collective operators B_{ij} satisfy the same commutation relations as the old operators A_{ij}. Using (3.4), (3.13) and (3.14), we can express A_{ij} in terms of B_{ij} as

$$\begin{aligned}
A_{11} &= \frac{1}{2}g_1^2 B_{11} + \frac{1}{\sqrt{2}}g_1 g_2 B_{12} + \frac{1}{2}g_1^2 B_{13} \\
&\quad + \frac{1}{\sqrt{2}}g_1 g_2 B_{21} + g_2^2 B_{22} + \frac{1}{\sqrt{2}}g_1 g_2 B_{23} \\
&\quad + \frac{1}{2}g_1^2 B_{31} + \frac{1}{\sqrt{2}}g_1 g_2 B_{32} + \frac{1}{2}g_1^2 B_{33}, \\
A_{12} &= \frac{1}{2}g_1 B_{11} - \frac{1}{2}g_1 B_{13} + \frac{1}{\sqrt{2}}g_2 B_{21} \\
&\quad - \frac{1}{\sqrt{2}}g_2 B_{23} + \frac{1}{2}g_1 B_{31} - \frac{1}{2}g_1 B_{33}, \\
A_{13} &= \frac{1}{2}g_1 g_2 B_{11} - \frac{1}{\sqrt{2}}g_1^2 B_{12} + \frac{1}{2}g_1 g_2 B_{13} \\
&\quad + \frac{1}{\sqrt{2}}g_2^2 B_{21} - g_1 g_2 B_{22} + \frac{1}{\sqrt{2}}g_2^2 B_{23} \\
&\quad + \frac{1}{2}g_1 g_2 B_{31} - \frac{1}{\sqrt{2}}g_1^2 B_{32} + \frac{1}{2}g_1 g_2 B_{33}, \\
A_{22} &= \frac{1}{2}B_{11} - \frac{1}{2}B_{13} - \frac{1}{2}B_{31} + \frac{1}{2}B_{33}, \\
A_{23} &= \frac{1}{2}g_2 B_{11} - \frac{1}{\sqrt{2}}g_1 B_{12} + \frac{1}{2}g_2 B_{13}
\end{aligned} \tag{3.15}$$

$$
\begin{aligned}
& -\frac{1}{2}g_2B_{31} + \frac{1}{\sqrt{2}}g_1B_{32} - \frac{1}{2}g_2B_{33}, \\
A_{33} &= \frac{1}{2}g_2^2B_{11} - \frac{1}{\sqrt{2}}g_1g_2B_{12} + \frac{1}{2}g_2^2B_{13} \\
& -\frac{1}{\sqrt{2}}g_1g_2B_{21} + g_1^2B_{22} - \frac{1}{\sqrt{2}}g_1g_2B_{23} \\
& +\frac{1}{2}g_2^2B_{31} - \frac{1}{\sqrt{2}}g_1g_2B_{32} + \frac{1}{2}g_2^2B_{33}.
\end{aligned}
$$

It is important to point out here that under the Hamiltonian H_0, the operators B_{ij} evolve in time as

$$B_{ij}(t) = B_{ij}(0)exp[i(\lambda_i - \lambda_j)t]. \tag{3.16}$$

The diagonal operators B_{ii} then vary slowly with time, while the off-diagonal operators $B_{ij}(i \neq j)$ oscillate rapidly with time. Hence, under intense field conditions, that is, when $G >> N\gamma_1, N\gamma_2, N\nu_1, N\nu_2, N\mu_1, N\mu_2$, we can insert the relations (3.15) into Eq. (3.6) and neglect the rapidly oscillating terms (secular approximation [46]) to arrive at the approximate master equation

$$
\begin{aligned}
\frac{\partial\rho}{\partial t} = & -iG[F,\rho] \\
& -\frac{\gamma_1(1+\bar{n}_1)}{2}\Big\{[\frac{1}{2}g_1^2(F^2 + B_{13}B_{31} + B_{31}B_{13}) + g_2^2(B_{12}B_{21} + B_{32}B_{23})]\rho \\
& \quad +HC - [g_1^2(F\rho F + B_{13}\rho B_{31} + B_{31}\rho B_{13}) + 2g_2^2(B_{21}\rho B_{12} + B_{23}\rho B_{32})]\Big\} \\
& -\frac{\gamma_2(1+\bar{n}_2)}{2}\Big\{[\frac{1}{2}g_2^2(F^2 + B_{13}B_{31} + B_{31}B_{13}) + g_1^2(B_{12}B_{21} + B_{32}B_{23})]\rho \\
& \quad +HC - [g_2^2(F\rho F + B_{13}\rho B_{31} + B_{31}\rho B_{13}) + 2g_1^2(B_{21}\rho B_{12} + B_{23}\rho B_{32})]\Big\} \\
& -\frac{(\nu_1+\gamma_1\bar{n}_1)}{2}\Big\{[\frac{1}{2}g_1^2(F^2 + B_{31}B_{13} + B_{13}B_{31}) + g_2^2(B_{21}B_{12} + B_{23}B_{32})]\rho \\
& \quad +HC - [g_1^2(F\rho F + B_{31}\rho B_{13} + B_{13}\rho B_{31}) + 2g_2^2(B_{12}\rho B_{21} + B_{32}\rho B_{23})]\Big\} \\
& -\frac{(\nu_2+\gamma_2\bar{n}_2)}{2}\Big\{[\frac{1}{2}g_2^2(F^2 + B_{31}B_{13} + B_{13}B_{31}) + g_1^2(B_{21}B_{12} + B_{23}B_{32})]\rho \\
& \quad +HC - [g_2^2(F\rho F + B_{31}\rho B_{13} + B_{13}\rho B_{31}) + 2g_1^2(B_{12}\rho B_{21} + B_{32}\rho B_{23})]\Big\}
\end{aligned}
$$

$$
\begin{aligned}
&-\frac{\mu_1}{2}\Big\{[\frac{1}{2}g_1^2g_2^2(E-2B_{22})^2+\frac{1}{2}g_1^2g_2^2(B_{13}B_{31}+B_{31}B_{13})+g_1^4(B_{21}B_{12}+B_{23}B_{32}) \\
&\qquad +g_2^4(B_{12}B_{21}+B_{32}B_{23})]\rho+HC-[g_1^2g_2^2(E-2B_{22})\rho(E-2B_{22}) \\
&\qquad +g_1^2g_2^2(B_{13}\rho B_{31}+B_{31}\rho B_{13})+2g_1^4(B_{12}\rho B_{21}+B_{32}\rho B_{23}) \\
&\qquad +2g_2^4(B_{21}\rho B_{12}+B_{23}\rho B_{32})]\Big\} \\
&-\frac{\mu_2}{2}\Big\{[\frac{1}{2}g_1^2g_2^2(E-2B_{22})^2+\frac{1}{2}g_1^2g_2^2(B_{31}B_{13}+B_{13}B_{31})+g_1^4(B_{12}B_{21}+B_{32}B_{23}) \\
&\qquad +g_2^4(B_{21}B_{12}+B_{23}B_{32})]\rho+HC-[g_1^2g_2^2(E-2B_{22})\rho(E-2B_{22}) \\
&\qquad +g_1^2g_2^2(B_{31}\rho B_{13}+B_{13}\rho B_{31})+2g_1^4(B_{21}\rho B_{12}+B_{23}\rho B_{32}) \\
&\qquad +2g_2^4(B_{12}\rho B_{21}+B_{32}\rho B_{23})]\Big\}, \qquad (3.17)
\end{aligned}
$$

where E and F are defined by

$$E = B_{33} + B_{11}, \qquad (3.18)$$

and

$$F = B_{33} - B_{11}. \qquad (3.19)$$

Eq. (3.17) can be expressed in the compact form

$$
\begin{aligned}
\frac{\partial\rho}{\partial t} &= -iG[F,\rho] \\
&\quad -\frac{x_1}{2}[(B_{12}B_{21}\rho+\rho B_{12}B_{21}-2B_{21}\rho B_{12})+(B_{32}B_{23}\rho+\rho B_{32}B_{23}-2B_{23}\rho B_{32})] \\
&\quad -\frac{x_2}{2}[(B_{21}B_{12}\rho+\rho B_{21}B_{12}-2B_{12}\rho B_{21})+(B_{23}B_{32}\rho+\rho B_{23}B_{32}-2B_{32}\rho B_{23})] \\
&\quad -\frac{x_{34}}{4}[(B_{13}B_{31}\rho+\rho B_{13}B_{31}-2B_{31}\rho B_{13})+(B_{31}B_{13}\rho+\rho B_{31}B_{13}-2B_{13}\rho B_{31})] \\
&\quad -\frac{x_3}{4}(F^2\rho+\rho F^2-2F\rho F) \\
&\quad -\frac{x_4}{4}[(E-2B_{22})^2\rho+\rho(E-2B_{22})^2-2(E-2B_{22})\rho(E-2B_{22})], \qquad (3.20)
\end{aligned}
$$

where

$$\begin{aligned}
x_1 &= \gamma_1(1+\bar{n}_1)g_2^2+\gamma_2(1+\bar{n}_2)g_1^2+\mu_1 g_2^4+\mu_2 g_1^4, &(3.21)\\
x_2 &= (\nu_1+\gamma_1\bar{n}_1)g_2^2+(\nu_2+\gamma_2\bar{n}_2)g_1^2+\mu_1 g_1^4+\mu_2 g_2^4, &(3.22)\\
x_3 &= \gamma_1(1+\bar{n}_1)g_1^2+\gamma_2(1+\bar{n}_2)g_2^2\\
&\quad +(\nu_1+\gamma_1\bar{n}_1)g_1^2+(\nu_2+\gamma_2\bar{n}_2)g_2^2, &(3.23)\\
x_4 &= (\mu_1+\mu_2)g_1^2 g_2^2, &(3.24)\\
x_{34} &= x_3+x_4. &(3.25)
\end{aligned}$$

We shall now set out to derive the equations of motion for the expectation values $\langle B_{ij}\rangle$. By using

$$\begin{aligned}
\langle B_{ij}\rangle &= Tr(\rho B_{ij})\\
&= \sum_{k=1}^{N} Tr\rho|\beta_i\rangle_k\,{}_k\langle\beta_j|\\
&= \sum_{k=1}^{N} {}_k\langle\beta_j|\rho|\beta_i\rangle_k
\end{aligned}$$

and expression (3.14) as well as the commutation relations for the operators B_{ij}, the coupling equations of motion for $\langle B_{ij}(t)\rangle$ can be obtained from Eq. (3.20) in the form

$$\begin{aligned}
\frac{d}{dt}\langle E(t)\rangle &= 2x_2N-x_5\langle E(t)\rangle-x_0\langle B_{22}(t)E(t)\rangle, &(3.26)\\
\frac{d}{dt}\langle F(t)\rangle &= -x_6\langle F(t)\rangle-x_0\langle B_{22}(t)F(t)\rangle, &(3.27)\\
\frac{d}{dt}\langle B_{12}(t)\rangle &= -(x_7+iG)\langle B_{12}(t)\rangle\\
&\quad -\frac{1}{2}x_0(\langle B_{22}(t)B_{12}(t)\rangle-\langle E(t)B_{12}(t)\rangle), &(3.28)\\
\frac{d}{dt}\langle B_{32}(t)\rangle &= -(x_7-iG)\langle B_{32}(t)\rangle\\
&\quad -\frac{1}{2}x_0(\langle B_{22}(t)B_{32}(t)\rangle-\langle E(t)B_{32}(t)\rangle), &(3.29)\\
\frac{d}{dt}\langle B_{13}(t)\rangle &= -(x_8+2iG)\langle B_{13}(t)\rangle-x_0\langle B_{22}(t)B_{13}(t)\rangle, &(3.30)
\end{aligned}$$

where

$$x_0 = x_1 - x_2, \tag{3.31}$$
$$x_5 = x_1 + 2x_2, \tag{3.32}$$
$$x_6 = x_1 + x_3 + x_4, \tag{3.33}$$
$$x_7 = x_1 + \frac{1}{2}(x_2 + x_{34}) + 2x_4, \tag{3.34}$$
$$x_8 = x_1 + x_3 + \frac{1}{2}x_{34}. \tag{3.35}$$

The coupling equations of motion (3.26-3.30) will be used later to study the steady-state spectral distributions of the biophoton field and the dynamic evolutions of the system.

Now we shall solve the approximate master equation (3.20). For this purpose, let us first introduce the collective states $|m,n\rangle$, which are eigenstates of the operators $\hat{N} = A_{11} + A_{22} + A_{33} = B_{11} + B_{22} + B_{33}$, E and F,

$$\hat{N}|m,n\rangle = N|m,n\rangle, \tag{3.36}$$
$$E|m,n\rangle = m|m,n\rangle, \tag{3.37}$$
$$F|m,n\rangle = n|m,n\rangle. \tag{3.38}$$

Thus, in the states $|m,n\rangle$, $\frac{1}{2}(m-n)$ molecules are in the dressed state $|\beta_1\rangle$, $(N-m)$ molecules in the state $|\beta_2\rangle$, and $\frac{1}{2}(m+n)$ molecules in the state $|\beta_3\rangle$, i.e.

$$B_{11}|m,n\rangle = \frac{1}{2}(m-n)|m,n\rangle, \tag{3.39}$$
$$B_{22}|m,n\rangle = (N-m)|m,n\rangle, \tag{3.40}$$
$$B_{33}|m,n\rangle = \frac{1}{2}(m+n)|m,n\rangle. \tag{3.41}$$

In addition, since the operators B_{ij} are functional to create a molecule in state $|\beta_i\rangle$ and annihilate a molecule in state $|\beta_j\rangle$, so that

$$B_{12}|m,n\rangle = \sqrt{(N-m)[\frac{1}{2}(m-n)+1]}\,|m+1,n-1\rangle, \tag{3.42}$$

$$
\begin{aligned}
B_{21}|m,n\rangle &= \sqrt{\frac{1}{2}(m-n)(N-m+1)}\,|m-1,n+1\rangle, &(3.43)\\
B_{23}|m,n\rangle &= \sqrt{\frac{1}{2}(m+n)(N-m+1)}\,|m-1,n-1\rangle, &(3.44)\\
B_{32}|m,n\rangle &= \sqrt{(N-m)[\frac{1}{2}(m+n)+1]}\,|m+1,n+1\rangle, &(3.45)\\
B_{13}|m,n\rangle &= \sqrt{\frac{1}{2}(m+n)[\frac{1}{2}(m-n)+1]}\,|m,n-2\rangle, &(3.46)\\
B_{31}|m,n\rangle &= \sqrt{\frac{1}{2}(m-n)[\frac{1}{2}(m+n)+1]}\,|m,n+2\rangle. &(3.47)
\end{aligned}
$$

In terms of the collective states, we can express a formal solution of the master equation (3.20) as

$$
\rho = \sum_{m=0}^{N} \sum_{n=-m,-m+2,\cdots}^{m} P_{m,n}|m,n\rangle\langle n,m|, \qquad (3.48)
$$

Note that for a given N, the allowed values of m,n are $m = 0,1,2,\cdots,N$, while $n = -m,-m+2,\cdots,m-2,m$. Inserting (3.48) into the master equation (3.20), we obtain the master equation for the probability $P_{m,n}$ in the form

$$
\begin{aligned}
\frac{\partial P_{m,n}}{\partial t} = \; & (N-m)[\frac{1}{2}(m-n)+1](x_1 P_{m+1,n-1} - x_2 P_{m,n})\\
& -\frac{1}{2}(m-n)(N-m+1)(x_1 P_{m,n} - x_2 P_{m-1,n+1})\\
& +(N-m)[\frac{1}{2}(m+n)+1](x_1 P_{m+1,n+1} - x_2 P_{m,n})\\
& -\frac{1}{2}(m+n)(N-m+1)(x_1 P_{m,n} - x_2 P_{m-1,n-1})\\
& +\frac{x_{34}}{4}(m+n)[\frac{1}{2}(m-n)+1](P_{m,n-2} - P_{m,n})\\
& -\frac{x_{34}}{4}(m-n)[\frac{1}{2}(m+n)+1](P_{m,n} - P_{m,n+2}). \qquad (3.49)
\end{aligned}
$$

The steady-state solution of Eq. (3.49) can be obtained under the steady-state condition $\frac{\partial P_{m,n}}{\partial t} = 0$, by using the detailed balancing principle, in the form

$$
P_{m,n} = x^{-m} P_{0,0}\,, \qquad (3.50)
$$

where

$$x = \frac{x_1}{x_2}, \tag{3.51}$$

and $P_{0,0}$ can be determined by the conservation condition

$$\sum_{m=0}^{N} \sum_{n=-m,-m+2,\cdots}^{m} P_{m,n} = 1 \,. \tag{3.52}$$

It is referred to the normalization constant

$$\begin{aligned} C &\equiv \frac{1}{P_{0,0}} \\ &= \sum_{m=0}^{N} (m+1) x^{-m} \\ &= \frac{x^{N+2} - (N+2)x + (N+1)}{(x-1)^2 x^N} . \end{aligned} \tag{3.53}$$

Substituting (3.50) as well as (3.53) into (3.48), we obtain the steady-state density operator in the form

$$\rho_S = \frac{1}{C} \sum_{m=0}^{N} \sum_{n=-m,-m+2,\cdots}^{m} x^{-m} |m,n\rangle\langle n,m| . \tag{3.54}$$

The analytical expressions of the steady-state expectation values for all the quantities to be investigated in the next sections are based on the steady-state solution (3.54). We can calculate, for example, the statistical moments of the operators E as follows:

$$\begin{aligned} \langle E^k \rangle_S &= Tr(\rho_S E^k) \\ &= \frac{1}{C} \sum_{m,n} \langle m,n| \sum_{m',n'} x^{-m'} |m',n'\rangle\langle n',m'|E^k|n,m\rangle \\ &= \frac{1}{C} \sum_{m=0}^{N} m^k (m+1) x^{-m} . \end{aligned} \tag{3.55}$$

In particular, one can find from (3.15) and (3.55) that

$$\langle E\rangle_S = \frac{2x^{N+2}-(N+1)(N+2)x^2+2N(N+2)x-N(N+1)}{(x-1)[x^{N+2}-(N+2)x+(N+1)]}, \quad (3.56)$$

$$\begin{aligned}\langle E^2\rangle_S = & [2x^{N+3}+4x^{N+2}-(N+1)^2(N+2)x^3+(N+2)(3N^2+3N-2)x^2 \\ & -N(3N^2+6N-1)x+N^2(N+1)] \\ & \cdot\frac{1}{(x-1)^2[x^{N+2}-(N+2)x+(N+1)]}. \end{aligned} \quad (3.57)$$

Also, we can express

$$\langle E^aF^b\rangle_S = \frac{1}{C}\sum_{m=0}^{N} m^a x^{-m} \sum_{n=-m,-m+2,\cdots}^{m} n^b \; (a,b=1,\cdots,4) \quad (3.58)$$

in terms of the statistical moments $\langle E^k\rangle_S$ $(k=1,\cdots,4)$ as

$$\langle E^aF^b\rangle_S = 0 \; (b \,\text{is odd}), \quad (3.59)$$

$$\langle F^2\rangle_S = \frac{1}{3}(\langle E^2\rangle_S + 2\langle E\rangle_S), \quad (3.60)$$

$$\langle EF^2\rangle_S = \frac{1}{3}(\langle E^3\rangle_S + 2\langle E^2\rangle_S), \quad (3.61)$$

$$\langle E^2F^2\rangle_S = \frac{1}{3}(\langle E^4\rangle_S + 2\langle E^3\rangle_S), \quad (3.62)$$

$$\langle F^4\rangle_S = \frac{1}{15}(3\langle E^4\rangle_S + 12\langle E^3\rangle_S + 8\langle E^2\rangle_S - 8\langle E\rangle_S). \quad (3.63)$$

We can also note here some explicit results which are used to calculate the limiting values for all the quantities to be considered in the next sections:

1. for $N=1$ and any x:

$$\langle E^k\rangle_S = \frac{2}{x+2}; \quad (3.64)$$

2. for $x=1$ and any N:

$$
\begin{aligned}
C &= \frac{1}{2}(N+1)(N+2), && (3.65)\\
\langle E\rangle_S &= \frac{2}{3}N, && (3.66)\\
\langle E^2\rangle_S &= \frac{1}{6}N(3N+1), && (3.67)\\
\langle E^3\rangle_S &= \frac{1}{30}N(12N^2+9N-1), && (3.68)\\
\langle E^4\rangle_S &= \frac{1}{30}N(10N^3+12N^2-N-1); && (3.69)
\end{aligned}
$$

3. for $x > 1$ and $N \to \infty$:

$$
\begin{aligned}
\lim_{N\to\infty} C &= \frac{x^2}{(x-1)^2}, && (3.70)\\
\lim_{N\to\infty} \langle E\rangle_S &= \frac{2}{x-1}, && (3.71)\\
\lim_{N\to\infty} \langle E^2\rangle_S &= \frac{2(x+2)}{(x-1)^2}, && (3.72)\\
\lim_{N\to\infty} \langle E^3\rangle_S &= \frac{2(4x^2+7x+1)}{(x-1)^3}, && (3.73)\\
\lim_{N\to\infty} \langle E^4\rangle_S &= \frac{2(x^3+18x^2+33x+8)}{(x-1)^4}; && (3.74)
\end{aligned}
$$

4. for any k and $N \to \infty$:

$$
\lim_{N\to\infty} \frac{\langle E^k\rangle_S}{N^k} = \begin{cases} 1 & , \quad x<1 \\ \frac{2}{k+2} & , \quad x=1 \\ 0 & , \quad x>1 \end{cases} . \qquad (3.75)
$$

One can see from (3.75) a discontinuous limiting behaviour of the expectation values $\frac{\langle E^k\rangle_S}{N^k}$ as $N \to \infty$, which will result in the corresponding discontinous behaviour of all quantities to be discussed in the subsequent sections.

3.3 Excited States of Biomolecules

In this section we discuss the steady-state behaviour of biomolecular populations (in the three levels $|\alpha_i\rangle (i = 1, 2, 3)$) and their fluctuations. All these quantities are functions of x and also dependent on the values of N, g_1 and g_2. We shall also examine the limiting behaviour as $N \to \infty$ and explain the implications of the results obtained from the analytical and numerical methods.

The steady-state expectation values of the diagonal operators A_{ii} yield the populations in the states $|\alpha_i\rangle$. One can find from (3.15) that

$$\begin{aligned} \langle A_{11}\rangle_S &= Ng_2^2 + (\frac{1}{2}g_1^2 - g_2^2)\langle E\rangle_S, && (3.76)\\ \langle A_{22}\rangle_S &= \frac{1}{2}\langle E\rangle_S, && (3.77)\\ \langle A_{33}\rangle_S &= Ng_1^2 + (\frac{1}{2}g_2^2 - g_1^2)\langle E\rangle_S, && (3.78) \end{aligned}$$

where $\langle E\rangle_S$ is expressed by (3.56).

It is of interest to derive the critical behaviour of $\langle A_{ii}\rangle_S$ at $x = 1$ and the limiting behaviour of $\frac{\langle A_{ii}\rangle_S}{N}$ as $N \to \infty$. The resulting expressions read as

$$\langle A_{11}\rangle_S = \langle A_{22}\rangle_S = \langle A_{33}\rangle_S = \frac{N}{3} \qquad (3.79)$$

for $x = 1$ and any value of N, and

$$\begin{aligned} \lim_{N\to\infty} \frac{\langle A_{11}\rangle_S}{N} &= \begin{cases} \frac{1}{2}g_1^2 & , \quad x < 1 \\ g_2^2 & , \quad x > 1 \end{cases}, && (3.80)\\ \lim_{N\to\infty} \frac{\langle A_{22}\rangle_S}{N} &= \begin{cases} \frac{1}{2} & , \quad x < 1 \\ 0 & , \quad x > 1 \end{cases}, && (3.81)\\ \lim_{N\to\infty} \frac{\langle A_{33}\rangle_S}{N} &= \begin{cases} \frac{1}{2}g_2^2 & , \quad x < 1 \\ g_1^2 & , \quad x > 1 \end{cases}. && (3.82) \end{aligned}$$

The plots of the fractional populations as functions of x for some finite values of N are shown in Fig. 3.2, where the parameters g_1 and g_2 have taken the experimental data [33,42]. The dashed curves in these figures show the limiting behaviour as $N \to \infty$. In addition, the schematic diagrams of populations of molecules in the three levels for $N \to \infty$ are shown in Fig. 3.3.

It may be of some interest to examine the behaviour of the fluctuations in the biomolecular populations. The expression for the fluctuations in the ith level is given by

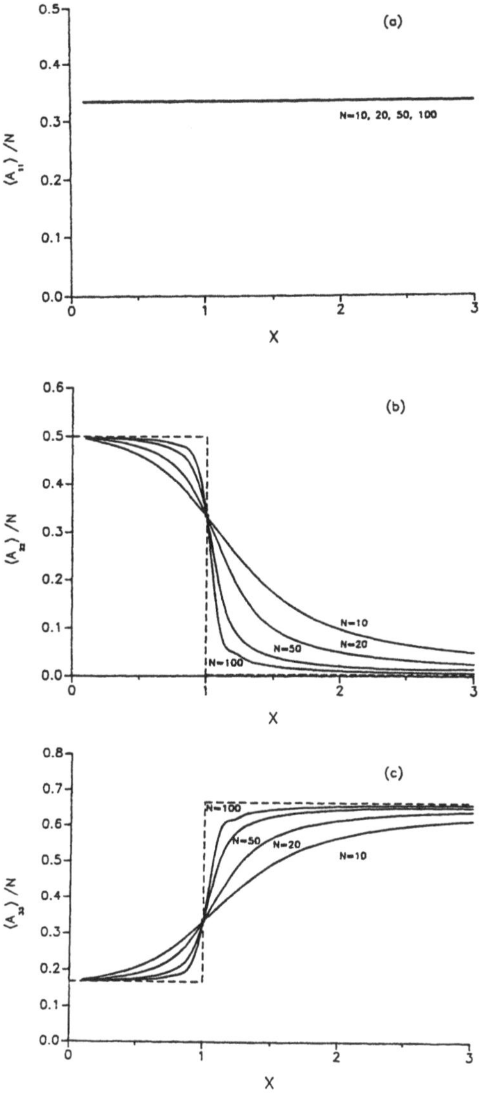

Figure 3.2: *Steady-state fractional populations of biomolecules in the three levels as functions of the parameter x for some values of N, where $g_1^2 = \frac{2}{3}$ and $g_2^2 = \frac{1}{3}$ are the experimental data. The dashed curves represent the limiting behaviour as $N \to \infty$.*

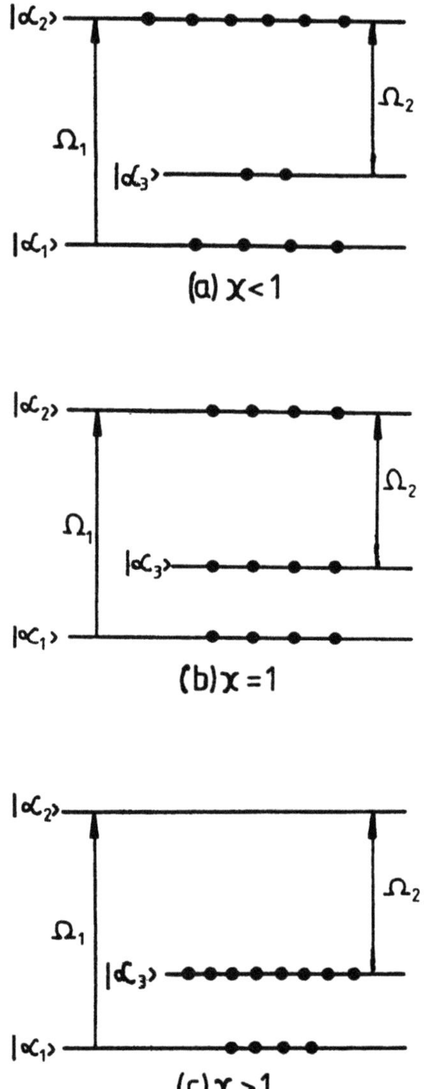

Figure 3.3: *Schematic diagrams of steady-state populations of biomolecules in the three levels for* $N \to \infty$.

$$\sigma_i^2 = \langle A_{ii}^2 \rangle_S - \langle A_{ii} \rangle_S^2. \tag{3.83}$$

Using relations (3.15) and (3.60), we get

$$\begin{aligned}
\sigma_1^2 &= \left(\frac{1}{3}g_1^4 - 2g_1^2g_2^2 + g_2^4\right)\langle E^2\rangle_S - \left(\frac{1}{2}g_1^2 - g_2^2\right)^2 \langle E\rangle_S^2 \\
&\quad + \left[\frac{1}{6}g_1^4 + g_1^2g_2^2\left(N - \frac{1}{2}\right)\right]\langle E\rangle_S + g_1^2g_2^2N, \qquad (3.84) \\
\sigma_2^2 &= \frac{1}{3}\langle E^2\rangle_S - \frac{1}{4}\langle E\rangle_S^2 + \frac{1}{6}\langle E\rangle_S, \qquad (3.85) \\
\sigma_3^2 &= \left(\frac{1}{3}g_2^4 - 2g_1^2g_2^2 + g_1^4\right)\langle E^2\rangle_S - \left(\frac{1}{2}g_2^2 - g_1^2\right)^2 \langle E\rangle_S^2 \\
&\quad + \left[\frac{1}{6}g_2^4 + g_1^2g_2^2\left(N - \frac{1}{2}\right)\right]\langle E\rangle_S + g_1^2g_2^2N. \qquad (3.86)
\end{aligned}$$

We can now take the limit $N \to \infty$ and show that

$$\lim_{N\to\infty} \frac{\sigma_1^2}{N^2} = \begin{cases} \frac{1}{12}g_1^4 & , \quad x < 1 \\ \frac{1}{18} & , \quad x = 1 \\ 0 & , \quad x > 1 \end{cases} , \tag{3.87}$$

$$\lim_{N\to\infty} \frac{\sigma_2^2}{N^2} = \begin{cases} \frac{1}{12} & , \quad x < 1 \\ \frac{1}{18} & , \quad x = 1 \\ 0 & , \quad x > 1 \end{cases} , \tag{3.88}$$

$$\lim_{N\to\infty} \frac{\sigma_3^2}{N^2} = \begin{cases} \frac{1}{12}g_2^4 & , \quad x < 1 \\ \frac{1}{18} & , \quad x = 1 \\ 0 & , \quad x > 1 \end{cases} . \tag{3.89}$$

The curves in Fig. 3.4 show the behaviour of the relative fluctuations $\frac{\sigma_i^2}{N^2} (i = 1, 2, 3)$ with respect to the parameter x for some finite values of N as well as for $N \to \infty$.

From the analytical and numerical results obtained in this section, one can understand the properties of the steady-state populations of biomolecules in the various states and their fluctuations as follows:

1. One may consider a given x as a working point of the system. There are the quasi-stationary excited states of biomolecules in the region $x < 1$, in

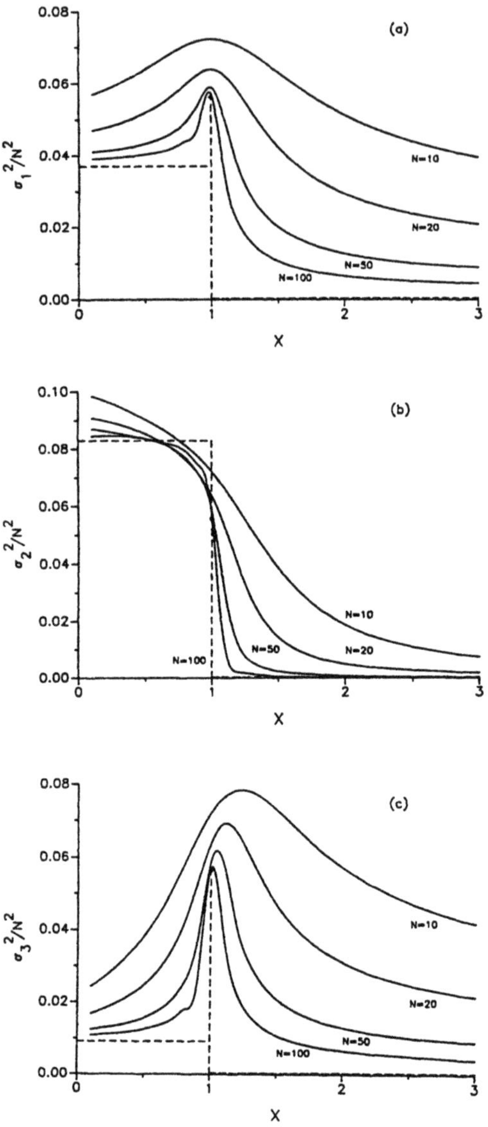

Figure 3.4: *Relative fluctuations in the steady-state populations of biomolecules in the three levels as functions of x. Data same as in Fig. 3.2.*

particular, for the infinite N the fractional populations in the excited state $|\alpha_2\rangle$ are always $\frac{1}{2}$ and the relative fluctuations $\frac{\sigma_2^2}{N^2}$ are finite and non-zero, which is characteristic of cooperativity of biological systems as open systems far away from thermal equilibrium. Therefore, the region $x < 1$ may be regarded as a "cooperative region".

2. In the region $x > 1$, both the populations in the excited state $|\alpha_2\rangle$ and the fluctuations of populations in the three states tend to zero for the infinite N, which characterizes the molecules as a group of individuals without correlation to each other. Therefore, this region may be considered as an "individual region". However, one can see from Fig. 3.3(c) that there is the inversion distribution of molecules between states $|\alpha_2\rangle$ and $|\alpha_3\rangle$, which seems to imply a presence of vibrational energy of the lattice system of biomolecules [39,41].

3. The fractional populations in all states and the relative fluctuations show, in the limit $N \to \infty$, a discontinous behaviour through $x = 1$, which displays a nonequilibrium first-order phase transition at the critical point $x = 1$. Also, one can see from (3.79) that for all values of N the biomolecular populations are equally divided among the three states at the critical point, which agrees with Popp's f_ν=constant-rule [32,39] (also see Chapter 1) describing dexterously the essential view of living state [47]. Evidently, the point $x = 1$ belongs to the cooperative region.

It is well known that biological systems work within a very narrow range of variation of the parameters that characterize them and their function [48]. According to this point of view, the living systems under description by our model may be expected to work within a very narrow region of variation of the parameter x around the threshold $x = 1$. Such a working state of living systems, which is reminiscent of lasers in threshold operation, can make a cooperative biophoton emission stabilized, as will be seen in next section.

3.4 Emission Intensities

The expression for the intensity of a radiation field can be given, in general, by [45]

$$I = \langle \mathcal{E}^{(-)} \mathcal{E}^{(+)} \rangle, \tag{3.90}$$

where $\mathcal{E}^{(\pm)}$ are the positive and negative frequency parts of the amplitude of the radiation field at the detector, which are functions of space and time. In the far-field limit, $\mathcal{E}^{(\pm)}$ can be expressed, for three-level systems, in terms of the molecular operators as [45]

$$\mathcal{E}^{(\pm)}(\vec{r},t) = E^{(\pm)}(\vec{r},t) + \psi(\vec{r})A_{32}\left(t - \frac{r}{c}\right), \tag{3.91}$$

where $E^{(\pm)}(\vec{r},t)$ correspond to the pumping field and $\psi(\vec{r})$ is a geometrical factor of the system.

In the following calculations of biophoton-emission intensities, using (3.90) and (3.91), we can drop $E^{(\pm)}(\vec{r},t)$ in (3.91) which do not contribute to the biophoton field; therefore, the steady-state intensities can be expressed by $I_2 \equiv \langle A_{23}A_{32}\rangle_S$, apart from an unimportant spatial factor. The evaluation of I_2, using (3.15), yields the analytical expression

$$I_2 = \frac{1}{6}\left\{[3(N+1)g_1^2 + 2g_2^2]\langle E\rangle_S - (3g_1^2 - g_2^2)\langle E^2\rangle_S\right\}. \tag{3.92}$$

It is clear from (3.92) and (3.75) that

$$\lim_{N\to\infty}\frac{I_2}{N^2} = \begin{cases} \frac{1}{6}g_2^2 & , \quad x < 1 \\ \frac{1}{12} & , \quad x = 1 \\ 0 & , \quad x > 1 \end{cases}. \tag{3.93}$$

The relative intensities $\frac{I_2}{N^2}$ as functions of the parameter x for some finite values of N are shown in Fig. 3.5. The dotted curve in the figure describes the limiting behaviour as $N \to \infty$. One can find from (3.93) that in the cooperative region $x \leq 1$ the intensities are proportional to N^2, exhibiting an aspect of cooperativity of biophoton emission. This arises solely from the phase-locking in the sequences of biomolecules, which is analogous to multi-mode lasers with N phase-locked modes, exhibiting an intensity proportional to N^2. Above the threshold $x = 1$, the intensities vary as N and the system behaves as a group of individuals radiating independently of each other. One can see again the different behaviour of the molecules in the cooperative region $x \leq 1$ and in the individual region $x > 1$. Also, one can see from Fig. 3.5 that for the finite N, $\frac{I_2}{N^2}$ pass through some maximum values around $x = 1$, which is due to $\frac{I_2}{N^2}$ taking, as $N \to \infty$, maximum value $\frac{1}{12}$ ($> \frac{1}{6}g_2^2$) at $x = 1$ in the phase transition from the cooperative region to the individual region. This may be an interesting phenomenon to be examined experimentally.

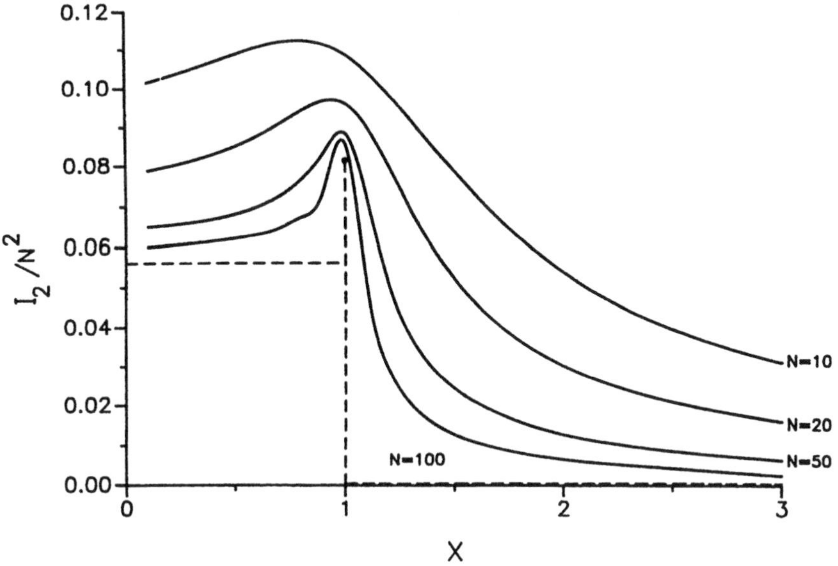

Figure 3.5: *Steady-state relative intensities of biophoton emission as functions of x. Data same as in Fig. 3.2.*

The dependence of emission intensity on N^2 has been indicated experimentally in drosophola embryos (see Chapter 11).

3.5 Intensity Correlations

The coherence of the biophoton field as a typical multi-mode field [32] can be understood from the normalized second-order intensity correlation functions defined by[27]

$$g_{ij}^{(2)}(\tau) = \frac{\langle : \hat{I}_i(t+\tau)\hat{I}_j(t) : \rangle}{\langle \hat{I}_i(t+\tau)\rangle\langle \hat{I}_j(t)\rangle}, \tag{3.94}$$

where :: represents the normal order of intensity operators $\hat{I}_i(t)$ and $\hat{I}_j(t)(i,j=1,2)$.

In the far-field limit, we can express the normalized second-order intensity correlation functions of the biophoton field in terms of the molecular operators A_{ij} [45]. In particular, we are interested in the steady-state correlation functions $g_{ij}^{(2)}(0)$, involving

$$g_{11}^{(2)}(0) = \frac{\langle A_{21}A_{21}A_{12}A_{12}\rangle_S}{\langle A_{21}A_{12}\rangle_S^2}, \tag{3.95}$$

$$g_{22}^{(2)}(0) = \frac{\langle A_{23}A_{23}A_{32}A_{32}\rangle_S}{\langle A_{23}A_{32}\rangle_S^2}, \tag{3.96}$$

$$g_{12}^{(2)}(0) = \frac{\langle A_{21}A_{23}A_{32}A_{12}\rangle_S}{\langle A_{21}A_{12}\rangle_S\langle A_{23}A_{32}\rangle_S}, \tag{3.97}$$

$$g_{21}^{(2)}(0) = \frac{\langle A_{23}A_{21}A_{12}A_{32}\rangle_S}{\langle A_{23}A_{32}\rangle_S\langle A_{21}A_{12}\rangle_S}, \tag{3.98}$$

The quantity $g_{12}^{(2)}(0)$, for example, is a measure of the probability of detecting simultaneously one photon from transition $|\alpha_2\rangle \to |\alpha_1\rangle$ and another photon from transition $|\alpha_2\rangle \to |\alpha_3\rangle$.

By using (3.15) and (3.59-3.63), these intensity correlation functions can be expressed, in terms of the statistical moments $\langle E^k\rangle_S$ $(k = 1, \cdots, 4)$, in the form

$$\begin{aligned} g_{22}^{(2)}(0) = \frac{1}{30I_2^2}\{&-2[5(N+1)(N+2)g_1^4 + 10(N+1)g_1^2g_2^2 + 3g_2^4]\langle E\rangle_S \\ &+[10(N^2+5N+5)g_1^4 + 10(N+3)g_1^2g_2^2 + g_2^4]\langle E^2\rangle_S \\ &-2[10(N+2)g_1^4 - 5Ng_1^2g_2^2 - 2g_2^4]\langle E^3\rangle_S \\ &+[10g_1^4 - 10g_1^2g_2^2 + g_2^4]\langle E^4\rangle_S\}, \end{aligned} \tag{3.99}$$

$$\begin{aligned} g_{21}^{(2)}(0) = \frac{1}{60I_1I_2}\{&[5(3N+2)g_1^4 - 2(10N^2+5N+1)g_1^2g_2^2 \\ &-10Ng_2^4 - 20(N+1)g_1^2 - 10g_2^2]\langle E\rangle_S \\ &+[-5(N+3)g_1^4 + 2(10N^2+45N+21)g_1^2g_2^2 \\ &+5(N+4)g_2^4 + 5(N+5)g_1^2 - 5g_2^2]\langle E^2\rangle_S \\ &+[5(N+2)g_1^4 - 2(25N+36)g_1^2g_2^2 + 5Ng_2^4 - 5g_1^2]\langle E^3\rangle_S \\ &+[-5g_1^4 + 32g_1^2g_2^2 - 5g_2^4]\langle E^4\rangle_S\}, \end{aligned} \tag{3.100}$$

where I_2 is expressed by (3.92) and $I_1 \equiv \langle A_{21}A_{12}\rangle$ is given by exchanging g_1 and g_2 in (3.92). Also, $g_{11}^{(2)}(0)$ can be given by exchanging g_1 and g_2 in (3.99) and then $g_{12}^{(2)}(0)$ in (3.97) is exactly equal to $g_{21}^{(2)}(0)$ in (3.98) because A_{21} and A_{23} are commutable.

Using Eqs. (3.64) and (3.75), the limiting values of these intensity correlations can be obtained:

$$g_{11}^{(2)}(0) = g_{22}^{(2)}(0) = g_{12}^{(2)}(0) = g_{21}^{(2)}(0) = 0 \quad (N=1); \tag{3.101}$$

$$\lim_{N\to\infty} g_{11}^{(2)}(0) = \lim_{N\to\infty} g_{22}^{(2)}(0) = \begin{cases} 1.2 & , \quad x<1 \\ 1.6 & , \quad x=1 \\ 2.0 & , \quad x>1 \end{cases} , \tag{3.102}$$

$$\lim_{N\to\infty} g_{12}^{(2)}(0) = \lim_{N\to\infty} g_{21}^{(2)}(0) = \begin{cases} 1.2 & , \quad x<1 \\ 0.8 & , \quad x=1 \\ 2.0 & , \quad x>1 \end{cases} . \tag{3.103}$$

The plots of $g_{22}^{(2)}(0)$ as functions of x for some finite values of N are shown in Fig. 3.6, where Eq. (3.55) has been used to calculate numerically the statistical moments $\langle E^k\rangle_S (k=1,\cdots,4)$. The dashed curve in the figure shows the behaviour as $N\to\infty$. For the case of small N (for example, $N\le 2$) and in the region around $x=1$, the emission field is of antibunching property, corresponding to $g_{22}^{(2)}(0) < 1$, which is a kind of nonclassical coherence effect, like sub-Poissonian photon statistics and squeezed state [49–56]. However, for the case of large N, the emission field is partially coherent ($g_{22}^{(2)}(0) < 2$). In the limit $N\to\infty$, the field has quite a high degree of coherence with $g_{22}^{(2)}(0) = 1.2$ for $x<1$, which is consistent with cooperative behaviour in this region. On the other hand, in the region $x>1$, the limiting value of $g_{22}^{(2)}(0) = 2.0$ as $N\to\infty$ makes the emission field fully incoherent, which is a manifestation of the fact that the molecules tend to radiate independently in the individual region, as is also evident from the limiting behaviour of the intensity $\frac{I_2}{N^2}$ discussed in Sec. 3.4.

It may be noted here that for the finite N, the different curves of $g_{22}^{(2)}(0)$ pass through some minimum values around $x=1$, indicating an increase in coherence there. This behaviour is caused mathematically by some critical maxima of $\frac{I_2}{N^2}$ around $x=1$. Essentially, it seems to indicate a novel feature of biophoton emission.

The intensity correlations $g_{21}^{(2)}(0)$ as function of x are shown in Fig. 3.7. Note that the antibunching behaviour is shown in the region around $x=1$ for any finite value of N. As $N\to\infty$, the region over which the antibunching is exhibited collapses to a single critical point $x=1$ (dashed curve in Fig. 3.7). The antibunching effects in the biophoton emission from some biological samples, such as acetabularia, gonyaulax, and firefly, have been observed recently (see Chapter 1).

It should be explained here that the different regimes of the biophoton field, that is, chaotic regime or coherent regime (including classical and nonclassical ones), may play different roles in life processes [39]. The chaotic state may govern, for example, the

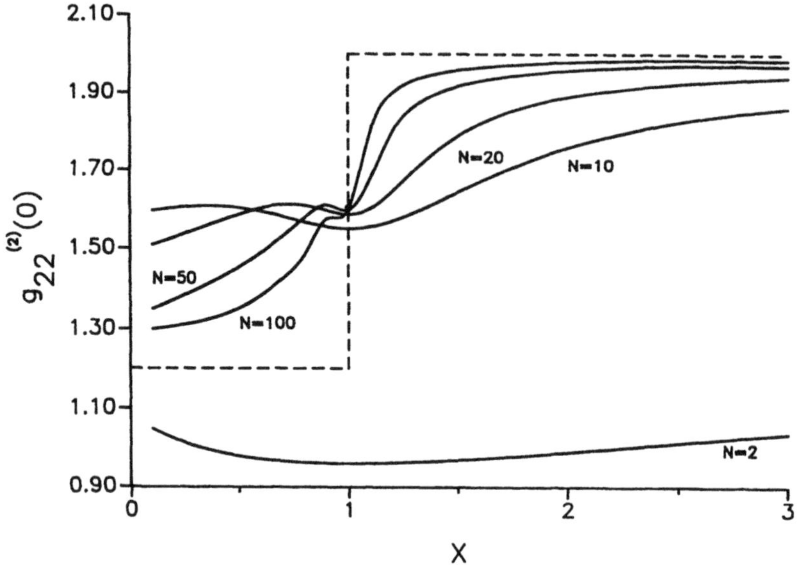

Figure 3.6: *Steady-state normalized second-order intensity correlations* $g_{22}^{(2)}(0)$ *as functions of* x. *Data same as in Fig. 3.2.*

individual regime of cell division, which appears to provide the necessary condition of intercellular decoupling. On the other hand, the coherent regime is responsible for cooperative processes, e.g., cell differentiation, memory functions, and intercellular communication; in particular, nonclassical biophotons with very high signal-to-noise ratio [57] may be expected to play a specially important role in the bio-communication both within one biological organism and between biological individuals belonging to the same kind.

3.6 Spectral Distributions

The steady-state spectrum of the biophoton emission may be given by the Fourier transform of steady-state the two-time correlation function $\langle A_{23}(\tau)A_{32}\rangle_S$ [44],

$$S(\omega-\Omega_2) = \gamma_2\int_0^\infty d\tau\langle A_{23}(\tau)A_{32}\rangle_S e^{-i(\omega-\Omega_2)\tau} + HC. \tag{3.104}$$

Using the secular approximation and relations (3.15), one can obtain

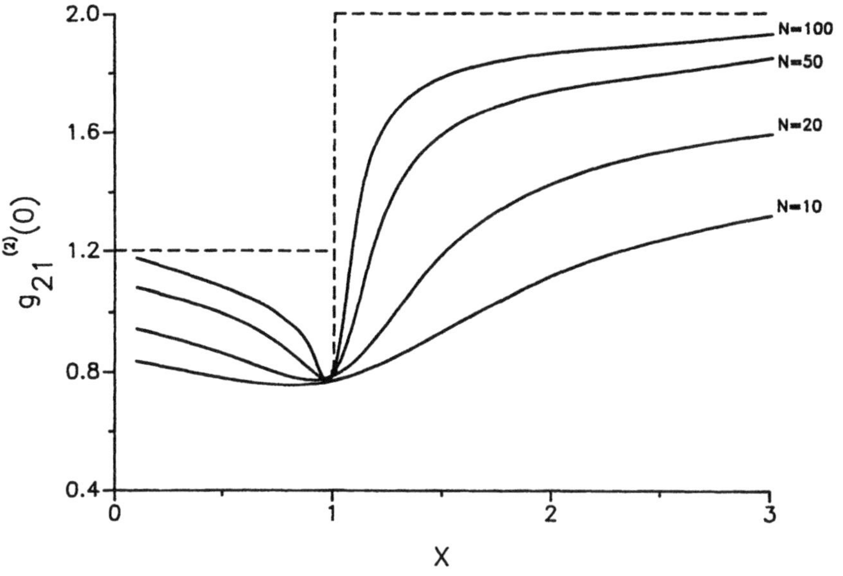

Figure 3.7: *Steady-state normalized second-order intensity correlations* $g_{21}^{(2)}(0)$ *as functions of* x. *Data same as in Fig. 3.2.*

$$\begin{aligned}\langle A_{23}(\tau)A_{32}\rangle_S &= \frac{1}{4}g_2^2\langle F(\tau)F\rangle_S + \frac{1}{2}g_1^2[\langle B_{12}(\tau)B_{21}\rangle_S + \langle B_{32}(\tau)B_{23}\rangle_S] \\ &\quad + \frac{1}{4}g_2^2[\langle B_{13}(\tau)B_{31}\rangle_S + \langle B_{31}(\tau)B_{13}\rangle_S]. \qquad (3.105)\end{aligned}$$

In order to find the expectation values appearing on the right-hand side of Eq. (3.105), we apply the quantum regression theorem [58] to Eqs. (3.27-3.30) and then write down the equations of motion for $\langle B_{ij}(\tau)B_{ji}\rangle_S$ in the form

$$\begin{aligned}\frac{d}{d\tau}\langle F(\tau)F\rangle_S &= -x_6\langle F(\tau)F\rangle_S - x_0(N - \langle E\rangle_S)\langle F(\tau)F\rangle_S, \\ \frac{d}{d\tau}\langle B_{12}(\tau)B_{21}\rangle_S &= -(x_7 + iG)\langle B_{12}(\tau)B_{21}\rangle_S \\ &\quad -x_0\left(\frac{N}{2} - \langle E\rangle_S\right)\langle B_{12}(\tau)B_{21}\rangle_S, \\ \frac{d}{d\tau}\langle B_{32}(\tau)B_{23}\rangle_S &= -(x_7 - iG)\langle B_{32}(\tau)B_{23}\rangle_S\end{aligned}$$

$$-x_0\left(\frac{N}{2}-\langle E\rangle_S\right)\langle B_{32}(\tau)B_{23}\rangle_S, \quad (3.106)$$

$$\begin{aligned}
\frac{d}{d\tau}\langle B_{13}(\tau)B_{31}\rangle_S &= -(x_8+2iG)\langle B_{13}(\tau)B_{31}\rangle_S \\
&\quad -x_0(N-\langle E\rangle_S)\langle B_{13}(\tau)B_{31}\rangle_S, \\
\frac{d}{d\tau}\langle B_{31}(\tau)B_{13}\rangle_S &= -(x_8-2iG)\langle B_{31}(\tau)B_{13}\rangle_S \\
&\quad -x_0(N-\langle E\rangle_S)\langle B_{31}(\tau)B_{13}\rangle_S,
\end{aligned}$$

where, following the treatment of Compagno and Persico [59], we have factorised:

$$\begin{aligned}
\langle B_{22}(\tau)B_{ij}(\tau)B_{ji}\rangle_S &\approx (N-\langle E\rangle_S)\langle B_{ij}(\tau)B_{ji}\rangle_S, \\
\langle E(\tau)B_{ij}(\tau)B_{ji}\rangle_S &\approx \langle E\rangle_S\langle B_{ij}(\tau)B_{ji}\rangle_S. \quad (3.107)
\end{aligned}$$

The solutions of Eqs. (3.106) are given by

$$\langle F(\tau)F\rangle_S = \langle F^2\rangle_S e^{-W_0\tau}, \quad (3.108)$$
$$\langle B_{12}(\tau)B_{21}\rangle_S = \langle B_{12}B_{21}\rangle_S e^{-(W_1+iG)\tau}, \quad (3.109)$$
$$\langle B_{32}(\tau)B_{23}\rangle_S = \langle B_{32}B_{23}\rangle_S e^{-(W_1-iG)\tau}, \quad (3.110)$$
$$\langle B_{13}(\tau)B_{31}\rangle_S = \langle B_{13}B_{31}\rangle_S e^{-(W_2+i2G)\tau}, \quad (3.111)$$
$$\langle B_{31}(\tau)B_{13}\rangle_S = \langle B_{31}B_{13}\rangle_S e^{-(W_2-i2G)\tau}, \quad (3.112)$$

where

$$W_0 = x_6+x_0(N-\langle E\rangle_S), \quad (3.113)$$
$$W_1 = x_7+x_0\left(\frac{N}{2}-\langle E\rangle_S\right), \quad (3.114)$$
$$W_2 = x_8+x_0(N-\langle E\rangle_S). \quad (3.115)$$

Substituting (3.108-3.112) into (3.105) and making use of

$$\begin{aligned}
\langle F^2\rangle_S &= 2\langle B_{13}B_{31}\rangle_S = 2\langle B_{31}B_{13}\rangle_S \\
&= \frac{2}{3}\langle E\rangle_S+\frac{1}{3}\langle E^2\rangle_S, \quad (3.116) \\
\langle B_{12}B_{21}\rangle_S &= \langle B_{32}B_{23}\rangle_S \\
&= \frac{1}{2}[(N+1)\langle E\rangle_S-\langle E^2\rangle_S], \quad (3.117)
\end{aligned}$$

we get

$$\langle A_{23}(\tau)A_{32}\rangle_S = \frac{1}{2}Y_0 e^{-W_0\tau} + Y_1 cosG\tau e^{-W_1\tau} + Y_2 cos2G\tau e^{-W_2\tau}, \tag{3.118}$$

where

$$\begin{aligned} Y_0 &= \frac{1}{6}g_2^2(2\langle E\rangle_S + \langle E^2\rangle_S), \\ Y_1 &= \frac{1}{2}g_1^2[(N+1)\langle E\rangle_S - \langle E^2\rangle_S], \\ Y_2 &= \frac{1}{2}Y_0. \end{aligned} \tag{3.119}$$

Note that Eq. (3.118) will reduce to (3.92) when $\tau = 0$. Substituting expression (3.118) into Eq. (3.104) and performing the integral, we obtain

$$\begin{aligned} \frac{S(\omega - \Omega_2)}{\gamma_2} &= \frac{W_0Y_0}{W_0^2 + (\omega - \Omega_2)^2} + \frac{W_1Y_1}{W_1^2 + [(\omega - \Omega_2) - G]^2} \\ &+ \frac{W_1Y_1}{W_1^2 + [(\omega - \Omega_2) + G]^2} + \frac{W_2Y_2}{W_2^2 + [(\omega - \Omega_2) - 2G]^2} \\ &+ \frac{W_2Y_2}{W_2^2 + [(\omega - \Omega_2) + 2G]^2}, \end{aligned} \tag{3.120}$$

In general, expression (3.120) yields a symmetrical structure with the five peaks located at the frequencies $\omega - \Omega_2 = 0, \pm G, \pm 2G$. The peak widths are, respectively, W_0, W_1 and W_2, while their heights are determined by $\frac{Y_0}{W_0}$, $\frac{Y_1}{W_1}$ and $\frac{Y_2}{W_2}$, respectively. This spectral distribution exhibits the dependence of biophoton emission on all the parameters under consideration through W and Y. Eq. (3.120) with $N = 1$ reduces to the single-molecule spectrum [44].

It may be of some interest to examine the limiting behaviour as $N \to \infty$. Using (3.113-3.115) , (3.120) and (3.75), widths and relative heigths of those peaks can be found, which are shown in Table 3.1. There are three peaks located at the frequencies $\omega - \Omega_2 = 0, \pm 2G$ in the region $x < 1$ and five peaks at the threshold $x = 1$. On the other hand, there are no peaks in the individual region $x > 1$, indicating a disappearing spectrum. This behaviour is not unexpected. As has been discussed earlier, both the population of biomolecules in the excited state $|\alpha_2\rangle$ and the relative intensity of the biophoton emission disappear in this region.

Table 3.1: The widths (W) and relative heigths ($y \equiv \frac{Y}{N^2}$) of the spectral peaks of the biophoton emission in limit $N \longrightarrow \infty$.

	W_0	W_1	W_2	y_0	y_1	y_2
$x < 1$	x_6		x_8	$\frac{g_2^2}{6}$	0	$\frac{g_2^2}{12}$
$x = 1$	x_6	x_7	x_8	$\frac{g_2^2}{12}$	$\frac{g_1^2}{12}$	$\frac{g_2^2}{24}$
$x > 1$				0	0	0

Let us discuss, for the finite values of N, the dependence of widths and heights of the spectral peaks in (3.120) on N. It is simple to show that in the threshold $x = 1$, the peak height of each spectrum line in (3.120) varies as N^2, whereas the width of each line is the same as in the single-molecule spectrum [44]. The picture changes for the case of $x \neq 1$. For the case of $x > 1$, $N >> 1$, so that $x^N >> N^3$, the widths of all spectrum lines vary with N. The heights of the three lines located at the frequencies $0, \pm 2G$ are independent of N, but the heights of the two lines located at the frequencies $\pm G$ are proportional to N. For the case where $x < 1$ and $N >> 1$, so that $x^N << N^{-1}$, the widths of the three lines located at the frequencies $0, \pm 2G$ are independent of N and their peak heights vary as N^2, while the widths and heights of the two lines located at the frequencies $\pm G$ are proportional to N.

In order to show the detailed behaviour of the spectral structure in (3.120), it is necessary to make some numerical treatments. For this purpose, we write it in the dimensionless form

$$\begin{aligned} \frac{S(\Delta)}{N^2} &= \frac{w_0 y_0}{w_0^2 + \Delta^2} + \frac{w_1 y_1}{w_1^2 + (\Delta - \delta)^2} + \frac{w_1 y_1}{w_1^2 + (\Delta + \delta)^2} \\ &+ \frac{w_2 y_2}{w_2^2 + (\Delta - 2\delta)^2} + \frac{w_2 y_2}{w_2^2 + (\Delta + 2\delta)^2}, \end{aligned} \tag{3.121}$$

where

$$\Delta = \frac{\omega - \Omega_2}{\gamma_2}, \tag{3.122}$$

$$\delta = \frac{G}{\gamma_2}, \tag{3.123}$$

$$w = \frac{W}{\gamma_2}, \tag{3.124}$$

$$y = \frac{Y}{N^2}. \tag{3.125}$$

The spectral distributions from (3.121) are shown in Fig. 3.8 with respect to the parameter Δ for different values of $X_0 \equiv \frac{x_0}{\gamma_2}$. Also, the five lines in (3.121) may collapse into a single line when, for example, the width of the line located at the frequency $\Delta = 0$ is large compared to the separation between the two lines located at the frequencies $\Delta = \pm 2\delta$. Similary, they may also collapse into three lines when, for example, $2\delta < w_0 < 4\delta$. These can be demonstrated if one examines Fig. 3.9.

Some spectral distributions with five peaks or three peaks, which display a certain degree of symmetry, have been observed experimentally in the biophoton emission from cress seeds (see Chapter 16).

3.7 Dynamic Evolution of the System

The biological system, as a typical open system far away from thermal equilibrium, always exchanges energy and/or matter with the external world. This makes it work, in general, in a nonequilibrium stationary state with respect to a relatively long time scale, which is reminiscent of lasers in continuous-wave operation [60,61]. However, when the biological system is excited by a certain kind of reaction, it should exhibit a separation from the stationary state and display the dynamic evolution. Therefore, it is necessary to examine the time dependence of the system after excitation. By such an examination, it will be seen that biophoton emission in the region $x < 1$ is of long-range and long-time cooperative properties.

It is clear that the time evolution of a population of molecules in the excited state $|\alpha_2\rangle$ can be characterized, under the secular approximation, by

$$\langle A_{22}(t)\rangle = \frac{1}{2}\langle E(t)\rangle \equiv Q(t), \tag{3.126}$$

and then the corresponding emission rate is expressed by [6]

$$R(t) = -\frac{d}{dt}Q(t). \tag{3.127}$$

In order to establish a differential equation for $Q(t)$, we start with the coupling equation of motion (3.26) and take an approximation:

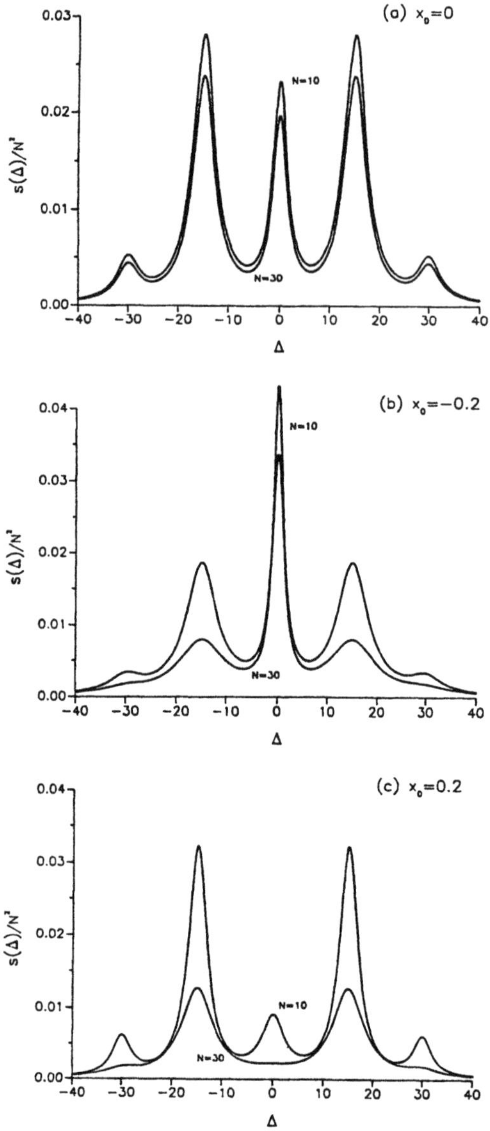

Figure 3.8: *Steady-state dimensionless spectral distributions of biophoton emission with* $g_1^2 = \frac{2}{3}$, $g_2^2 = \frac{1}{3}$, $\delta = 15$, *and* $\gamma_1 = \mu_2 = \nu_1 = \nu_2 = \bar{n}_1 = \bar{n}_2 = 0$, *(a)* $X_0 = 0$, *(b)* $X_0 = -0.2$, *(c)* $X_0 = 0.2$.

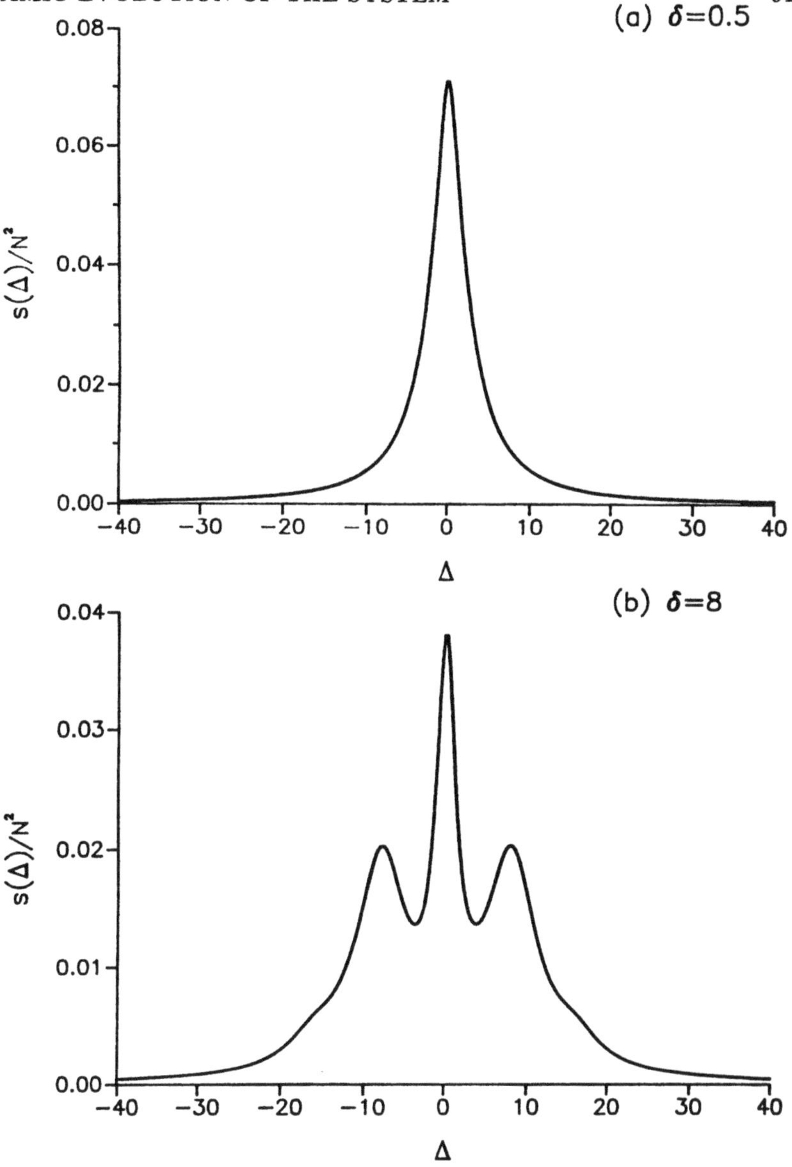

Figure 3.9: *Steady-state dimensionless spectral distributions of biophoton emission for* $N = 10$, $g_1^2 = \frac{2}{3}$, $g_2^2 = \frac{1}{3}$, $X_0 = -0.2$, *and* $\gamma_1 = \mu_2 = \nu_1 = \nu_2 = \bar{n}_1 = \bar{n}_2 = 0$, *(a) single line for* $\delta = 0.5$, *(b) three lines for* $\delta = 8$.

$$\langle E^2(t)\rangle \approx \alpha\langle E(t)\rangle^2 + \beta\langle E(t)\rangle + \gamma(t), \tag{3.128}$$

where α and β are two constants and $\gamma(t)$ may be an unknown function of time. Note that the approximation (3.128) follows from a consideration on (3.85). We then obtain from (3.26)

$$\frac{d}{dt}Q(t) = x_2N + \gamma(t) - (x_0N + x_5 - 2x_0\beta)Q(t) + 2x_0\alpha Q^2(t). \tag{3.129}$$

Eq. (3.129) can be used to describe generally the dynamic evolution of the system. However, in the present paper, we put emphasis on the case of $\alpha = 1$, $\beta = 0$, and $\gamma(t) = 0$, so that Eq. (3.129) reduces to

$$\frac{d}{dt}Q(t) = x_2N - (x_0N + x_5)Q(t) + 2x_0Q^2(t). \tag{3.130}$$

Let us solve equation (3.130) in three cases, that is, $x_0 = 0$, $x_0 < 0$ and $x_0 > 0$, as follows:

I. $x_0 = 0$ (i.e. $x = 1$),

Eq. (3.130) reduces to

$$\frac{d}{dt}Q(t) = x_1N - 3x_1Q(t). \tag{3.131}$$

It is simple to write its solution as

$$Q_F(t) = \frac{1}{3}N + \left(Q_F(0) - \frac{1}{3}N\right)e^{-3x_1t}, \tag{3.132}$$

and the corresponding emission rate as

$$R_F(t) = 3x_1\left(Q_F(0) - \frac{1}{3}N\right)e^{-3x_1t}, \tag{3.133}$$

where $Q_F(0)$ is the population of molecules in excited state $|\alpha_2\rangle$ at $t = 0$ which corresponds to the time when the excitation just goes away. It is clear that for any initial population $Q_F(0)$ and any value of N, $Q_F(t)$ displays an exponential decay until $Q_F(\infty) = \frac{1}{3}N$, which is consistent with the corresponding steady-state result shown in Eq. (3.79) and in Fig. 3.3(b). Also, $R_F(t)$ represents an exponential emission rate from $Q_F(0)$ excited molecules. One can see that it is possible to display an exponential decay in the cooperative region.

II. $x_0 < 0$ **(i.e.** $x < 1$**),**

Eq. (3.130) reduces to

$$\frac{d}{dt}Q(t) = x_2 N + (|x_0|N - x_5)Q(t) - 2|x_0|Q^2(t), \tag{3.134}$$

which has the solution

$$Q_{SR}(t) = \frac{1}{2}\left[\mathcal{X}_N + \mathcal{Y}_N \frac{1 + \mathcal{D}\, coth\left(\frac{t}{\mathcal{T}_R}\right)}{\mathcal{D} + coth\left(\frac{t}{\mathcal{T}_R}\right)}\right], \tag{3.135}$$

where

$$\mathcal{X}_N = \frac{1}{2}\left(N - \frac{x_5}{|x_0|}\right), \tag{3.136}$$

$$\mathcal{Y}_N = \left(\mathcal{X}_N^2 + \frac{2x_2 N}{|x_0|}\right)^{\frac{1}{2}}, \tag{3.137}$$

$$\mathcal{T}_R = \frac{1}{|x_0|\mathcal{Y}_N}, \tag{3.138}$$

$$\mathcal{D} = \frac{2Q_{SR}(0) - \mathcal{X}_N}{\mathcal{Y}_N}. \tag{3.139}$$

The corresponding emission rate is given, from (3.127) and (3.135), by

$$R_{SR}(t) = \frac{1}{2}(\mathcal{D}^2 - 1)|x_0|\mathcal{Y}_N^2 \frac{csch^2\left(\frac{t}{\mathcal{T}_R}\right)}{\left[\mathcal{D} + coth\left(\frac{t}{\mathcal{T}_R}\right)\right]^2}. \tag{3.140}$$

We will discuss (3.135) and (3.140) in three cases as follows:

(i) $|\mathcal{D}| > 1$,

Eqs. (3.135) and (3.140) reduce to

$$Q_{SRE}(t) = \frac{1}{2}\left[\mathcal{X}_N + \mathcal{Y}_N coth\left(\frac{t+\mathcal{T}_{DE}}{\mathcal{T}_R}\right)\right], \tag{3.141}$$

$$R_{SRE}(t) = \frac{1}{2}|x_0|\mathcal{Y}_N^2 csch^2\left(\frac{t+\mathcal{T}_{DE}}{\mathcal{T}_R}\right), \tag{3.142}$$

where

$$\mathcal{T}_{DE} = \mathcal{T}_R \; arccoth\mathcal{D}. \tag{3.143}$$

(ii) $\mathcal{D} = \pm 1$,

From (3.135) and (3.140), one can obtain

$$Q_{SRS}(t) = \frac{1}{2}(\mathcal{X}_N \pm \mathcal{Y}_N), \tag{3.144}$$

$$R_{SRS}(t) = 0. \tag{3.145}$$

(iii) $|\mathcal{D}| < 1$,

Eqs. (3.135) and (3.140) reduce to

$$Q_{SRA}(t) = \frac{1}{2}\left[\mathcal{X}_N + \mathcal{Y}_N tanh\left(\frac{t+\mathcal{T}_{DA}}{\mathcal{T}_R}\right)\right], \tag{3.146}$$

$$R_{SRA}(t) = -\frac{1}{2}|x_0|\mathcal{Y}_N^2 sech^2\left(\frac{t+\mathcal{T}_{DA}}{\mathcal{T}_R}\right), \tag{3.147}$$

where

$$\mathcal{T}_{DA} = \mathcal{T}_R \; arctanh\,\mathcal{D}\,. \tag{3.148}$$

The results obtained for $x_0 < 0$ are now discussed as follows:

1. It is clear from (3.141) that $Q_{SRE}(t)$ takes the maximum $Q_{SRE}(0)$ at $t = 0$, then it decreases with t, and finally exhibiting a non-vanishing minimum $Q_{SRE}(\infty) = \frac{1}{2}(\mathcal{X}_N + \mathcal{Y}_N)$ as $t \to \infty$. One can obtain $\frac{Q_{SRE}(\infty)}{N} = \frac{1}{2}$ as $N \to \infty$, which is consistent with corresponding steady-state results shown in Eq. (3.75) and in Fig. 3.3(a). This means that the system by relaxation is able to return in a relatively long time to its original stationary state, in other words, it may commit the original correlated state to its memory.

2. $Q_{SRE}(t)$ in (3.141) can be expanded, the first term is

$$Q_{SRE}(t) = \frac{\frac{1}{2|x_0|}}{t + \mathcal{T}_{DE}} + \frac{1}{2}\mathcal{X}_N, \tag{3.149}$$

   which displays an expected hyperbolic decay [32,34,35,36,39]. According to Chapter 2, such a hyperbolic decay may be a sufficient condition for a fully coherent ergodic field remaining coherent in time evolution.

3. $R_{SRE}(t)$ in (3.142) displays a dynamic cooperative emission with an intensity proportional to N^2 at $t = 0$, which, in fact, represents a typical superradiant emission [13], as will be discussed in more detail later.

4. There is the discontinous behaviour of both the fractional population $\frac{Q_{SRE}(t)}{N}$ and the relative emission rate $\frac{R_{SRE}(t)}{|x_0|N^2}$ in the limit $\mathcal{T}_R \to 0$ (when, for example, $N \to \infty$ and $|x_0| \neq 0$), which are caused by the discontinuity of the hyperbolic functions of themselves, appearing in (3.141) and (3.142). We can find out the limiting values (for example, taking $\frac{Q_{SRE}(0)}{N} = 1$) as

$$\lim_{\mathcal{T}_R \to 0} \frac{Q_{SRE}(t)}{N} = \begin{cases} 1 & , \quad t = 0 \\ \frac{1}{2} & , \quad t > 0 \end{cases}, \tag{3.150}$$

$$\lim_{\mathcal{T}_R \to 0} \frac{R_{SRE}(t)}{|x_0|N^2} = \begin{cases} 1 & , \quad t = 0 \\ 0 & , \quad t > 0 \end{cases} \tag{3.151}$$

5. $Q_{SRS}(t)$ in (3.144) and $R_{SRS}(t)$ in (3.145) yield a constant population of molecules in the excited state and a vanishing emission rate, respectively, which indicates a biophoton storage process.

6. $Q_{SRA}(t)$ in Eq. (3.146) yields an increasing population of molecules in the excited state and $R_{SRA}(t)$ in Eq. (3.147) displays a cooperative photon-absorption process.

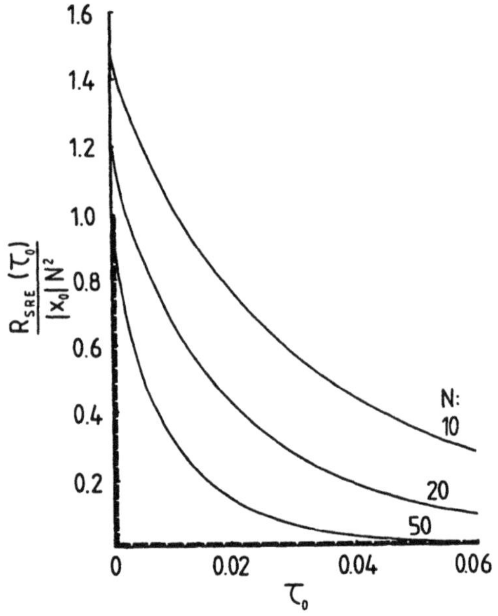

Figure 3.10: *Superradiance relative-emission-rates from Eq. (3.142) as a function of the time parameter* $\tau_0 \equiv |x_0|t$, *with* $\frac{x_1}{|x_0|} = 2$ *and* $\frac{Q_{SRE}(0)}{N} = 1$. *The dashed curve represents the limiting behaviour as* $T_R \to 0$.

The curves in Fig. 3.10 show the variations of the relative emission rates with time, using formula (3.142).

III. $x_0 > 0$ **(i.e.** $x > 1$**),**

Eq. (3.130) reduces to

$$\frac{d}{dt}Q(t) = x_2 N - (|x_0|N + x_5)Q(t) + 2|x_0|Q^2(t), \tag{3.152}$$

which has the solution

$$Q_{SF}(t) = \frac{1}{2}\left[X_N + Y_N \frac{D - tanh\left(\frac{t}{T_R}\right)}{1 - Dtanh\left(\frac{t}{T_R}\right)}\right], \tag{3.153}$$

where

$$X_N = \frac{1}{2}\left(N + \frac{x_5}{|x_0|}\right), \tag{3.154}$$

$$Y_N = \left(X_N^2 - \frac{2x_2 N}{|x_0|}\right)^{\frac{1}{2}}, \tag{3.155}$$

$$T_R = \frac{1}{|x_0|Y_N}, \tag{3.156}$$

$$D = \frac{2Q_{SF}(0) - X_N}{Y_N}. \tag{3.157}$$

The corresponding emission rate is given by

$$R_{SF}(t) = \frac{1}{2}(1 - D^2)|x_0|Y_N^2 \frac{sech^2\left(\frac{t}{T_R}\right)}{\left[1 - D\, tanh\left(\frac{t}{T_R}\right)\right]^2}. \tag{3.158}$$

We shall discuss (3.153) and (3.158) in three cases as follows:

(i) $|D| < 1$,

Eqs. (3.153) and (3.158) reduce to

$$Q_{SFE}(t) = \frac{1}{2}\left[X_N - Y_N tanh\left(\frac{t - T_{DE}}{T_R}\right)\right], \tag{3.159}$$

$$R_{SFE}(t) = \frac{1}{2}|x_0|Y_N^2 sech^2\left(\frac{t - T_{DE}}{T_R}\right), \tag{3.160}$$

where

$$T_{DE} = T_R\ arctanh\, D\,. \tag{3.161}$$

(ii) $D = \pm 1$,

From (3.153) and (3.158), one can obtain

$$Q_{SFS}(t) = \frac{1}{2}(X_N \pm Y_N), \tag{3.162}$$

$$R_{SFS}(t) = 0. \tag{3.163}$$

(iii) $|D| > 1$,

Eqs. (3.153) and (3.158) reduce to

$$Q_{SFA}(t) = \frac{1}{2}\left[X_N - Y_N coth\left(\frac{t - T_{DA}}{T_R}\right)\right], \quad (3.164)$$

$$R_{SFA}(t) = -\frac{1}{2}|x_0|Y_N^2 csch^2\left(\frac{t - T_{DA}}{T_R}\right), \quad (3.165)$$

where

$$T_{DA} = T_R \; arccoth \; D\,. \quad (3.166)$$

For the above-mentioned results about $x_0 > 1$, we can discuss as follows:

1. Expanding $Q_{SFE}(t)$ in (3.159) and taking the first term, we get

$$Q_{SFE}(t) = -\frac{1}{2}|x_0|Y_N^2 t + \frac{1}{2}(X_N + |x_0|T_{DE}Y_N^2), \quad (3.167)$$

which displays a linear decay. It is not strange to obtain such a result for the individual region, but it seems to be surprising that (3.160) displays a cooperative emission with an intensity proportional to N^2 at $t = T_{DE}$.

2. However, one should note the condition $|D| < 1$ under which (3.160) is valid. It implies that in this case the value of the initial population can take, at most, about $\frac{N}{2}$, i.e. $Q_{SFE}(0) \leq \frac{N}{2}$. On the other hand, the condition $|\mathcal{D}| > 1$ allows equation (3.142) with $Q_{SRE}(0) \leq N$. It is obvious that the former is just a short-range cooperative effect, while the latter may be a long-range one.

3. Actually, Eq. (3.160) represents a curve of typical superfluorescence [7,8], being another cooperative emission similar to superradiance. But, the distinction between the two phenomena is rather striking and it has to do with the way in which the initial state is prepared [13]. For superfluorescence the phases of the dipole moments belonging to different molecules are random at time $t = 0$, the molecules start radiating independently as in normal fluorescence with an intensity proportional to N. In fact, it can be seen from (3.133) and (3.160) that the initial emission rate of superfluorescence is exactly equal to that of ordinary fluorescence, i.e.

$$R_{SFE}(0) = R_F(0) = \frac{1}{2} x_1 N, \tag{3.168}$$

for $Q_{SFE}(0) = Q_F(0) = \frac{N}{2}$, where

$$D = \frac{\mathcal{X}_N}{\left(\mathcal{X}_N^2 + \frac{x_1 N}{|x_0|}\right)^{\frac{1}{2}}} < 1 \,. \tag{3.169}$$

On the other hand, in the case of superradiance, the dipole moments belonging to different molecules are initially in phase, so that a macroscopic polarization is present at $t = 0$, which would give a radiation field amplitude proportional to N and hence a radiation intensity proportional to N^2. In fact, it can be found from (3.142) that

$$R_{SRE}(0) = |x_0| N^2 + (x_1 + x_2) N, \tag{3.170}$$

for $Q_{SRE}(0) = N$, where

$$\mathcal{D} = \frac{N + X_N}{\left(X_N^2 - \frac{x_1 N}{|x_0|}\right)^{\frac{1}{2}}} > 1 \,. \tag{3.171}$$

4. The distinction of the time evolutions of the emission processes between the two phenomena is also rather striking. For superradiance the maximum emission rate stands at $t = 0$, having a radiation lifetime determined by $\mathcal{T}_R$ shown in (3.138). In the case of superfluorescence, following initial ordinary spontaneous emission, a correlation between the molecules may develop in an interesting way, as follows. When the spontaneous emission starts, the quantum fluctuations of the radiation field of the vacuum act on independent molecules. From this action, the dipole-dipole correlations gradually emerge through a kind of symmetry breaking phenomenon [12]. For a delay time T_{DE} shown in (3.161), the system reaches an optimum correlated state in which the emission rate is proportional to N^2.

5. These distinctions between superradiance and superfluorescence originate essentially from the big difference between the two relaxation laws shown in (3.149) and (3.167), one hyperbolic and another linear.

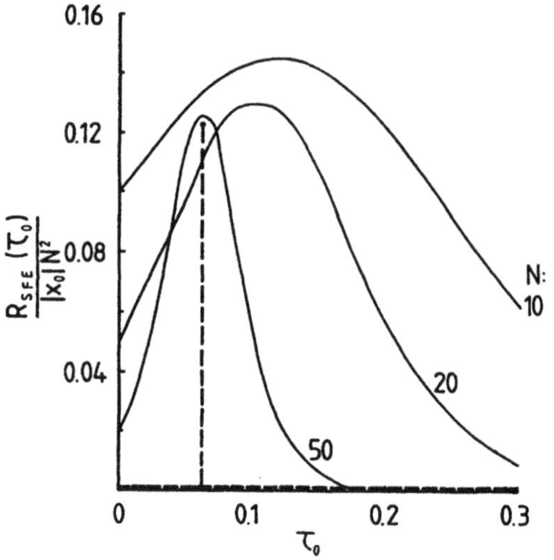

Figure 3.11: *Superfluorenscence relative-emission-rates from Eq. (3.160) as a function of the time parameter* $\tau_0 \equiv |x_0|t$, *with* $\frac{x_1}{|x_0|} = 2$ *and* $\frac{Q_{SRE}(0)}{N} = \frac{1}{2}$. *The dashed curve represents the limiting behaviour as* $T_R \to 0$.

The curves of emission rate for superfluorescence are shown in Fig. 3.11, using formula (3.160). Fortunately, three different spontaneous emission-processes can be shown simultaneously in our problems, i.e., superradiance with hyperbolic relaxation, superfluorescence with linear relaxation and normal fluorescence with exponential relaxation.

The analytical and numerical results obtained in this section as well as those in Sec. 3.4 show an imaginable physical picture of biophoton emission from collective biomolecules working in the cooperative region $x < 1$. When the biomolecules work in the stationary state, they always keep a phase correlation to each other and undergo continuously cooperative emission. After excitation, they display a hyperbolic relaxation and a superradiant emission, finally returning to the original correlated state. It can be seen that such a cooperative effect can occur in a long range (between N molecules) and in a long time (over the steady-state and the decay periods).

On the other hand, in the individual region, even if a cooperative process appears, it occurs in a short range (only between $\frac{N}{2}$ molecules) and in a short time (only over the decay period), that will finally lead the system into the originally non-correlated state.

It should be remarked here that all experimental observations of cooperative emission have been of the superfluorescence-type [5,9,10,11,26] since the first experimental demonstration of cooperative emission, which was made in optically pumped HF gas by Skribanowitz et al. [5]. We have not yet found a report on the experimental demonstration of superradiance from cooperative systems since Dicke first introduced the concept of collective spontaneous emission [4]. In a non-living matter, it seems to be extremely difficult to prepare an initial state in which a phase correlation between the dipole moments belonging to different molecules is already present. It may be inferred that such an initial state exists only in biological systems. So, superradiance may be considered as a phenomenon peculiar to living systems, perhaps.

3.8 Comparison with Experiments

In the previous sections, the theoretical results of biophoton emission have been presented. Correctness of those results must be tested by comparison with the corresponding experimental observations of the biophoton emission from biological samples.

Firstly, there is experimental evidence of N^2-dependence of emission intensity (see Chapter 11), supporting the theoretical results shown in Eq. (3.92).

Secondly, according to the theoretical results shown in Eqs. (3.99-3.101) and (3.103), the biophoton emission may be of antibunching property, which agrees with the recent experimental observations (see Chapter 1).

Also, there are some experimental results (see Chapter 16), which are consistent with the theoretical prediction about the steady-state spectral distributions with five peaks or three peaks (see Eq. (3.120) or Eq. (3.121)).

In addition, in this section, we will make a comparison of the theoretical results about the dynamic evolution of the system with the corresponding experimental observations. In Sec. 3.7 we have obtained some expressions describing the decay of the three-level system. Now, let us consider the decay from a practical biological system after excitation. Due to our experience there is good reason to believe that:

(i) There are two possible parts of biomolecules within one biological sample, one of which exhibits the photon emission after excitation, while, at the same time, another part may display more or less photon absorption.

(ii) There are various possible types of the biomolecules within each of the two parts, which may have different values of parameters, such as $\mathcal{D}, \mathcal{Y}_N$ and $\mathcal{T}_R$ for the part displaying photon emission and D, Y_N, and T_R for the part displaying photon absorption.

(iii) There are three possible working states for any biomolecular type, which may belong to the regions $x < 1$, $x = 1$, or $x > 1$.

Therefore, in general, a practical biological sample after excitation may display a decay in superposition over two biomolecular parts and all types within them as well as over the regions $x < 1$, $x = 1$, and $x > 1$. Thus, we may express, from (3.133), (3.142), (3.147), (3.160) and (3.165), the photon emission rate of a practical biological system, in the form

$$
\begin{aligned}
I(t) \; = \; & \sum_{i=1}^{M_R} R_i csch^2\left(\frac{t+\mathcal{T}_{R_i}}{\mathcal{T}_{R_i}}\right) - \sum_{i=1}^{M_A} A_i sech^2\left(\frac{t+T_{A_i}}{T_{A_i}}\right) \\
& + \sum_{i=1}^{M_r} r_i \, sech^2\left(\frac{t-\mathcal{T}_{r_i}}{\mathcal{T}_{r_i}}\right) - \sum_{i=1}^{M_a} a_i \, csch^2\left(\frac{t-T_{a_i}}{T_{a_i}}\right) \\
& + \sum_{i=1}^{M_E} E_i e^{-\frac{t}{\mathcal{T}_{E_i}}} + S,
\end{aligned}
\tag{3.172}
$$

where R, E, r-terms describe the cooperative photon emissions from the regions $x < 1$, $x = 1$ and $x > 1$, respectively, and A, a-terms describe the cooperative photon absorptions from the regions $x < 1$ and $x > 1$, respectively. $M_R, M_E, M_r(M_A, M_a)$ are the corresponding numbers of types of biomolecules with photon emissions (absorptions) and S is the steady-state emission rate. Eq. (3.172) looks very complicated indeed!

However, for an **active** living system within which a strong correlation between almost all biomolecules occurs, it may be sufficient to take into account only the contributions from the cooperative region $x < 1$ and, in this case, the biomolecules within the system may be considered to belong to the same type with respect to biophoton emission, so Eq. (3.172) can be much simplified as

$$
I(t) \; = \; R \, csch^2\left(\frac{t+\mathcal{T}_R}{\mathcal{T}_R}\right) - A \, sech^2\left(\frac{t+T_A}{T_A}\right) + S \, . \tag{3.173}
$$

The second term in Eq. (3.173) even may be neglected in case of weak absorption, so that

$$I(t) = R\,csch^2\left(\frac{t+\mathcal{T}_R}{\mathcal{T}_R}\right) + S\,. \tag{3.174}$$

Expanding Eq. (3.174), the first term leads to

$$I(t) \approx \frac{R\mathcal{T}_R^2}{(t+\mathcal{T}_R)^2} + S\,, \tag{3.175}$$

which displays a hyperbolic decay.

Fig. 3.12 show the theoretical curves on decay law from Eq. (3.173) in comparison with the corresponding experimental results from several biological samples after excitation by white-light-illumination. One can see that the actually measured results follow clearly this decay law.

Note that in Eq. (3.173) (as well as in Eq. (3.174)), the parameter $\mathcal{T}_R$ plays the role of lifetime of cooperative decay of the system after excitation, which may be understood as a "coherence" time [39].

It is of interest to examine an order of magnitude of $\mathcal{T}_R$ from Fig. 3.12. One can find that the three measured samples display the same order of magnitude of lifetime, which is about

$$\mathcal{T}_R \sim 1s\,. \tag{3.176}$$

Furthermore, an order of magnitude of the parameter $|x_0|$ can be estimated as follows. We consider, for example, the decay of rye grains shown in Fig. 3.12(a). The order of magnitude of the initial emission rate is given by

$$|x_0|\mathcal{Y}_N^2 = \frac{\mathcal{Y}_N}{\mathcal{T}_R} \sim 10^3 s^{-1}\,. \tag{3.177}$$

The use of (3.176) and (3.177) yields the order of magnitude of $|x_0|$ as

$$|x_0| \sim 10^{-3} s^{-1}\,. \tag{3.178}$$

It is clear that the other two measured samples shown in Fig. 3.12(b,c) have smaller values of $|x_0|$.

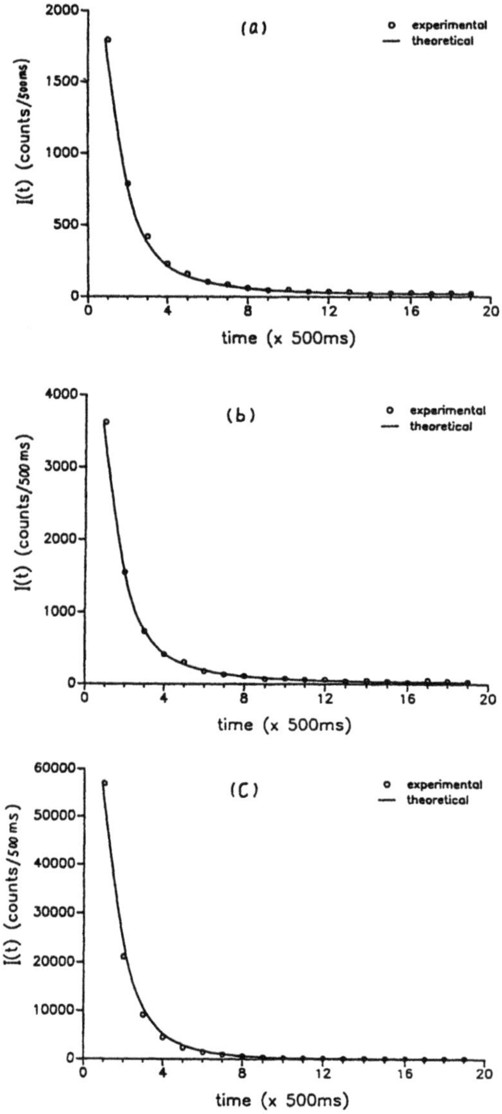

Figure 3.12: *Comparison of the biophoton decay behaviour between theoretical formulation Eq. (3.173) and the experimental results from rye grains (a), oat grains (b), and wheat grains (c), which are excited initially by while-light-illumination. The measured intensities range from about 200–800nm.*

One can see from such a small value of $|x_0|$ that living systems do work within a very narrow region of variation of the parameter x around $x = 1$. In other words, such a small value of $|x_0|$ is responsible that cooperative decay of living systems after excitation displays a low intensity and a long lifetime. This is very different from those of a normal cooperative decay from the two-level atomic system, where no parameter $|x_0|$ is introduced and the peaked power of radiation and its lifetime are typically of orders 1 kW and $1ns$, respectively [11].

3.9 Conclusions

In this paper we have presented the quantum theory of biophoton emission in terms of cooperativity. We have examined analytically and numerically the steady-state behaviour of the populations of biomolecules in various levels and their fluctuations as well as the steady-state properties of the biophoton field, including emission intensities, intensity correlations and spectral distributions. In addition, we have studied the dynamic evolution of the system after excitation, involving the relaxation of the system and the biophoton emission rate. Finally, we have given the comparison between some theoretical results and corresponding experimental observations.

It is clear from our results that the system is governed by a nonequilibrium first-order phase transition with respect to the parameter x given, from (3.51), (3.21) and (3.22), by

$$x = \frac{\gamma_1(1+\bar{n}_1)g_2^2 + \gamma_2(1+\bar{n}_2)g_1^2 + \mu_1 g_2^4 + \mu_2 g_1^4}{(\nu_1 + \gamma_1\bar{n}_1)g_2^2 + (\nu_2 + \gamma_2\bar{n}_2)g_1^2 + \mu_1 g_1^4 + \mu_2 g_2^4}. \tag{3.179}$$

The phase transition occurs at the critical point $x = 1$, which marks a transition of the working point of the system from the cooperative region to the individual region, or from an "ordered state" to a "chaotic state" [1]. The molecules corresponding to the individual region $x > 1$ behave as a group of individuals without correlation to each other. On the other hand, a collection of biomolecules working in the cooperative region displays long-range and long-time cooperative effects with respect to all their aspects:

1. As the biophoton emission takes place continously, the biomolecules can form quasi-stationary excited states.

2. The fluctuations of the steady-state biomolecular populations in the excited states are of the order of N^2, where N is the number of biomolecules under consideration.

3. The emission intensities are proportional to N^2 rather than N, representing a spontaneous phase-locking in the sequences of biomolecules.

4. The biophoton field exhibits, in different cases, partial coherence, complete coherence or nonclassical coherence, alternatively.

5. The spectral distributions result in a peaked structure, exhibiting a dependence of the heigths of some spectrum lines on N^2.

6. In particular, the relaxation dynamics of the system after excitation is characterized by a hyperbolic decay and a superradiant emission.

7. Through relaxation, the system is able to return in a relatively long time to its original correlated state, i.e., it may commit the normal working state to its memory.

The parameter x plays such a dominant role that it is necessary to go further into the expression (3.179). It is clear that x is related to all factors of the system under consideration, i.e., $g_1, g_2, \bar{n}_1, \bar{n}_2, \gamma_1, \gamma_2, \nu_1, \nu_2, \mu_1$ and μ_2, which determine collectively the working point x of the system. It is well known that very weak external influences are sufficient for significant changes of the intensity or the spectrum of biophoton emission. Also, all biological processes, e.g., cell cycle phase, growth, differentiation, etc., have sensitive influence on the biophoton emission. In our point of view, all these influences may be interpreted as changes of the parameters on the right-hand side of Eq. (3.179) and finally leading to a change of the working point x of the system. For example, one can see that an increase of temperature T of the system leads to increases of $\bar{n}_1$ in (3.7) and $\bar{n}_2$ in (3.8), the value of x, in cooperative region $x < 1$, increases in turn, so that the emission intensity increases. If the working point x changes so drastically that it rises above the threshold $x = 1$, the system will go into the individual region.

In order to understand the threshold in terms of those few parameters, which may play more important roles compared to other ones, we take $g_1^2 = \frac{2}{3}$, $g_2^2 = \frac{1}{3}$ (from the experimental data [42,33]) and assume $\gamma_2 >> \gamma_1$, $\mu_1 >> \mu_2$, $\nu_1 = \nu_2 = 0$, $\bar{n}_1 = \bar{n}_2 = 0$, then obtaining from (3.179)

$$x \approx \frac{3}{2}\frac{\gamma_2}{\mu_1} + \frac{1}{4}. \tag{3.180}$$

It is clear from (3.180) that apart from γ_2, the nonradiative rate μ_1, which has much to do with the vibration of the lattice system of biomolecules, can play a more important role, where condition $\mu_1 \geq 2\gamma_2$ is sufficient for retaining the system in the cooperative region $x \leq 1$. A theoretical work, taking account of the influence of time-dependent vibration of the lattice system of biomolecules on the biophoton emission, is in preparation.

3.10 Acknowledgments

The author is enormously grateful to Professor F.A. Popp for his valuable discussions and significant suggestions. The author expresses his thanks to Professor R.K. Mishra, Professor K.H. Li, Professor M.W. Ho, Professor K. Zioutas, Dr. W.P. Mei, Messrs. M. Bischof, R. Neurohr, K. Lambing, M. Galle for their useful discussions and Professor J.Y. Zhang for his interest in this work. Many thanks go to Mr. J.B. Li, Miss J. Deny and Miss C. Benkel for their help with numerical work and to Mr. M. Bischof for proofreading.

3.11 References

1. I. Prigogine and I. Stengers, *Order Out of Chaos* (Bantam Books Inc., 1984).

2. H. Haken (ed.), *Cooperative Effects: Progress in Synergetics* (North-Holland, American Elsevier, 1974).

3. C.M. Bowden, D.W. Howgate, and H.R. Robl (eds.), *Cooperative Effects in Matter and Radiation* (Plenum Press, New York and London, 1977).

4. R.H. Dicke, Coherence in spontaneous radiation processes. *Phys. Rev.* **93** (1954) p. 99-110.

5. N. Skribanowitz, I.P. Herman, J.C. MacGillivray, and M.S. Feld, Observation of Dicke superradiance in optically pumped HF gas. *Phys. Rev. Lett.* **30** (1973) p. 309-312.

6. L. Allen and J.H. Eberly, *Optical Resonance and Two-Level Atoms* (Wiley, New York, 1975).

7. R. Bonifacio and L.A. Lugiato, Cooperative radiation processes in two-level systems: Superfluorescence. *Phys. Rev. A* **11** (1975) p. 1507-1521.

8. R. Bonifacio and L.A. Lugiato, Cooperative radiation processes in two-level systems: Superfluorescence II. *Phys. Rev. A* **12** (1975) p. 587-598.

9. H.M. Gibbs, Q.H.F. Viehen, and H.M.J. Hikspoors, Single-pulse superfluorescence in cesium. *Phys. Rev. Lett.* **39** (1977) p. 547-550.

10. M.S. Feld and J.C. MacGillivray, Superradiance, in *Coherent Nonlinear Optics, Vol. 21 of Topics in Current Physics*, eds. M.S. Feld and V.S. Letokhov (Springer, Berlin, 1980) p. 7-57.

11. Q.H.F. Vrehen and H.M. Gibbs, Superfluorescence experiments, in *Dissipative Systems in Quantum Optics, Vol. 27 of Topics in Current Physics*, ed. R. Bonifacio (Springer, Berlin 1982) p. 111-147.

12. M. Gross and S. Haroche, Superradiance: An essay on the theory of collective spontaneous emission. *Phys. Rep.* **93** (1982) p. 301-396.

13. O. Svelto, *Principles of Lasers, 2nd edition* (Plenum Press, New York, London, 1982).

14. R.K. Bullough, Conceptual models of co-operative processes in nonlinear electrodynamics, in *Nonlinear Electrodynamics in Biological Systems*, eds. W.R. Adey and A.F. Lawrence (Plenum Press, New York and London, 1984) p. 347-392.

15. F. Haake and R. Reibold, and F.P. Mattar, Quantum fluctuations in two color superfluorescence from three-level systems. *J. Opt. Soc. Am. B* **1** (1984) p. 547-548.

16. N.N. Bogolubov, Jr., T. Quang, and A.S. Shumovsky, Double optical resonance in a system of atoms. *Phys. Lett. A* **112** (1985) p. 323-326.

17. Q. Gu, Superradiant beats and their spectroscopic applications, in *Trends in Quantum Electronics*, ed. S.M. Bochum (European Physical Society, 1985) p. 359-361.

18. Q. Gu, A simple method for describing superradiance. *Chin. Phys. Lasers (USA)* **13** (1986) p. 855-858.

19. Q. Gu, Quantum beats in superradiance. *Chin. Phys. Lasers (USA)* **14** (1987) p. 709-714.

20. Q. Gu, Some nonlinear effects in superradiance (review). *Chin. J. Quantum Electron.* **4** (1987) p. 107-121.

21. Q. Gu, Squeezing is superradiance. *Chin. Phys. Lasers (USA)* **15** (1988) p. 218-221.

22. Q. Gu, On superradiance. *Physics* **18** (1989) p. 15-20.

23. S.V. Lawande and B.N. Jagatap, Cooperative effects in a system of strongly driven three-level atoms. *Phys. Rev. A* **39** (1989) p. 683-693.

24. J. Knoester and S. Mukamel, Intermolecular forces, spontaneous emission, and superradiance in a dielectric medium: Polariton-mediated interactions. *Phys. Rev. A* **40** (1989) p. 7065-7080.

25. T. Quang, Squeezing via a two-photon transition in a system of driven three-level atoms. *Phys. Rev. A* **43** (1991) p. 2561-2564.

26. T. Becker and R.-H. Rinkleff, Superfluorescent transitions between high-lying levels in an external electric field. *Phys. Rev. A* **44** (1991) p. 1806-1816.

27. T.A.B. Kennedy, Quantum theory of cross-phase-modulational instability: Twin-beam correlations in a $\chi^{(3)}$ process. *Phys. Rev. A* **44** (1991) p. 2113-2123.

28. S. Haroche and D. Kleppner, Cavity quantum electrodynamics. *Phys. Today* **42** (1989) p. 24-30.

29. D.A. Miller, Useful perspective on the relation between biological and physical descriptions of phenomena. *J. theor. Biol.* **152** (1991) p. 341-355.

30. A. Szent-Györgyi, *Introduction to A Submolecular Biology* (Academic Press A, New York, 1960).

31. H. Fröhlich, Long-range coherence and energy storage in biological systems. *Int. J. Quantum Chem.* **2** (1968) p. 641-649.

32. F.A. Popp, B. Ruth, W. Bahr, J. Boehm, P. Grass, G. Grolig, M. Rattemeyer, H.G. Schmidt, and P. Wulle, Emission of visible and ultraviolet radiation by active biological system. *Colletive Phenomena* **3** (1981) p. 187-213.

33. K.H. Li, F.A. Popp, W. Nagl, and H. Klima, Indications of optical coherence in biological systems and its possible significance, in *Coherent Excitations in Biological Systems*, eds. H. Fröhlich and F. Kremer (Springer, Heidelberg, Berlin, New York, 1983) p. 117-122.

34. F.A. Popp, On the coherence of ultraweak photon emission from living tissues, in *Disequilibrium and Self-Organization*, ed. C.W. Kilmister (Reidel, Dordrecht, 1986) p. 207-230.

35. W. Nagl and F.A. Popp, Opposite long-range interactions between normal and malignant cells, in *Energy Transfer Dynamics*, eds. T.W. Barret and H.A. Pohl (Springer, Heidelberg, Berlin, New York, 1987) p. 248-256.

36. F.A. Popp, G. Cilento, W.B. Chwirot, M. Galle, A.A. Gurwitsch, H. Inaba, K.H. Li, W.P. Mei, W. Nagl, R. Neurohr, D.H.J. Schamhart, J. Slawinski, and R. van Wijk, Biophoton emission (multi-author review). *Experientia* **44** (1988) p. 543-600.

37. Q. Gu, The ultraweak photon emission from biological systems (review). *Chin. J. Quantum Electron.* **5** (1988) p. 97 - 112.

38 Q. Gu, Ultraweak photon emission from living systems. *Science (China)* **41** (1989) p. 35-41.

39. F.A. Popp, Photon storage in biological systems, in *Electromagnetic Bio-Information, 2nd edition*, eds. F.A. Popp, U. Warnke, H.L. Koenig, and W. Peschka (Urban & Schwarzenberg, Muenchen, Wien, Baltimore, 1989) p. 144-167.

40. F.A. Popp, *Biologie de la Lumiere* (Marc Pietteur, Editeur, 1989) p. 113-123.

41. K.H. Li and F.A. Popp, Dynamics of DNA excited states, in *Molecular and Biological Physics of Living Systems*, ed. R.K. Mishra (Dordrecht, Boston, London, 1990) p. 31-52.

42. M. Gueron, J. Eisinger, and A.A. Lamola, Excited states of nucleic acids, in *Basic Principles in Nucleic Acid Chemistry*, ed. Paul O.P. Tso (Academic Press, New York, London, 1974) p. 311-398.

43. J.J. McCullough, Photoadditions of aromatic compounds. *Chem. Rev.* **87** (1987) p. 811-860.

44. G.S. Agarwal and S.S. Jha, Theory of resonant Raman scattering of intense optical waves. *J. Phys. B* **12** (1979) p. 2655-2671.

45. G.S. Agarwal, Quantum statistical theories of spontaneous emission and their relation to other approaches, in *Quantum Optics, Vol. 70 of Springer Tracts in Modern Physics* (Springer-Verlag, Berlin, Heidelberg, New York, 1974) p. 1-129.

46. G.S. Agarwal, L.M. Narducci, D.H. Feng, and R. Gilmore, Intensity correlations of a cooperative system. *Phys. Rev. Lett.* **42** (1979) p. 1260-1263.

47. R.K. Mishra (ed.), *The Living State-II* (World Scientific, 1985).

48. E. Balanovski and P. Beaconsfield, Solitonlike excitations in biological systems. *Phys. Rev. A* **32** (1985) p. 3059-3064.

49. P. Tombesi and E.R. Pike (eds.), *Squeezed and Nonclassical Light* (Plenum Press, New York and London, 1989).

50. Q. Gu, Photon antibunching in light from a single-mode laser. *Chin. Phys. Lasers (USA)* **15** (1988) p. 615-619.

51. Q. Gu, Squeezing in a two-photon Jaynes-Cummings model. *Acta Phys. Sin.* **37** (1988) p. 751-759.

52. Q. Gu, The quantum statistical properties of the Jaynes-Cummings model. *Acta Phys. Sin.* **38** (1989) p. 737-744.

53. Q. Gu, Sub-Poissonian photon statistics in the degenerate two-photon laser. *J. Shenzhen Univ. (Science & Engineering)* **6** (1989) p. 1-9.

54. Q. Gu, Photon statistics for laser with a weak spontaneous emission. *J. Shenzhen Univ. (Science & Engineering)* **7** (1990) p. 11-20.

55. Q. Gu, Squeezing in One-Atom Maser. *Chin. J. Lasers* **17** (1990) p. 352-356.

56. Q. Gu, Novel features in photon statistics from a single-mode laser. *Science in Chin. A* **33** (1990) p. 1460-1468.

57. H.P. Yuen, Nonclassical light, in *Photons and Quantum Fluctuations*, eds. E.R. Pike and H. Walther (Adam Hilger, Bristol and Philadelphia, 1988) p. 1-9.

58. M. Lax, Quantum Noise XI: Multitime correspondence between quantum and classical stochastic processes. *Phys. Rev.* **172** (1968) p. 350-361.

59. G. Compagno and F. Persico, Theory of collective resonance fluorescence in strong driving fields. *Phys. Rev. A* **25** (1982) p. 3138-3168.

60. Q. Gu, Coherent radiation and the ultraweak photon emission from living systems. *Physics* **18** (1989) p. 235-242.

61. Q. Gu, Analogy between life and the laser. *Chin. J. Infrared and Millimeter (USA)* **8** (1989) p. 83-89.

Chapter 4

Coherence in Physics and Biology

Ke-hsueh Li

Abstract

In this paper we will show first that the term "coherence" in quantum theory has to be considered as a fundamental physical concept for all matter waves, as important as Heisenberg's Uncertainty Principle. Then we discuss cooperative radiation phenomena and the possibility of communication in living systems by the use of some extreme optical effects of quantum nature, which can be displayed only by a photon emission with very weak intensity.

4.1 Introduction

Because of the achievements of modern science almost nobody believes that the living processes ought to be attributed to some mystical forces. "It would be very singular that all nature, all the planets, should obey eternal laws, and that there should be a little animal, five feet high, who, in contempt of these laws could act as he pleased, solely according to his caprice." This remark from Voltaire in his "Ignorant Philosopher" epitomizes the healthy scepticism of the humanist movement[1].

Bohr[2] thought that "we all agree with Newton that the real basis of science is that Nature under the same conditions will always exhibit the same regularities. Therefore, if we were able to push the analysis of the mechanism of living organisms as far as that of atomic phenomena, we should scarcely expect to find any features differing from the properties of inorganic matter".

Despite the fact that the living phenomena are displayed macroscopically, the main biological processes are carried out microscopically. Even before the discovery of the wonderful structure of deoxyribonucleic acid (DNA), Schrödinger [3] had already roughly described the hereditary mechanism of "chromosome fibres" and mutations of living molecules in the language of quantum theory.

So the question at issue is not whether physics can explain some features of the functions of living organisms, which it clearly can, but whether some fundamental traits are still being overlooked, ignored or even missed in the analysis of living or natural phenomena on the basis of physical experience.

The surprisingly essential feature of a biological process seems to be that, in spite of the vast numbers of molecules involved, it is ordered in time and space. The molecules taking part in the process are cooperatively following lawful behaviour of matter and collectively displaying new features which are missing from those of individual molecules. This new collective ordering of the structure of living matter must avoid the decay into the disordered region of the equilibrium state.

According to the second law of thermodynamics "every process, event, happening - call it what you will; in a word, everything that is going on in Nature means an increase of the entropy of the part of the world where it is going on. Thus a living organism continually increases its entropy - or, as you may say, produces positive entropy - and thus tends to approach the dangerous state of maximum entropy, which is death. It can only keep aloof from it, i.e. alive, by continually drawing from its environment negative entropy" [3].

We try to show that even from the viewpoint of physics, which aims to invoke general principles of the motion of matter in the space-time frame of our real world, the biological phenomena are not too complicated, and can be explained as features of open systems far away from thermal equilibrium at macroscopic as well as at microscopic levels.

At the microscopic level the life process is sustained mainly by the photosynthesis of important biomolecules from simple inorganic substances in the presence of sunlight. For instance, plants absorb photons of visible light from the sun's radiation and thereby convert carbon dioxide CO_2 and water H_2O into oxygen O_2 and the simple carbohydrate $C_6H_{12}O_6$. If the photosynthesis occurs in the structurally highly organized chloroplasts, as has been generally accepted [4], then the process of photosynthesis can only manifest itself within the coherence volume of the sunlight, as can be shown by the following consideration:

The angular diameter 2α which the sun's disc substends on the surface of the earth is about $0°32' \sim 0.0093$ radian. The diameter d of the area of coherence is approximately $0.16\bar{\lambda}/0.0047 \sim 34\bar{\lambda}$ [5]. Taking the mean wavelength of light $\bar{\lambda}$ as 5.5×10^{-5} cm gives $d \sim 20\mu m$. At the same time the coherence length of visible light is about 3m as it should be greatly reduced on account of a number of unknown factors, whereas the size of chloroplast is only about $1 \sim 2\mu m$.

On the other hand, we know that the primary electromagnetic interaction between dipole or electric charges and e.m. waves concerns the amplitude (strength of the field, which is a vector) but not intensity (energy of the field, which is a scalar). The interaction part of the Hamiltonian can be represented simply by [6]

$$H_I = e\vec{D} \cdot \vec{E},$$

where $e\vec{D}$ is the effective total dipole of a molecule, and $\vec{E}$ represents the strength of field. For the case of black-body radiation we can represent $\vec{E}$ as a Fourier series in the form [7]

$$\vec{E} = \sum_n \left[\vec{A}_n sin\left(\frac{2\pi nt}{T}\right) + \vec{B}_n cos\left(\frac{2\pi nt}{T}\right)\right],$$

where T is the period considered for observation and the coefficients $\vec{A}_n$ and $\vec{B}_n$ are statistically independent of each other. To prove the statistical independence of the different modes of the radiation field the assumption of randomly distributed phases of the different oscillators is sufficient.

We know that the field $\vec{E}$ can actually exist only within the coherent space-time, while beyond the coherent space-time $\vec{E}$ must effectively be zero due to randomly distributed phases. That is to say that the electromagnetic interaction H_I can occur only within the coherent space-time while beyond that the effect of electromagnetic interactions is statistically not measureable, or $H_I \approx 0$. Thus we see that the coherence of fields plays a very important role for the interaction, and hence for photochemical reactions to occur at all. In fact we find that this statement is always true in the quantum theory for interactions of all sorts of particles or matter waves. But, for biological systems the electromagnetic interaction is particularly important since biological giant macro-molecules contain tens to hundreds of thousands of atoms assembled into complex geometrical configurations. The shape and stability of such molecules (which actually are ordered electromagnetic structures) are dependent on the delicate balance of electromagnetic interactions between neighbouring atoms.

In Feynman diagrams the electromagnetic interaction (or e.m. force) can be described by virtual photons exchanged between particles. The existence of virtual photons would be limited by the time- and space-interval of the uncertainty relation. The time- and space-scope of interactions is limited by

$$\Delta t = \frac{\hbar}{\Delta E}.$$

E is the energy of a virtual photon and the range of interaction forces is

$$R = \Delta x = c\Delta t = \frac{c\hbar}{\Delta E} = \frac{c\hbar}{h\Delta\nu} = \frac{1}{2\pi}\frac{\bar{\lambda}^2}{\Delta\lambda} = L,$$

where L is the coherence length of photons [8]. If we take $\bar{\lambda} = 5.5 \times 10^{-5}$ cm and $\Delta\lambda = 1.18 \times 10^{-4} \dot{A}$ (the natural line broadening), we have

$$L \approx 4m,$$

in consistence with that roughly estimated above.

This scheme for electromagnetic interaction can be shown to be a general one for all other elementary particles. The uncertainty relation of Heisenberg is the only fundamental condition to determine the space-time region of interactions between particles and fields. The important fact which may be traced back to the original papers of Heisenberg [9] is that the uncertainty relation is only an alternative approach to coherence properties of fields and particles.

The interference between different probability amplitudes, hence the coherence properties of probability packets constitute the essential point of quantum theory [10].

4.2 Uncertainty Principle and Coherence

The uncertainty relations for systems of material particles follow from the wave-corpuscle duality of matter. Any attempt to reveal one of the two aspects requires a modification of the experimental set-up which destroys any possibility of observing the other aspect.

The dynamic state of a classical particle is defined at every instant by specifying precisely its position $\vec{r}(x, y, z)$ and its momentum $\vec{p}(p_x, p_y, p_z)$. On the contrary, since the wave function $\psi(\vec{r})$ has a certain spatial extension, one cannot attribute a precise position to a quantum particle; one can only define the probability of finding the particle in a given region of space when one carries out a measurement of its position. The same consideration is also valid for the observation of momentum. The wave function actually represents the probability amplitude including all information about the system under consideration.

The probability amplitude is the basic element or alternative (in the sense of Feynman) for analysing the interference and interaction between different particles and fields in quantum mechanics. Thus we may remind of Feynman's formulation of the uncertainty principle [11]: *Any determination of the alternative taken by a process capable of following more than one alternative destroys the interference between alternatives.* Heisenberg's original statement of the uncertainty principle was not given in the form we have cited here.

We will see next how closely the uncertainty relation connects to the coherence consideration of the wave function:

4.2.1 Some elementary relations

For a one-dimensional problem, which can easily be extended to many-dimensional cases, from Heisenberg's original paper we have [10]

$$\Delta x \, \Delta p \geq \hbar \tag{4.1}$$

with the de Broglie relation

$$\Delta p = \hbar \Delta k = \hbar \Delta \frac{2\pi}{\lambda} = h \frac{\Delta \lambda}{\lambda_o^2}. \tag{4.2}$$

Here λ_o denotes the mean value of the wavelength under consideration.

From (4.2) and (4.1) we have

$$\Delta x \geq \frac{1}{2\pi} \frac{\lambda_o^2}{\Delta \lambda} = L, \tag{4.3}$$

where L is the coherence length of the matter wave [5,12].

With the exception on the right hand side with an equality sign, the relation (4.3) had already been deduced by Heisenberg himself.

Just as momentum is a wave number and cannot be localized in space, so energy is a frequency and cannot be localized in time. Thus there exists, as suggested by the principle of relativity, a time-energy uncertainty relation

$$\Delta t \, \Delta E \geq \hbar, \tag{4.4}$$

where $\Delta E = \Delta \hbar\omega$ according to Planck's relation.

Because of

$$\Delta \hbar\omega = \hbar \left(2\pi\Delta \frac{v}{\lambda}\right) = \hbar v \left(2\pi \frac{\Delta\lambda}{\lambda_o^2}\right) = \hbar \frac{v}{L}, \tag{4.5}$$

we have from (4.4)

$$\Delta t \geq \frac{L}{v} = \tau, \tag{4.6}$$

where $v = c$ for electromagnetic waves, and τ represents the coherence time.

We must be aware, however, that the interpretation of equation (4.4) is quite different. In the position-momentum uncertainty relation, the position and momentum variables play exactly symmetrical roles: they are both conjugate operators and can be measured at a given time t. The statistical distributions of the results of a measurement and consequently the uncertainties $\Delta x_i \Delta p_i$ are all derivable from the value of the wave function at that time. In relation (4.4), on the other hand, energy and time play fundamentally different roles: the energy is a dynamic variable of the system, whereas the time is a parameter. The relation (4.4) connects the uncertainty ΔE in the value taken by this dynamic variable to a time interval Δt characteristic of the rate of change of the system [13].

Now, let us consider a one-dimensional free particle which may be represented by a monochromatic plane wave

$$e^{i(kx - \omega t)}$$

with well-defined momentum $\hbar\vec{k}$ and energy $\hbar\omega$. By superposition of plane waves, one can, by interferences, get a wave packet, the length of which can be denoted by Δx or L of the wave train and its group velocity by v. It travels at velocity v along the x-axis; however, the instant at which it passes a given point of the axis is not determined precisely, but carries an uncertainty

$$\Delta t = \frac{\Delta x}{v} = \frac{L}{v} = \tau. \tag{4.7}$$

Furthermore, this wave train has a certain spread in momentum space, hence an uncertainty ΔE in the value of energy of the particle

$$\Delta E \simeq \frac{\partial E}{\partial p}\Delta p = v\Delta p. \tag{4.8}$$

Three important points must be mentioned here: one is that all the inequalities discussed above must be replaced by equalities, if we are engaged with coherent states, i.e. the state of minimum uncertainty; secondly, the mean lifetime τ of a radioactive system is represented by the decay time of a nonstationary wave packet; it does not correspond to a well-defined value of the energy, but to an energy spectrum with a certain spread ΔE, usually called the level width. One must wait for a time of order τ to observe an appreciable change in the properties of the system. Consequently

$$\tau\Delta E = \hbar. \tag{4.9}$$

The third point is that, if we insert subnuclear masses and dimensions in (4.1) we find within the uncertainty region a velocity range $\frac{\Delta p}{m}$ which is by no means small compared to the speed of light [14]. This fact indicates that the phenomena displayed within the coherence volume ought to be considered as a whole and the "particle" cannot be localized.

4.2.2 Coherent structures

In classical optics, the coherence length L is equivalent to the breadth of the wave packet [15]. Similarly, in quantum mechanics the uncertainty region Δx in the position of a particle corresponds to the breadth of the wave function within which the matter is statistically distributed in the sense of Born [13]. However, the wave function as a solution of Schrödinger's equation must be confined by its boundary conditions which are particularly important for the building of interference structures of matter waves. In order to show the interference pattern building within the coherence length or the coherence volume to be a general fact, we rewrite, first of all, the general time-dependent Schrödinger equation

$$i\hbar\frac{\partial}{\partial t}\psi = \left(-\frac{\hbar^2}{2m}\nabla^2 + V\right)\psi \tag{4.10}$$

with the Hamiltonian:

$$H = -\frac{\hbar^2}{2m}\nabla^2 + V$$

in the form

$$\hbar^2\nabla^2\psi = 2m\left[V - i\hbar\frac{\partial}{\partial t}(ln\psi)\right]\psi \tag{4.11}$$

with the momentum operator

$$\vec{p} = \frac{\hbar}{i}\nabla \quad \text{and} \quad p^2 = -\hbar^2\nabla^2. \tag{4.12}$$

If we consider equation (4.11) as an eigenvalue problem, we then have the appropriate eigenvalue of the momentum operator

$$p = \sqrt{(-2m)\left[V - i\hbar\frac{\partial}{\partial t}(ln\psi)\right]}. \tag{4.13}$$

For the case when the solution ψ of the Schrödinger equation can be factorised into space- and time-parts, i.e. $\psi(x,t) = \varphi(x)f(t)$, the momentum eigenvalue (4.13) has the simple form [8,13]:

$$\mid p \mid = \sqrt{2m(E - V)}, \tag{4.14}$$

where E is the well-defined energy of the whole system in a certain state. We see that the very important information carried by the wave function actually arises from the boundary conditions, where it is represented by potential V and the total energy E of the system. The latter term can generally be taken as a constant.

On account of the relation

$$\Delta \mid p \mid = \sqrt{\langle \mid p \mid^2 \rangle - \langle \mid p \mid \rangle^2} \tag{4.15}$$

and the constraint $\mid \frac{\langle V \rangle}{E} \mid < 1$, we obtain from inequalities (4.1) and (4.3)

$$\Delta x \geq \hbar\sqrt{\frac{2E}{m}}\frac{1}{\Delta V} = L. \tag{4.16}$$

Taking for V a simple Coulomb potential of a hydrogen atom, we get [13]

$$\begin{aligned} \langle V\rangle &= \langle\frac{-ke^2}{r}\rangle = -\frac{ke^2}{n^2 a_0} \\ \langle V^2\rangle &= \frac{2k^2e^4}{(2l+1)n^3a_0^2}, \end{aligned} \tag{4.17}$$

where n is the principal quantum number and l is the angular momentum quantum number, a_0 the radius of the first Bohr orbit. If $\mid E \mid = \frac{mk^2e^4}{2\hbar^2}$ is constant, then the equation (4.16) becomes

$$\Delta x \geq \frac{n^2 a_0}{\sqrt{\frac{2n}{2l+1} - 1}} = L_{n,l}, \tag{4.18}$$

where $L_{n,l}$ denotes the coherence length of particles with different quantum numbers n and l. For minimum uncertainty states we have

$$\Delta x = L_{n,l}.$$

When the wave function $\psi(x,t)$ cannot be factorised, the problem becomes very complicated, but the fundamental correspondence between the coherence length and the breadth of the wave packet must remain.

We see that the atomic orbits represent the structures of electrons or electron waves within their coherence (uncertainty) volume. They are actually the different interference patterns of electron waves constrained by the boundary condition (4.14).

This well-known fact has been discussed in all elementary text books of quantum mechanics but it was not evidently shown that the structures are formed within the uncertainty (or coherence) volume. Interference is a general phenomenon of matter waves and coherence is an inherent feature of them.

Hereby we may illustrate this with another elegant example [16]: Let us consider molecular orbitals of the hydrogen molecule-ion. Its Hamiltonian is usually given by

$$H = -\frac{\hbar^2}{2m}\nabla^2 + \frac{e^2}{4\pi\epsilon_o}\left(-\frac{1}{r_a} - \frac{1}{r_b} + \frac{1}{R}\right), \tag{4.19}$$

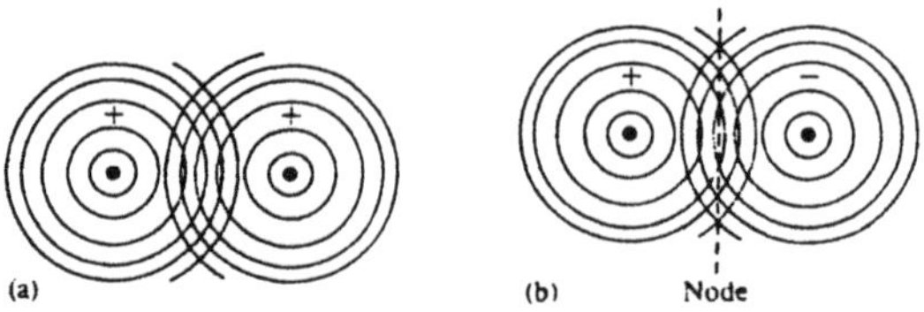

Figure 4.1: *(a) Constructive, (b) destructive overlap between 1s-orbitals.*

where r_a and r_b are the distances of the single electron from the two nuclei a and b, separated by R. The lowest energy solution, a bonding orbital, can be modelled as the linear combination $\psi_a + \psi_b$ of the hydrogen 1s-orbitals centred on each nucleus, see Fig. 4.1(a).

The antibonding orbital can also be modelled as $\psi_a - \psi_b$ by using the same two atomic orbitals, but this time superimposed with opposing amplitudes, as shown in Fig. 4.1(b). In the first case there is constructive interference where the two wave functions overlap and their amplitudes augment each other: this models the accumulation of amplitude in the internuclear region which we know is an important feature of the bonding process. In the antibonding combination the amplitudes tend to cancel each other and there is exact cancellation to form a nodal plane half way between the two nuclei, just as in the exact solution. The destructive interference where the two wave functions overlap with opposite phase models the annihilation of the amplitude in the internuclear region.

Now let us replace the electron by a photon and assume that the two atoms are sufficiently separated so that the overlap of electron wave functions is impossible. In this case the symmetric state describes the fact that if a photon is emitted by one atom, one must also be emitted by the other. That is just the most superradiant state, a cooperative coherent spontaneous radiation process. On the other hand, the antisymmetric state describes the fact that a photon released by one atom must be absorbed by the other and vice versa. That is only possible if one atom is initially excited and the other not and both the atoms must be within the uncertainty length (coherence length) of the photons. This latter case corresponds, therefore, to the most subradiant state of the two atom-system in which the photon can be trapped for a long time. All these cases have been discussed in detail by different authors since the fundamental pioneering work of Dicke [17].

Here we see that the symmetric state consists in the constructive interference of photons between two atoms and the antisymmetric state forms the destructive interference which causes an attractive force between two atoms due to the exchanges of photons. This force has been named the First-Order Dispersion Forces by Stephen [18].

By the discussions in this subsection we have aimed to show that there really are coherent structures within the uncertainty length. These structures are formed by the interference of matter waves constrained by boundary conditions. The existence of structures does not contradict the state indeterminacy of the uncertainty principle. The indeterminacy only actually means the nonlocalization of the particle in position and momentum compared with the localization in the classical sense. Due to the nonlocalization within the uncertainty length the particle loses its classical picture.

4.2.3 The uncertainty principle and biology

The title of this subsection is very necessary and relevant to the theme, because all fundamental and elementary processes in organisms proceed on microscopic, i.e. atomic and molecular levels as we mentioned in the introduction. In order to further illustrate the necessity for discussion of this theme in biology we give some examples:

A: The average kinetic energy of each degree of freedom of a molecular motion is about 0.6 $kcal/Mol$ at $25°C$ [19]. According to Boltzmann's law of energy distribution there are then enough molecules with several times the energy to break molecular bonds which are in the order of ca. 1 $kcal/Mol$ for dispersion forces and $10 \sim 20$ $kcal/Mol$ for ionic bonds. However, we may not assume that the stability of the molecular structure and of chemical reactions between molecules is destroyed by thermal agitations if the pulse widths of thermal fluctuations, no matter how strong, are smaller than the coherence times of molecular interaction forces, because instantaneous disturbances within a large uncertainty space-time cannot play a role in damaging the long term behaviour of molecular structures and chemical reactions.

On the contrary, the thermal agitations may even be considered as a favourable condition for biochemical processes in cells due to the fact that the motions of chemical complementary partners may help attain proper geometric positions for a biochemical reaction.

B: On the basis of the uncertainty principle we also understand why the enzyme molecules must have a high molecular weight (e.g. lysozyme has the molecular weight of ca. 14300):

In biochemical catalysis an enzyme must interact with a macromolecule precisely at a certain, determined position such that the uncertainty in position Δx may be reduced to a value as small as possible. But, the uncertainty principle forces Δp to take an increasingly larger value.

That does not mean, however, that the uncertainty velocity Δv must be large due to the relation: $\Delta p = m\Delta v$, where the molecular weight m of the enzymes is very high. For a large value of Δv there may be a possibility that the interacting molecules may break away from the coherence volume for a short time, and the stability of the biochemical reaction cannot therefore be maintained. By a simple calculation we can find the lower limit of the uncertainty region Δx for the lysozyme, with molecular weight of 14300 at room temperature, which is of the order of 10^{-3} Å.

C: It has been known for a long time that DNA molecules are luminescent materials[20]. The fluorescence spectra of native and denatured calf thymus DNA are distributed around the wavelength of 3500 Å. These spectra have been identified as excimers and exciplexes built by base-molecules due to the base stacking. We know that in the DNA molecule the distance between each two bases is ca. 3.4Å, thus each wavelength of the DNA fluorescence may cover about 1000 base-molecules along the DNA-skeleton.

One should know that the coherence length of the radiation is much longer than the wavelength. In this case we must surely have the collective (or cooperative) emission from DNA molecules, according to Dicke's theory [17]. The coherence lengths may be very long due to the cooperatively coherent emissions. The photons within the DNA skeleton may also be completely delocalized due to the fact that the molecular energy levels are relatively definite, i.e., $\Delta\omega$ are relatively smaller, and $\Delta x = c\Delta t$ is large, due to the uncertainty relation between $\Delta\omega$ and Δt.

The evidence for these statements comes from the fact that for a bacterium lacking the normal repair mechanism the formation of only 10 photochemical defects per DNA molecule (10^7 nucleotides) is fatal, with a 50 % probability[21]. The details of events initiated by the absorption of a photon by one of the bases and ending with the creation of a stable photochemical product on the same or a different base can be understood only with the aid of the knowledge of basic quantum theory and of the coherence effect of photons within DNA skeletons.

4.2.4 Minimum uncertainty and coherence

The subject of this subsection is well known in quantum optics. On account of its importance in the analysis of elementary processes by quantum mechanical methods we will present some new aspects which may be useful for understanding some biological phenomena on microscopic levels.

Using the inequality of arithmetic mean and geometric mean, for $\langle(\Delta\hat{p})^2\rangle$ and $\omega^2\langle(\Delta\hat{q})^2\rangle$ we have the following inequality [22]:

$$\frac{1}{2}\left[\langle(\Delta\hat{p})^2\rangle + \omega^2\langle(\Delta\hat{q})^2\rangle\right] \geq \omega\left[\langle(\Delta\hat{q})^2\rangle\langle(\Delta\hat{p})^2\rangle\right]^{1/2} \geq \frac{\hbar\omega}{2}, \tag{4.20}$$

where

$$\Delta\hat{p} = \hat{p} - \langle\hat{p}\rangle, \qquad \Delta\hat{q} = \hat{q} - \langle\hat{q}\rangle. \tag{4.21}$$

From the commutation rule

$$[\hat{q}, \hat{p}] = i\hbar, \tag{4.22}$$

we have

$$\left[\langle(\Delta\hat{q})^2\rangle\langle(\Delta\hat{p})^2\rangle\right]^{1/2} \geq |\langle\Delta\hat{q}\Delta\hat{p}\rangle| \geq \frac{1}{2}|\langle[\Delta\hat{q}, \Delta\hat{p}]\rangle| = \frac{1}{2}|\langle[\hat{q}, \hat{p}]\rangle| = \frac{1}{2}\hbar \tag{4.23}$$

If we use the eigenvalue properties of coherent states

$$\hat{a}\,|\alpha\rangle = \alpha\,|\alpha\rangle, \qquad \langle\alpha|\,\hat{a}^{+} = \langle\alpha|\,\alpha^{*}, \tag{4.24}$$

we obtain, from the definitions

$$\begin{aligned} \hat{q} &= \left(\frac{\hbar}{2\omega}\right)^{1/2}(\hat{a} + \hat{a}^{+}) \\ \hat{p} &= -i\left(\frac{\hbar\omega}{2}\right)^{1/2}(\hat{a} - \hat{a}^{+}), \end{aligned} \tag{4.25}$$

the other forms of Eq. (4.21):

$$\begin{aligned}\langle(\Delta\hat{q})^2\rangle &= \langle\hat{q}^2\rangle - \langle\hat{q}\rangle^2 = \frac{\hbar}{2\omega},\\ \langle(\Delta\hat{p})^2\rangle &= \langle\hat{p}^2\rangle - \langle\hat{p}\rangle^2 = \frac{\hbar\omega}{2}.\end{aligned} \tag{4.26}$$

Thus, for the coherent states the inequality (4.23) reduces to the equality

$$\langle(\Delta\hat{q})^2\rangle\langle(\Delta\hat{p})^2\rangle = \frac{\hbar^2}{4}. \tag{4.27}$$

On account of this relation for coherent states we may replace all inequalities of foregoing sections by equalities.

Of course, from Eq. (4.20) we also have the expression for coherent states:

$$\frac{1}{2}\left[\langle(\Delta\hat{p})^2\rangle + \omega^2\langle(\Delta\hat{q})^2\rangle\right] = \frac{1}{2}\left[\langle\hat{p}^2\rangle + \omega^2\langle\hat{q}^2\rangle\right] - \frac{1}{2}\left[\langle\hat{p}\rangle^2 + \omega^2\langle\hat{q}\rangle^2\right] = \frac{\hbar\omega}{2}. \tag{4.28}$$

This has been explained as the difference between the total energy of the field and that of its coherent part, so it is the incoherent energy of the field. Therefore it may be said that in the coherent state the incoherent energy takes on its minimum value $\frac{\hbar\omega}{2}$ in energy modes arising only from the zero-point fluctuations [23].

If we calculate the same quantity of Eq. (4.28), but, in the Fock state instead of the coherent state, then we get the total energy of the field in the form

$$\frac{1}{2}\left[\langle\hat{p}^2\rangle + \omega^2\langle\hat{q}^2\rangle\right]_{Fock} = \hbar\omega(n + \frac{1}{2}). \tag{4.29}$$

By assuming that the total field energy remains constant, independent of the representation of mathematical pictures, then from Eqs. (4.28) and (4.29) we have the field energy of the coherent part:

$$\frac{1}{2}\left[\langle\hat{p}\rangle^2 + \omega^2\langle\hat{q}\rangle^2\right] = n\hbar\omega, \tag{4.30}$$

which is the excited energy of the nth state of a harmonic oscillator with the frequency ω.

Now we consider the variance of photon numbers, the mean-square deviation of photon number:

$$\langle \Delta n^2 \rangle \equiv \langle n^2 \rangle - \langle n \rangle^2, \tag{4.31}$$

where $n = a^+ a$.

For the coherent state we have

$$\langle \Delta n^2 \rangle \equiv \langle a^+ a \, a^+ a \rangle - \langle a^+ a \rangle^2 = \langle n \rangle, \tag{4.32}$$

which obeys the well known Poisson distribution [24], but, in normal order

$$\langle \Delta n^2 \rangle = 0. \tag{4.33}$$

For the Fock-state we have

$$\langle \Delta n^2 \rangle_{Fock} \equiv \langle a^+ a \, a^+ a \rangle_{Fock} - \langle a^+ a \rangle^2_{Fock} = 0\,, \tag{4.34}$$

but, in normal order we get

$$\langle \Delta n^2 \rangle_{Fock} \equiv -\langle n \rangle_{Fock}. \tag{4.35}$$

For the Fock-state the photon number is completely fixed, so that the mean-square deviation is zero (see Eq. (4.34)). But, that Eq. (4.35) should be negative in the normal ordering is difficult to understand. The normal ordering of operators is important for evaluating the correlation function in second order. In fact, this result indicates the possibility of an occurence of the antibunching effect in the Fock-state.

To dodge this difficulty we prefer to introduce the quantity r to denote the relative excess coincidence counting rate which has been proposed by Paul [58]

$$r = \frac{\langle \Delta n^2 \rangle}{\langle n \rangle^2} = \frac{\langle n^2 \rangle - \langle n \rangle^2}{\langle n \rangle^2} \tag{4.36}$$

in accordance with its classical representation

$$r = \frac{\overline{I^2} - \bar{I}^2}{\bar{I}^2} \equiv \frac{(\Delta I)^2}{\bar{I}^2} \tag{4.37}$$

as a convenient measure of the strength of the bunching effect, the criterion for photon bunching to occur thus being

$$r > 0. \tag{4.38}$$

On the other hand, it will be identified as photon antibunching when the inverse of (4.38) is true.

We see that the quantity r in Eq. (4.36) in normal order for the Fock-state is just

$$r = -\frac{\langle n \rangle_{Fock}}{\langle n \rangle^2_{Fock}} = -\frac{1}{\langle n \rangle_{Fock}}, \tag{4.39}$$

which is incompatible with the classical wave picture represented by Eq. (4.37).

If we discuss the above problem consistently in the coherent field, then the field coherence will not be destroyed no matter which representation has been used, because the nature of the field cannot be dependent on its mathematical descriptions.

4.3 Cooperative Phenomena of Radiation

It was first pointed out by Dicke [17] that the usual model for the description of the spontaneous radiation of a gas is wrong in principle due to overlooking the fact that all the molecules are at least interacting with a common radiation field and hence cannot be treated as independent. We know from the discussions in previous sections that two radiating atoms can be extremely far apart, by many wavelengths, but, as long as they remain within the coherence length (uncertainty length) of the radiation they still exhibit some correlation effect. Is it allowed, therefore, to ask the question if, in such a radiating system, one atom by the process of radiation could "know" about the existence of another one in its neighbourhood.

The two (or more) atoms can also be considered to build the boundary conditions for the formation of interference patterns of the emitted photons in the coherence volume. The orderliness of interference space-time patterns should be a prerequisite for the communication linkage among individual atoms which actually build a complex cooperative system which has to be considered as a whole.

Such physical phenomena of very simple systems may be the fundamental basis for the complex displays of electromagnetic waves in biological systems [25,26].

4.3.1 Superradiance

The term "superradiance" was coined by Dicke in his celebrated 1954 paper. He quantum-mechanically discussed in it the cooperation of N two-level atoms by processes of spontaneous emission. Since that time similar problems, especially those for extended samples have been intensively and sucessfully discussed. The basic principles of cooperativity become apparent in the following simple discussion:

We identify N two-level identical atoms with the subsripts $1, 2, \cdots, i, \cdots, N$. The upper and lower states of each atom are called $|e\rangle$ and $|g\rangle$ respectively and are separated by an energy interval $\hbar\omega_o$ ($\lambda = \frac{2\pi c}{\omega_o}$ is the wavelength of the emitted radiation). The raising and lowering off-diagonal operators $D_i^{\pm}$ are defined by

$$D_i^+ = |e\rangle_i\, {}_i\langle g|, \quad D_i^- = |g\rangle_i\, {}_i\langle e| \tag{4.40}$$

and the diagonal operator D_i^3 by

$$D_i^3 = \frac{1}{2}\left(|e\rangle_i\, {}_i\langle e| - |g\rangle_i\, {}_i\langle g|\right). \tag{4.41}$$

The $D_i^{\pm}$ and D_i^3 operators act only in the i-th atom subspace and obey the well-known commutation rules:

$$[D_i^3, D_j^{\pm}] = \pm\delta_{ij} D_i^{\pm}; \quad [D_i^+, D_j^-] = 2\delta_{ij} D_i^3. \tag{4.42}$$

These operators can be used to define the electric dipole operator of the atom:

$$d_i = (D_i^+ + D_i^-)l\epsilon, \tag{4.43}$$

where ϵ is the normalized vector defining the polarization of the $|e\rangle - |g\rangle$ transition and l is the electric dipole matrix element of the transition, which is assumed to be real without loss of generality with a convenient choice of phases of the eigenstates $|e\rangle$ and $|g\rangle$.

Since the atoms are located in a volume which is assumed for a moment to be small compared to λ, it is natural to suppose that the N radiators behave as a point-like dipole resulting from the sum of the N individual dipoles:

$$d = \sum_i d_i = l\epsilon \sum_i (D_i^+ + D_i^-). \tag{4.44}$$

Note that the above assumption is not necessary for the physical systems discussed but only for simplifying the mathematical procedure.

Now recalling that the rate of photon emission for a single atom [27] is

$$\rho_i = \Gamma\langle D_i^+ D_i^- \rangle, \tag{4.45}$$

where Γ is the atomic natural linewidth corresponding to the $\mid e > \rightarrow \mid g >$ transition and $\langle\ \rangle$ denotes a quantum mechanical average, we quite naturally generalize this expression to get the N-atom point-like system radiation rate:

$$\rho_N = \Gamma\langle D^+ D^- \rangle. \tag{4.46}$$

For a $|JM\rangle$ state, it is easy to evaluate the expectation value of (4.46), as the $D^\pm$ operators are analogous to the transverse components $J^\pm$ of an angular momentum $J = \frac{N}{2}$. One immediately obtains [28]:

$$\rho_N = \Gamma(J+M)(J-M+1). \tag{4.47}$$

In a symmetrical state, the expectation value of any $D_i^+\ D_j^-$ products with $j \neq i$ does not depend upon the i, j couple so that one can write:

$$\langle\left(\sum_i D_i^+\right)\left(\sum_j D_j^-\right)\rangle = N(N-1)\langle D_i^+ D_j^-\rangle + \sum_i \langle D_i^+ D_i \rangle, \tag{4.48}$$

which, according to equation

$$\langle JM|\sum_i D_i^+ D_i^- |JM\rangle = \Gamma(J+M) \tag{4.49}$$

and Eq. (4.47), yields in $|JM\rangle$ states

$$\langle JM|D_i^+ D_j^-|JM\rangle = \frac{\Gamma(J^2-M^2)}{N(N-1)}. \tag{4.50}$$

That is the correlation function between every two atoms i and j, which measures the cooperation between them. If $M=0$, we have the best correlated system which has the highest rate of emission from Eq. (4.47)

$$\rho_N = \Gamma\frac{N}{2}\left(\frac{N}{2}+1\right), \tag{4.51}$$

being proportional to N^2. In this case the occupation number of the excited state is equal to that of the ground state. That is the case for $f_\nu = const.$, as has been discussed by Popp [29].

If we identify that the dipole $D_i^{\pm}$ is classically proportional to the amplitude of the electromagnetic field $E^{\pm}$ [30], then the terms (4.50) really represent the interference effect of the radiation field, which is also a measure of the coherence of the radiation field.

Paul [31] states: "In fact, Eq. (92) in Dicke's paper, when specialized to the case of two initially excited atoms ($n = 2$, $m_0 = 1$ in Dicke's notation) located at definite positions $\vec{r}_1$ and $\vec{r}_2$, respectively, reads

$$W(\vec{k}_2) = 2W_0(\vec{k}_2)\left|\frac{e^{i\Delta\vec{k}\,\vec{r}_1} + e^{i\Delta\vec{k}\,\vec{r}_2}}{2}\right|, \quad (\Delta k = \vec{k}_2 - \vec{k}_1), \tag{4.52}$$

and hence is easily rewritten as

$$W(\vec{k}_2) = 2W_0(\vec{k}_2)\{1 + cos[\Delta\vec{k}(\vec{r}_2 - \vec{r}_1)]\}. \tag{4.53}$$

The physical meaning of this result is as follows: It is assumed that the emission of a photon in the direction $\vec{k}_1$ has been observed. The quantity $W(\vec{k}_2)$ is the probability (per unit time and unit solid angle) that the second photon will be emitted in the direction $\vec{k}_2$ and $W_0(\vec{k}_2)$ is the corresponding radiation probability for a single isolated excited atom". We see in Eq. (4.53) that the probability $W(\vec{k}_2)$ becomes largest if $\Delta\vec{k} \perp (\vec{r}_2 - \vec{r}_1)$, i.e. $\vec{k}_1$ and $\vec{k}_2$ must be as nearly as possible parallel to the axis linking two atoms. When there are more atoms building a one-dimensional curve, then the photons emitted ought to move along the curve instead of the straight line. This is surely a solid pysical basis for cooperative emission from DNA molecules where the photons are completely delocalized along the DNA circular structure as Popp [32] has postulated.

4.3.2 Mathematical excursion

For the readers who are dealing with the problems of cooperative phenomena of radiation for the first time we propose that they omit this subsection during a first reading. Despite the fact that the physical principles of superradiance and subradiance, which we will discuss in the next subsection, are very clear and obvious, some detailed features of them may not become clear without the aid of mathematical descriptions.

A) The fundamental equation

In order to ensure, as far as possible, mathematical accuracy we confine our problem to only two atoms, by using a projector method. We will see how a photon (only one) will be emitted by the two atoms cooperatively, compared to the emission of individual atoms.

We start with the general time-dependent Schrödinger equation:

$$i\hbar\frac{\partial}{\partial t}|\psi(t)\rangle = H(t)|\psi(t)\rangle \tag{4.54}$$

with

$$H(t) = H_0 + H_r + H_{in}, \tag{4.55}$$

where

$$H_0 = \frac{\hat{p}_1^2}{2m} + \frac{\hat{p}_2^2}{2m} + e(\hat{V}(\vec{r}_1) + \hat{V}(\vec{r}_2)) \tag{4.56}$$

is the Hamiltonian for atom 1 at $\vec{r}_1$ with momentum operator $\hat{p}_1$ and atom 2 at $\vec{r}_2$ with $\hat{p}_2$;

$$H_r = \sum_k \hbar\omega_k(\hat{a}_{kl}^{+}\hat{a}_{lk}) \tag{4.57}$$

is that for an electromagnetic field with different k modes and l polarizations;

$$H_{in} = -(\hat{d}_1\hat{E}_1 + \hat{d}_2\hat{E}_2) \tag{4.58}$$

represents the interaction Hamiltonian for the dipole of atom 1 at field position $\vec{r}_1$ as well as for the dipole of atom 2 at field position $\vec{r}_2$ respectively.

Based on Louisell's approximation [33] we can express d and E in the simple form without loss of generality

$$\hat{d} = \hat{d}e^{-i\omega t}, \quad \hat{E} = \hat{E}e^{-i[(kr-\omega t)+\Phi(t)]}, \tag{4.59}$$

where $\Phi(t)$ is a random phase factor.

For our problem we take the following initial state

$$|i\rangle = \alpha|g,e;0\rangle \pm \beta|e,g;0\rangle. \tag{4.60}$$

Here we have α, β as normalization constants, g and e indicate the ground and excited state for two atoms, respectively. The letter 0 indicates the vacuum state of the electromagnetic field. The plus sign $+$, and minus sign $-$ represent the symmetric and antisymmetric state with respect to the exchange of two atoms.

The final state must be

$$|g,g;k\rangle \tag{4.61}$$

with $k = (\vec{k}, \omega)$.

That indicates that both the atoms come to their ground states due to releasing the photon into a state k. The Hilbert space of the problem is built by the state vectors of $|e,g;0\rangle$, $|g,e;0\rangle$ and $|g,g;k\rangle$ for a possible k. The subspaces of interest are (4.60) and (4.61).

Now we define the following projectors:

$$P = |i\rangle\langle i| \;\; and \;\; (1-P) = |ggk\rangle\langle ggk| \tag{4.62}$$

The projection of these operators on the initial state gives

$$P|\psi(0)\rangle = |i\rangle \;\; and \;\; (1-P)|\psi(0)\rangle = |0\rangle \tag{4.63}$$

with

$$P^2 = P \;\; and \;\; (1-P)^2 = (1-P). \tag{4.64}$$

Taking the projection with the same projectors on the Schrödinger equation (4.54), we get

$$i\hbar\frac{\partial}{\partial t}|\psi_i(t)\rangle = PHP|\psi_i(t)\rangle + PH(1-P)|\psi_k(t)\rangle, \qquad (4.65)$$

$$i\hbar\frac{\partial}{\partial t}|\psi_k(t)\rangle = (1-P)H(1-P)|\psi_k(t)\rangle + (1-P)HP|\psi_i(t)\rangle \qquad (4.66)$$

with

$$P|\psi(t)\rangle = |\psi_i(t)\rangle; \; (1-P)|\psi(t)\rangle = |\psi_k(t)\rangle \qquad (4.67)$$

and

$$\begin{aligned} PHP &= |i\rangle(\alpha^*\langle ge0| \pm \beta^*\langle eg0|)(H_0 + H_r + H_{in})(\alpha|ge0\rangle \pm \beta|eg0\rangle)\langle i| \\ &= |i\rangle[\alpha^2(W_1 + W_1^*) + \beta^2(W_1^* + W_2)]\langle i| = |i\rangle(\alpha^2+\beta^2)\bar{W}\langle i|, \end{aligned} \qquad (4.68)$$

where the relation (4.60) has been accounted for.

With the assumption that there is no direct interaction between two identical atoms we have,

$W = 0$ as the energy of atomic ground state;

$W^* = \bar{W}$ as the energy of excited state.

$$\begin{aligned} (1-P)H(1-P) &= |ggk\rangle\langle ggk|(H_0 + H_r + H_{in})|ggk\rangle\langle ggk| \\ &= |ggk\rangle U_k \langle ggk|, & (4.69) \\ PH(1-P) &= |i\rangle(\alpha^*\langle ge0| \pm \beta^*\langle eg0|)(H_0 + H_r + H_{in})|ggk\rangle\langle ggk| \\ &= |i\rangle(\alpha^* d_2^* E_2^* \pm \beta^* d_1^* E_1^*)\langle ggk|, & (4.70) \\ (1-P)HP &= |ggk\rangle\langle ggk|(H_0 + H_r + H_{in})(\alpha|ge0\rangle \pm \beta|eg0\rangle)|i\rangle \\ &= |ggk\rangle(\alpha d_2 E_2 \pm \beta d_1 E_1)\langle i|. & (4.71) \end{aligned}$$

By inserting all these relations (4.67- 4.71) into (4.65) and (4.66), we get

$$\begin{aligned} i\hbar\frac{\partial}{\partial t}|\psi_i(t)\rangle &= (\alpha^2+\beta^2)\bar{W}|\psi_i\rangle \\ &\quad + |i\rangle(\alpha^* d_2^* E_2^* \pm \beta^* d_1^* E_1^*)\langle ggk|\psi_k\rangle, & (4.72) \\ i\hbar\frac{\partial}{\partial t}|\psi_k(t)\rangle &= U_k|\psi_k\rangle + |ggk\rangle(\alpha d_2 E_2 \pm \beta d_1 E_1)\langle i|\psi_i\rangle. & (4.73) \end{aligned}$$

On account of initial condition (4.63) we obtain the solution of Eq. (4.73) in the simple form:

$$|\psi_k\rangle = -\frac{i}{\hbar}e^{-\frac{i}{\hbar}U_k t}\int_0^t |ggk\rangle(\alpha d_2 E_2 \pm \beta d_1 E_1)e^{\frac{i}{\hbar}U_k t'} \cdot \langle i|\psi_i(t')\rangle dt'. \quad (4.74)$$

By inserting (4.74) into (4.72) we have

$$i\hbar\frac{\partial}{\partial t}|\psi_i\rangle = (\alpha^2+\beta^2)\bar{W}|\psi_i\rangle - \frac{i}{\hbar}\int_0^t |i\rangle(\alpha^* d_2^* E_2^* \pm \beta^* d_1^* E_1^*)\langle ggk|ggk\rangle (\alpha d_2 E_2 \pm \beta d_1 E_1)e^{-\frac{i}{\hbar}(t-t')U_k}\langle i|\psi_i\rangle dt', \quad (4.75)$$

where

$$\langle ggk|ggk\rangle = \sum_{k,\omega}|C_{k,\omega}|^2$$

is the so called Parseval relation in Quantum Mechanics [13]. So we have to write Eq. (4.75) in the form

$$i\hbar\frac{\partial}{\partial t}|\psi_i\rangle = (\alpha^2+\beta^2)\bar{W}|\psi_i\rangle - \frac{i}{\hbar}\int_0^t \sum_{k\omega}|c_{k,\omega}|^2(\alpha d_2 E_2 \pm \beta d_1 E_1)^* (\alpha d_2 E_2 \pm \beta d_1 E_1)e^{-\frac{i}{\hbar}(t-t')U_k}|\psi_i(t')\rangle dt'. \quad (4.76)$$

Care must be taken to note that the first bracket depends on the time t and the last on t'.

Regarding the relation (4.59) the above equation can be written in the form

$$\begin{aligned}\frac{\partial}{\partial t}|\psi_i\rangle = & -\frac{i}{\hbar}(\alpha^2+\beta^2)\bar{W}|\psi_i\rangle \\ & -\frac{i}{\hbar}\int_0^t \sum_{k\omega}|c_{k,\omega}|^2\{\alpha^2|d(r_2)|^2E^*(r_2t)E(r_2t') \\ & +\beta^2|d(r_1)|^2E^*(r_1t)E(r_1t') \\ & \pm\alpha^*\beta d^*(r_2)d_0(r_1)E^*(r_2t)E(r_1t')e^{-i\vec{k}(r_2-r_1)} \\ & \pm c.c.\}e^{-i(\omega-\omega')(t-t')}e^{-\frac{i}{\hbar}(t-t')U_k}|\psi_i\rangle dt'. \quad (4.77)\end{aligned}$$

When we replace the summation over ω by an integral with respect to ω, i.e.

$$\sum_{\omega} \longrightarrow \int d\omega,$$

we get a δ-function of the kind

$$\int_{-\frac{\Delta\omega}{2}}^{+\frac{\Delta\omega}{2}} e^{-i(\omega-\omega')(t-t')} d\omega \;=\; 2\pi\delta(t-t'). \tag{4.78}$$

If it is assumed that $|c_{k,\omega}|$, E are nearly constant at $\omega - \omega' \simeq 0$ and $d(r_1) \simeq d(r_2)$, we can rewrite Eq. (4.77) in a new form:

$$\frac{\partial}{\partial t}|\psi_i\rangle \;=\; (-i\Omega - \Delta)|\psi_i\rangle, \tag{4.79}$$

where

$$\Omega \;=\; \frac{\alpha^2+\beta^2}{\hbar}\bar{W}, \tag{4.80}$$

$$\Delta \;=\; \frac{2\pi}{\hbar^2}d^2|c_{\bar{k}}|^2\{(\alpha^2 E(r_2)^2 + \beta^2 E(r_1)^2) \pm (\alpha^*\beta\langle E(r_2)E(r_1)\rangle)e^{-i\bar{k}(r_2-r_1)} + c.c\}. \tag{4.81}$$

Here we replace k by an average $\bar{k}$ because $k \| (r_2 - r_1)$ have main contributions to the integral due to the components $k \perp (r_2 - r_1)$ which belong to the field amplitudes $E \perp d$. In addition, only the terms for the case $\omega \;=\; c|k| \simeq \omega'$ contribute most to the integral.

B) Solutions of Eq. (4.79) and some consequences of it.

1. When we take $\alpha = \beta = \frac{1}{\sqrt{2}}$ in Eq. (4.60) and $\langle E(r_2)E(r_1)\rangle = \langle E(r_1)E(r_2)\rangle$ then we have

$$\Delta \;=\; \frac{\pi}{\hbar^2}d^2|c_{\bar{k}}|^2\{E^2(r_2) + E^2(r_1) \pm 2\langle E(r_2)E(r_1)\rangle cos\bar{k}(r_2 - r_1)\}. \tag{4.82}$$

We see that the terms in swept brackets represent an interference field of two atoms separated by the distance $r_2 - r_1$.

If we take this interference pattern as a stationary structure, then we can build a differential equation for the probability of the atomic subsystem following the projection (4.67) and Eq. (4.79):

$$\frac{\partial}{\partial t}\langle\psi_i|\hat{d}^2|\psi_i\rangle = -2\Delta\langle\psi_i|\hat{d}^2|\psi_i\rangle \tag{4.83}$$

or

$$\frac{\partial}{\partial t}d^2 = -\kappa d^4 \tag{4.84}$$

with

$$\kappa = \frac{2\pi}{\hbar^2}|c_{\bar{k}}|^2\{E^2(r_2) + E^2(r_1) \pm 2\langle E(r_2)E(r_1)\rangle cos\bar{k}(r_2 - r_1)\}, \tag{4.85}$$

where we used the relation $d^2 = \langle\psi_i|\hat{d}^2|\psi_i\rangle$. This approximation is reasonable because the two atoms are taken far enough apart so that their electron clouds cannot overlap.

Thus we get a hyperbolic decay law as we have repeatedly discussed elsewhere [34,35]:

$$d^2 = \frac{d_0^2}{1 + d_0^2\kappa(t - t_0)} \quad for\ t \geq t_0. \tag{4.86}$$

2. If we take the interaction between fields and matter as a random process, then we can consider Δ as a constant and the solution of (4.83) is a well known exponential function:

$$d^2 = d_0^2 e^{-2\Delta(t-t_0)}. \tag{4.87}$$

By expansion of (4.87) we have the term of the first order:

$$\begin{aligned} d^2 &\simeq d_0^2(1 - 2\Delta(t - t_0)), \\ \frac{d^2 - d_0^2}{(t - t_0)} &= -d_0^2\Delta = -\kappa d_0^2. \end{aligned} \tag{4.88}$$

Actually, (4.88) can be interpreted as the emission probability per unit time which is formally similar to the result of Dicke's Eq. (92) in his original paper, when specialized to the case of two atoms located at definite positions r_1 and r_2. The difference is that we only consider a one-photon problem which makes our problem obvious and simple. It shows that the Dicke's Eq. (92) is an effect of the first order of our treatment.

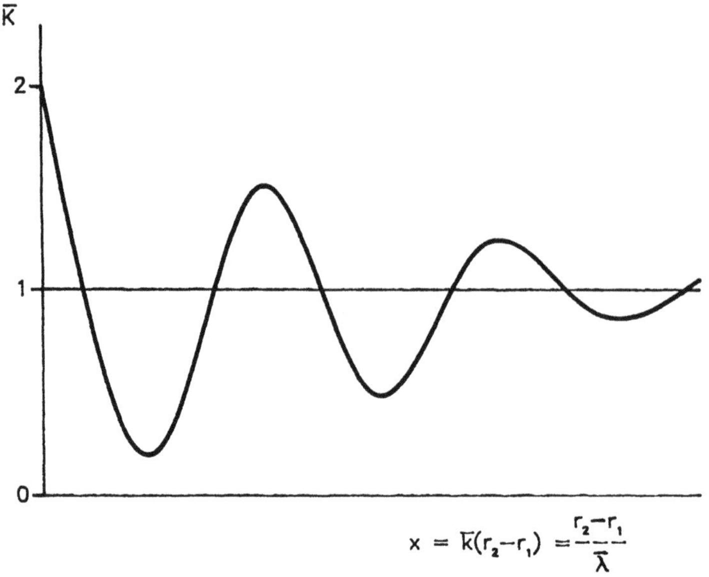

Figure 4.2: *Schematic diagram of the effect of correlation function (4.85) on k in Eq. (4.81). $\bar{k}$ denotes the normalized k.*

3. We see that the appearance of plus- and minus signs in the swept brackets in the expression of Δ comes from the symmetry considerations. But the symmetry features are dependent on different parameters $\bar{k}$ and $r_2 - r_1$ as boundary conditions or constraints of the problem, on which the constructive and destructive interference between two atoms, i.e. superradiance and subradiance, depend.

4. The correlation function in the expression (4.85) can be generally expressed in the form:

$$\langle E(r_2)E(r_1)\rangle = |E(r_1)^2|e^{-\frac{(r_2-r_1)}{L}}, \tag{4.89}$$

where L is the coherence length of the field amplitude E.

Hence we see that the interference term will disappear, if the distance between two atoms $(r_2 - r_1)$ is larger than the coherence length L. Here an effect of the distance between atoms, or concentration effect, is displayed (Fig. 4.2).

4.3.3 Subradiance

On account of its importance, especially in biology [36,37], we want to devote a subsection to the problem of subradiance.

From Eqs. (4.75), (4.77) and (4.79) we see that the system under discussion may emit hardly any or even no photons if $\Delta \simeq 0$ for some appropriate values of the parameters k and $(r_2 - r_1)$. In such cases we have an effect of photon-trapping within the system where the photon cannot be designated to a special atom. It is actually delocalized or socialized within the whole system. That may be learnt from the uncertainty principle discussed in Sec. 4.2 above.

When the atoms of the system are taken apart from each other beyond the coherence length (uncertainty length) of photons

$$\frac{(r_2 - r_1)}{L} > 1 ,$$

then the cooperative atomic system will decay to independently individual subsystems which radiate incoherently (see Eqs. (4.79), (4.81) and (4.89)).

For a many-atom (of number N) problem we have a similar picture: each arbitrary picked pair of neighbouring atoms can be viewed as a two-atom system discussed above. Hence we have $\frac{N}{2}$ atoms in excited states and $\frac{N}{2}$ atoms in the ground state for the ideal case where not only $M = 0$ but also $J = 0$ in the sense of Dicke's theory because each pair of $\frac{1}{2}$-spins are compensating to a zero-spin subsystem which no longer contribute to the total cooperative number in the sense of Dicke. Thus the total cooperative number is actually zero, when N is even. For this case we get just the subradiance system, which will not radiate according to the Eq. (4.47).

V. Ernst [38] has extended the Weisskopf and Wigner-theory of the natural line width to a case where many atoms at given positions interact with one common quantized radiation field. At time $t = 0$, one quantum of radiation energy is stored by the system, which however does not mean necessarily that exactly one of the atoms must be excited, according to the uncertainty principle. He obtained the following result: If his theory is applied to the lasing transition in ruby, we find that subradiance as well as superradiance is so strong that the process of natural spontaneous emission of a photon in the sense of the theory of Weisskopf and Wigner practically never occurs.

The subradiant state is a state of the two atoms in which the photon can be trapped for a long time. How long it is trapped is dependent on the ratios of different parameters:

$$\frac{(r_2 - r_1)}{\lambda} \quad and \quad \frac{(r_2 - r_1)}{L} \tag{4.90}$$

(see. Eqs. (4.79), (4.81), or (4.83,4.84)).

In principle we do not have a definite upper limit for the lifetime of the subradiant state, but the lifetime of the superradiant state has a lower boundary $\left(\frac{4}{\tau}\right)^{-1}$ due to the constructive interference of the fields, where τ is the lifetime of an individual atom.

As we have discussed elsewhere [39], the DNA-base molecules can radiate cooperatively by building so-called excimer or exciplex excited states of two neighbouring bases. If we take the wavelengths of DNA fluorescence of about 3500 $\dot{A}$ into consideration, as given by experiments of M. Gueron at al. [40], we can estimate the lifetime of the subradiant state of such excimers to be about 10^6 times that of the individual DNA bases, according to the ratio of

$$\frac{(r_2 - r_1)}{\lambda} \simeq 1000 \tag{4.91}$$

by the Eq. (3.3) in the Ernst paper where the second term can be neglected, because of $\lambda = 3500\ \dot{A}$, $(r_2 - r_1) = 3.4\ \dot{A}$, and the ratio of

$$\frac{(r_2 - r_1)}{L} \simeq 0. \tag{4.92}$$

It is usually assumed that an atom can be replaced by a tiny dipole. From classical electrodynamics we know that the magnetic induction in the near zone-approximation is a factor $(\vec{k}\vec{r})$ smaller than the electric field in the region where $\vec{k}\vec{r} << 1$. Thus the fields in the near zone are dominantly electric in nature. The dipole interaction is dependent on the orientation of the dipoles:

	p ↑↑ p	p ↑↓ p
force:	repulsive	attractive
interference:	constructive	destructive
cooper. radiation:	superradiant	subradiant

Early in 1964 M.J. Stephen [18] investigated quantum-mechanically the interaction energy between two atoms, one of which was excited. The form of the interaction depends on the relation between the distance R separating the atoms and the wavelength $2\pi k^{-1}$ of the transition from the excited state to the ground state. This state can well be represented by a superposition of states which are symmetric and antisymmetric with respect to the interchange of the two atoms. These states are analogous to the singlet and triplet states of the two spins. Only the symmetric state is capable of emitting a photon and decaying to the ground state of two unexcited atoms which is also symmetric (superradiance). In the antisymmetric state the photon is trapped between two atoms (subradiance) and leads to an interaction energy , which is attractive and called First-Order Dispersion Forces by Stephen.

4.3.4 Antibunching in biological systems

Antibunching and sub-Poissonian photon distribution (so-called non-classical effects) are statistical properties of light which have been the subjects of research in quantum optics for two decades. However, for electromagnetic processes in biological systems up to now this important theme has not attracted attention elsewhere [41].

In reality, some biophysical processes can much better fulfil the conditions producing non-classical phenomena of light (as electromagnetic waves). For example, Paul [42] has shown that any effort to measure the antibunching effect can only be successful at moderate photon numbers, say $\langle\langle n\rangle\rangle \leq 100$. The actual measurement of non-fluctuating photon numbers requires an electromagnetic wave train of pulses, well separated in space by equal distances cT, each of which contains precisely the same number of photons. The response time of the detector τ should obey the inequality $T > \tau \geq \Delta t$, where Δt is the pulse duration. If we measure the time-averaged coincidence counting rates $K(\tau)$ (the averaging procedure extending over a large number of single pulses), the antibunching effect would become manifest in a deficit of $K(0)$ compared to $K(mT)$, where m is any positive integer.

For biological objects which radiate very weakly ("ultra-weak" radiation) the highest photon number/$s \cdot cm^2$ registered amounts to about 10^5. Hence we have the photon number density:

$$\langle\langle n\rangle\rangle \;=\; 10^{-5} \sim 10^{-6}, \tag{4.93}$$

which is considerably less than the critical value $\langle\langle n\rangle\rangle \leq 100$ given by Paul. As the response time τ of the photomultiplier is about 10^{-9} s and the breadth of impulse registered is certainly smaller then 10^{-9} s, the distance between two pulses may conservatively be estimated using the assumption that each photon causes a pulse. Then for 10^5 photons in a second the distance between each pair of photons is of the order of $3 \times 10^5 cm$ or $T \simeq 10^{-5} sec$. Thus we have the condition $T > \tau \geq \Delta t$ as required for the emergence and possible detection of the antibunching phenomenon in biological systems.

Of course, the fundamental cause of antibunched photons emitted from biological tissues (called "biophotons") lies in the cooperative phenomena of the excited biomolecules as we discussed in the earlier subsections.

It is well known that quantum-mechanical cooperative phenomena have a tendency to reduce the fluctuations of the physical quantities (the order parameter). For instance, if a laser is operated below its threshold, the field is merely supported by stochastic processes (spontaneous emission noise). The inverse of the relaxation time of the field amplitude may be interpreted as the optical line width. By approaching the threshold, the system becomes more and more undamped. The optical line width decreases, which is a well-observed phenomenon in laser experiments.

When the threshold is passed, the field amplitude can be amplified coherently and at the same time the optical noise would be strongly suppressed. This coherence is only disturbed by small superimposed amplitude fluctuations caused by suppressed optical noise and by very small phase fluctuations. Similar phenomena are observed in the phase-transitions of ferromagnetism, superconductivity and others. As we said, the cooperative phenomenon has been reasonably defined by some authors[43] as an action that tends to reduce the fluctuation of the physical quantities (the order parameter). Therefore, if it acts ideally, the fluctuation of the observed physical quantity can approach its minimum value.

As we have also shown in subsection (4.2), an isolated excited atom radiates randomly within its lifetime and space-symmetrically with the same probability in all directions. But for two cooperative atoms which have an influence on each other only by means of the radiation field, the radiation of a photon from the system behaves entirely differently from that of an isolated atom. We know, as discussed in the subsections on superradiance and subradiance, that the field can no longer be random and hence there exists a basis for constructive or destructive interference between the two atoms. In other words, there appears a dynamic space-time structure (s. Eqs. (4.79) - (4.82)).

For destructive interference this structure is nearly stationary due to the long lifetime. In principle the lifetime of the photon remaining in the system can be indefinitely long, $\Delta t \rightarrow \infty$, and the photon can be considered as being stored in the system. This photon-storage effect, if it occurs in biological tissues, may be essential for living processes on the microscopical level [29]. Destructive interference means photon-storage or photon trapping and represents photon exchange between two atoms (see Sec. 4.2).

Because the photon carries information of one atom to the other due to interaction with them, the exchange of photons presents a real possibility for biological communication between tissues via electromagnetic fields. Photon-exchange as a photon storage-mechanism prolong the life time and lowers the intensity of photon emission. In particular, it changes a random radiation process of an isolated system into a regular one of a cooperative system. These are the reasons for the appearance of the antibunching effect in biological tissues. In order to seek the antibunching effect in physical systems, many authors have previously theoretically proposed methods of using destructive interference [42,44]. However, in biological systems the subradiance mechanism provides a good enough natural condition for the destructive interference of photon fields.

Photon antibunching is often defined by the condition [6,42,45]

$$g^{(2)} = \frac{\langle \hat{a}^+\hat{a}^+\hat{a}\hat{a}\rangle}{\langle \hat{a}^+ a\rangle^2} < 1, \tag{4.94}$$

where $\langle \hat{a}^+ a\rangle = \langle n\rangle$ and $\langle \hat{a}^+\hat{a}^+\hat{a}\hat{a}\rangle = \langle n^2\rangle$ are the first two lowest moments of $\langle n^k\rangle$. We have seen in Sec. 4.2 that the Eq. (4.94) cannot be satisfied by coherent states. This does not mean that the field of antibunched photons is incoherent. The antibunching phenomenon must be subjected to the constraint of the minimum uncertainty principle which may be a necessary condition for canonical coherent states. For concrete problems we must solve the corresponding dynamic equations of the operator $\hat{a}$ and the relevant distribution function in order to evaluate the expressions for moments. As an example Perina [24,46] has developed a calculation by using the Heisenberg-Langevin equation for operators $\hat{a}$, $\hat{a}^+$ and the Fokker-Planck equation for the probability distribution to treat optical parametric processes with classical pumping. He shows that antibunching and sub-Poissonian occur in some time interval of the dynamic process where

$$g^{(2)} - 1 < 0. \tag{4.95}$$

For the higher-order reduced factorial moments,

$$\frac{\langle n^k\rangle_N}{\langle n\rangle_N^k} - 1 < 0 \tag{4.96}$$

increasing k causes a downward trend to the limit value -1. Therefore, antibunching also shows itself in the higher-order moments, especially when the second order correlation function is too weak for judging any not distinct antibunching effects due to technical problems. Similar methods have been used in our laboratory for evaluating the antibunching phenomena of ultra-weak emission from biological samples [47].

4.4 Communication with Extremely Low Quantum Noise

It is very attractive for further exploration that some quantum features of electromagnetic fields lend themselves to an interesting application of optical communication below the vacuum noise. Especially, when the detector system is smaller in size than the coherence volume of electromagnetic waves then, due to the uncertainty principle, the detector cannot be considered as a point device. (This is just the case of biological tissues due to their ultraweak photon emission). Mandel [48] describes a situation where a single detector receives no intelligible information of a non-classical light-source, and yet two detectors working in coincidence can achieve almost error-free information of the source.

In fact, as we discussed in Sec. 4.2, the biological cell receives possibly the whole coherent pattern of the light field because it is an ideal coupled extended detector and is embedded in the field-structure as a member of the whole biological entity. We may not separate the biophoton field from the cell or other parts of the tissue just as we cannot separate one part of the cell from the other parts without destroying the living processes and living functions.

4.4.1 Squeezed state

We know that the coherent state is the minimum uncertainty state (4.27). From Eq. (4.25) we have

$$\begin{aligned} \hat{X} &= \frac{1}{2}(\hat{a}+\hat{a}^{+}) = \left(\frac{\omega}{2\hbar}\right)^{\frac{1}{2}}\hat{q} \\ \hat{Y} &= \frac{1}{2i}(\hat{a}-\hat{a}^{+}) = \left(\frac{1}{2\hbar\omega}\right)^{\frac{1}{2}}\hat{p}, \end{aligned} \tag{4.97}$$

which are the so-called quadrature operators satisfying the commutation relation

$$[\hat{X}, \hat{Y}] = \frac{i}{2} \tag{4.98}$$

and instead of (4.27) we have

$$\langle(\Delta X)^2\rangle\langle(\Delta Y)^2\rangle = \frac{1}{16}. \tag{4.99}$$

For coherent states (or for the special case of the vacuum state) the equality of Eq. (4.99) is valid and the variances of the quadrature operators are equal:

$$\langle(\Delta X)^2\rangle = \langle(\Delta Y)^2\rangle = \frac{1}{4}. \tag{4.100}$$

Considering Eq. (4.99) we have yet another alternative to choose different ΔX and ΔY while the equality of Eq. (4.99) remains valid; and this case has been named the squeezed state. The distinction is often academic, since the quantum squeezing occurs when the fluctuations in one field quadrature are reduced below the quantum limit for a coherent state at the expense of increased fluctuations in the other quadrature. Photon counting may be used to measure the squeezing in terms of sub-Poissonian counting statistics, characterized by the parameter [49]

$$\frac{\langle\Delta n^2\rangle}{\langle n\rangle} - 1. \tag{4.101}$$

Photon antibunching was first predicted in absorptive bistability by Casagrande and Lugiato [50]. Squeezing in optical bistability is closely related to the existence of photon antibunching. In general, the relationship between photon antibunching and squeezing may in short be stated: Particularly for short counting times photon antibunching implies the existence of squeezing and vice versa [51].

The earliest proposed application of light squeezing was the reduction of quantum noise in optical communications [52,53]. Loudon [54] describes the signal/noise ratio for balanced homodyne detection by the equation:

$$\frac{\langle m_{12}\rangle^2}{\langle(\Delta m_{12})^2\rangle} = \frac{4\mu\langle\hat{E}(x)\rangle^2}{1+\mu[4\langle(\Delta E(x))^2\rangle - 1]}, \tag{4.102}$$

where $\langle m_{12}\rangle$ is the photocount mean of the balanced homodyne detection and $\langle(\Delta m_{12})^2\rangle$ is its variance:

$$\begin{aligned}\langle m_{12}\rangle &= 2\mu|\alpha_L|\langle\hat{E}(x)\rangle \\ \langle(\Delta m_{12})^2\rangle &= \mu|\alpha_L|^2\{1+\mu[4\langle(\Delta E(x))^2\rangle - 1]\}\end{aligned}$$

with μ representing the efficiency coefficient, where for perfect detection $\mu = 1$. $|\alpha_L|$ is the strength of the local oscillator which is completely compensated in Eq. (4.102). $\hat{E}(x)$ is the operator of the field of an information-carrying signal.

For squeezed states one has

$$\langle(\Delta E(x))^2\rangle = \frac{1}{4}[e^{-2s}cos^2(x-\frac{1}{2}\Theta) + e^{2s}sin^2(x-\frac{1}{2}\Theta)]$$

and $x = \frac{1}{2}\pi + \Phi_L$ with s as the squeeze parameter.

Thus the Eq. (4.102) becomes

$$\frac{\langle m_{12}\rangle^2}{\langle(\Delta m_{12})^2\rangle} = \frac{4\mu|\alpha|^2cos(\frac{1}{2}\Theta - \Phi)}{1-\mu+\mu e^{-2s}} \tag{4.103}$$

for $\Phi_L = \frac{1}{2}\Theta + \frac{1}{2}\pi$.

The denominator is equal to unity for coherent signal light ($s = 0$), but it can in principle be made arbitrarily small for highly squeezed signal light ($s \to \infty$) with highly efficient photodetection ($\mu \to 1$). Thus greatly enhanced signal/noise ratios are possible, with consequent advantages for optical communication systems.

For biological systems both conditions $\mu \to 1$, and $s \to \infty$ are perhaps good enough for enhancing the signal/noise ratio. On account of the expression of the squeeze parameter s for optical fibre [54]:

$$s = \frac{4\pi\chi^{(3)}I_c}{\Omega_c}$$

with $\chi^{(3)}$ the nonlinear susceptibility, I_c the dimensionless pump intensity and

$$\Omega_c = \frac{2\pi c}{nl},$$

where l is the interaction length permitted by a fibre, and n is the refractive index defined by

$$n = \frac{c}{v}$$

with v the light velocity within the fibre.

We have to consider the DNA molecule as a biological optical fibre, within which the effective light velocity v may be considered to be very small due to photon trapping in it. The interaction length l ought to be the whole length of the DNA fibre. This fact may lead the squeeze parameter s to a high value while the biological detection system has always a perfect efficiency $\mu \to 1$.

4.4.2 Squeezing of a damped oscillator

The study of the harmonic oscillator is of great importance not only in Quantum Optics but also in the whole region of Quantum Theory, since the Hamiltonian of the type

$$H = \frac{1}{2m(t)}\hat{p}^2 + m(t)^2\omega(t)^2\hat{q}^2 \tag{4.104}$$

enters all problems involving quantized oscillations: one encounters it in Quantum Electrodynamics, and generally in Quantum Field Theory; one likewise encounters it in the theory of molecular and crystalline vibrations. Thus the study of it is also of profound value for understanding some fundamental problems of radiation processes in biological systems.

In the Heisenberg picture we have corresponding equations of motion for the Hamiltonian (4.104):

$$\frac{\partial \hat{q}}{\partial t} = \frac{\hat{p}}{m(t)}, \quad \frac{\partial \hat{p}}{\partial t} = -m(t)\omega(t)^2\hat{q}. \tag{4.105}$$

From these equations we obtain the second-order differential equation for the operator $q(t)$

$$\frac{\partial^2 \hat{q}}{\partial t^2} + \frac{\dot{m}(t)}{m(t)}\frac{\partial \hat{q}}{\partial t} + \omega(t)^2\hat{q} = 0. \tag{4.106}$$

From general solutions of (4.105) and (4.106) we obtain the well known relation:

$$[\hat{q}(t), \hat{p}(t)] = [\hat{q}(0), \hat{p}(0)] = i\hbar. \tag{4.107}$$

According to Yuen [52] the uncertainty in $\hat{q}$ and $\hat{p}$ for the squeezed state is given by

$$(\Delta q)^2 = \frac{\hbar}{2\omega}|\mu - \nu|^2, \quad (\Delta p)^2 = \frac{\hbar\omega}{2}|\mu + \nu|^2 \tag{4.108}$$

thus

$$(\Delta p)(\Delta q) = \frac{\hbar}{2}|\mu + \nu||\mu - \nu| \tag{4.109}$$

with $|\mu|^2 - |\nu|^2 = 1$.

Yuen's annihilation and creation operators are defined by

$$A = \mu\hat{a} + \nu\hat{a}^+, \tag{4.110}$$

$$A^+ = \mu^*\hat{a}^+ + \nu^*\hat{a}. \tag{4.111}$$

Corresponding operators have been constructed by A. Jannussis et al. [55]:

$$A = \frac{1}{\sqrt{\alpha\beta^* + \alpha^*\beta}}\left(\alpha\hat{q} + \frac{i\beta}{\hbar}\hat{p}\right), \tag{4.112}$$

$$A^+ = \frac{1}{\sqrt{\alpha\beta^* + \alpha^*\beta}}\left(\alpha^*\hat{q} - \frac{i\beta^*}{\hbar}\hat{p}\right). \tag{4.113}$$

The operators $\hat{q}, \hat{p}, A$ and A^+ satisfy simultaneously the following relations:

$$[\hat{q}, \hat{p}] = i\hbar, \ [A, A^+] = 1 \ and \ \alpha\beta^* + \alpha^*\beta > 0$$

with complex numbers:

$$\alpha = \alpha_1 + i\alpha_2, \ \beta = \beta_1 + i\beta_2.$$

On account of (4.112, 4.113) we have

$$(\Delta q)(\Delta p) = \frac{\hbar}{2}\sqrt{\frac{(\alpha_1^2+\alpha_2^2)(\beta_1^2+\beta_2^2)}{(\alpha_1\beta_1+\alpha_2\beta_2)^2}} \geq \frac{\hbar}{2}, \tag{4.114}$$

because $(\alpha_1^2+\alpha_2^2)(\beta_1^2+\beta_2^2) \geq (\alpha_1\beta_1+\alpha_2\beta_2)^2$.

From the above results we obtain the minimalization condition for the coherent states

$$\frac{\alpha_1}{\beta_1} = \frac{\alpha_2}{\beta_2} = constant.$$

For a harmonic oscillator with a time-dependent mass and frequency with the help of Hamiltonian (4.104) Jannussis et al. have shown that the exact coherent states of it are equivalent to the squeezed states [55], in accordance with (4.108) and (4.109).

We see a simplified case of them for a system with constant frequency and time-dependent mass in the forms

$$m(t) = m(0)\,exp(-\gamma t), \quad \omega(t) = \omega = const., \tag{4.115}$$

where γ is a corresponding friction coefficient.

For this case we have the general solutions of Eq. (4.106):

$$\begin{aligned} g_1(t) &= exp\left[\left(i\Omega - \frac{\gamma}{2}\right)t\right], \\ g_2(t) &= exp\left[-\left(i\Omega + \frac{\gamma}{2}\right)t\right], \end{aligned}$$

where $\Omega = \sqrt{\frac{\gamma^2}{4} - \omega^2} > 0$ corresponds to the strong friction case, and the corresponding values of the parameters $\mu(t)$, $\nu(t)$ are the following:

$$\begin{aligned} \mu(t) &= cosh\,\Omega t - i\frac{\omega}{\Omega}sinh\,\Omega t, \\ \nu(t) &= \frac{\gamma}{2\Omega}sinh\,\Omega t \end{aligned}$$

with condition

$$|\mu|^2 - |\nu|^2 = 1.$$

The relation (4.109) takes the form

$$(\Delta q)(\Delta p) = \frac{\hbar}{2}\sqrt{1 + \frac{\gamma^2\omega^2}{\Omega^4} sinh^4 \Omega t} \geq \frac{\hbar}{2}. \tag{4.116}$$

For the case $\Omega = \sqrt{\omega^2 - \frac{\gamma^2}{4}} > 0$, which corresponds to a weak friction, we obtain

$$\begin{aligned} \mu(t) &= cos\, \Omega t - i\frac{\omega}{\Omega} sin\, \Omega t, \\ \nu(t) &= \frac{\gamma}{2\Omega} sin\, \Omega t \end{aligned}$$

and the uncertainty relation takes the form

$$(\Delta q)(\Delta p) = \frac{\hbar}{2}\sqrt{1 + \frac{\gamma^2\omega^2}{\Omega^4} sin^4 \Omega t} \geq \frac{\hbar}{2}, \tag{4.117}$$

but for $\Omega t = \pi n$ $(n = 0, \pm 1, \pm 2, \cdots)$ we have the minimum value

$$(\Delta q)(\Delta p) = \frac{\hbar}{2}. \tag{4.118}$$

4.4.3 The hyperbolic decay law and squeezed states

Jannussis et al. have reasonably shown that the excited coherent states of a harmonic oscillator with time-dependent mass and frequency are equivalent to the squeezed states. In the last subsection we have exhibited a special example from Jannussis with conditions given in Eq. (4.115).

We know since ten years [34,35,56] that a damped harmonic oscillator in the coherent state cannot follow the special ansatz:

$$m(t) = m(0)\, exp(-\gamma t)$$

as given in (4.115), if it remains in a coherent interaction with its surrounding radiation field. The term $\frac{\dot{m}(t)}{m(t)}$ in Eq. (4.106) would be replaced by an indefinite function of time, which must be determined by the requirement of the frequency stability, i.e., $\omega =$ const. So that we have the following condition

$$\frac{\dot{m}(t)}{m(t)} = \lambda(t) \; and \; \omega = const \tag{4.119}$$

instead of Eq. (4.115).

With this condition the general solutions of Eq. (4.106) read

$$\begin{aligned} q_1(t) &= \frac{q(0)}{1+\lambda_0 t} e^{-i\omega t} \\ q_2(t) &= \frac{q(0)}{1+\lambda_0 t} e^{i\omega t} \end{aligned} \tag{4.120}$$

and their derivatives:

$$\begin{aligned} \dot{q}_1(t) &= \frac{-(\lambda_0 + i\omega(1+\lambda_0 t))}{(1+\lambda_0 t)^2} q(0) e^{-i\omega t} \\ \dot{q}_2(t) &= \frac{-(\lambda_0 - i\omega(1+\lambda_0 t))}{(1+\lambda_0 t)^2} q(0) e^{i\omega t} \end{aligned}$$

The initial values are:

$$q_1(0) = q_2(0) = q(0), \;\; \dot{q}_1(0) = -(\lambda_0 + i\omega_0) q(0), \;\; \dot{q}_2(0) = -(\lambda_0 - i\omega_0) q(0).$$

With these functions and values we obtain, from Eqs. (4.14),(4.15) of Jannussis et al.[55], the relations:

$$|\mu|^2 - |\nu|^2 = 1$$

and

$$|\mu + \nu||\mu - \nu| = \frac{\lambda_0}{\omega(1+\lambda_0 t)^2} \tag{4.121}$$

for $\cos\omega t = 0$ and $\sin\omega t = 1$, but,

$$|\mu + \nu||\mu - \nu| = 0 \tag{4.122}$$

for $cos\omega t = 1$ and $sin\omega t = 0$.

With these results of two ideal cases we see that the damped harmonic oscillator with constraints of (4.119) actually represents the squeezed states, Eq. (4.121) is always smaller than unity, if $\lambda_0 \leq \omega$ (that is the usual case for our discussions), but Eq. (4.121) $= 1$ only for $\lambda_0 = \omega$, $t = 0$.

In the case of ultraweak photon (biophoton) emission we always have $\lambda_0 << \omega$. λ_0 is the decay rate, the inverse value of the coherence time of the biophoton. In the model of Jannussis λ_0 represents the growth rate of the oscillator mass. Only for $\lambda_0 > \omega$ and only at the very beginning of the time development of the oscillation, equation (4.121) can exceed unity and with time it reduces rapidly to a value smaller than unity. Thus we have, in the case of a damped oscillator with hyperbolic decay, Eq. (4.120), a very ideal example of a squeezed state, too.

We are very inspired by the above discussions about the coherent states of oscillators with time-dependent mass and frequency, the so called squeezed states, as they may become the basis for some new understanding of a number of dynamic molecular processes in vivo, as for instance reproduction, differentiation of DNA molecules, and communication with electromagnetic fields under the limit of the quantum noise of vacuum.

4.5 References

1. D.E. Green and R.F. Goldberger, *Molecular Insights into the Living Processes* (Academic Press, New York-London, 1967).

2. N. Bohr, 1885-1962, *Der Kopenhagener Geist in der Physik* (Vieweg, 1985).

3. E. Schrödinger, *What is life?* (Cambridge University Press, 1955).

4. G. Renger, Photosynthese, in *Biophysik 2nd. ed.*, eds. W. Hoppe, W. Lohmann, H. Markl and H. Ziegler (Springer-Verlag, 1982) p. 532.

5. M. Born and E. Wolf, *Principles of Optics, 5th edition* (Pergamon Press, Oxford, 1975) p. 511.

6. R. Loudon, *The Quantum Theory of Light, 2nd ed.* (Clarendon Press, Oxford, 1986) p. 43.

7. D. Ter Haar, On the History of Photon Statistics, in *Quantum Optics*, ed. R.J. Glauber (Academic Press, New York-London, 1969) p. 1-14.

8. R.T. Weidner and R.L. Sells, *Elementare moderne Physik* (Friedr. Vieweg+Sohn, Braunschweig-Wiesbaden 1982) p. 419-436.

9. K.H. Li and F.A. Popp, Coherence and the EPR paradox, in this volume.

10. W. Heisenberg, *Physikalische Prinzipien der Quantentheorie* (Bibliograph. Institut, Mannheim, 1958).

11. R.P. Feynman and A.R. Hibbs, *Quantum Mechanics and Path Integrals* (Mc Graw-Hill Book Company, 1965).

12. M. Garbuny, *Optical Physics* (Academic Press, New York and London, 1965).

13. A. Messiah, *Quantum Mechanics vol. 1* (North-Holland Pub. Comp. Amsterdam, 1965) p. 135.

14. L. Streit, The Construction of Quantum Field Theories, in *The Uncertainty Principle and Foundations of Quantum Mechanics. A Fifty Years' Survey* eds. W.C. Prise and S.S. Chissick (John Wiley + Son, London, New York, Sydney, Toronto 1977) p. 349-350.

15. E. Goldin, *Waves and Photons, An Introduction to Quantum Optics* (John Wiley + Sons, Inc., 1982).

16. P.W. Atkins, *Molecular Quantum Mechanics 2nd ed.* (Oxford University Press, Oxford-New York, 1983) p. 250-283.

17. R.H. Dicke, *Phys. Rev.* **93** (1954) p. 99.

18. M.J. Stephen, *J. Chem. Phys.* **40** (1964) p. 669.

19. E. Mueller, Allgemeine Struktur- und Funktionsprinzipien der Zellen, in *Molekulare Biologie der Zelle* (Gustav Fischer Verlag, Stuttgart, 1973) p. 262.

20. J. Eisinger and K.G. Shulman, *Science* **161** (1968) p. 1311.

21. J.K. Setlow, *Compr. Biochem.* **27** (1967) 190.

22. H. v. Mangoldt und K. Knopp, *Einführung in höhere Mathematik, 1. Band, 10. Auflage* (S. Hirzel Verlag Stuttgart, 1956) p. 128.

23. M. Sargent, M.O. Scully and W.E. Lamb Jr., *Laser Physics* (Addison-Wesley, Reading, 1974) p. 410.

24. J. Perina, *Coherence of Light* (D. Reidel Publ. Comp. Dordrecht, Boston, Lancaster, 1985).

25. R. Neurohr and M. Galle, *Natur- und Ganzheits-Medizin* **4** (1991) p. 28-33.

26. M. Galle, R. Neurohr, G. Altmann, F.A. Popp and W. Nagl, *Experientia* **47** (1991) p. 457.

27. G.S. Agarwal, *Springer Tracts in Modern Physics* **70** (1974) p. 60.

28. A.R. Edmonds, *Drehimpulse in der Quantenmechanik* (B.F. Mannheim, 1964) p. 26.

29. F.A. Popp, *Biologie des Lichts* (Verlag Paul Parey, Berlin, Hamburg, 1984).

30. J.D. Jackson, *Classical Electrodynamics* (John Wiley + Sons, Inc., New York, London, 1962).

31. H. Paul, Interference between independent photons, *Rev. Mod. Phys.* **58** (1986) p. 209.

32. F.A. Popp and V.E. Strauss, *"So koennte Krebs entstehen"* (Fischer Taschenbuch Verlag, Frankfurt, 1979).

33. W.H. Louisell, *Quantum Statistical Properties of Radiation* (John Wiley + Sons, New York, Singapore, 1973).

34. K.H. Li and F.A. Popp, *Phys. Lett A* **93** (1983) p. 262.

35. F.A. Popp, *Proc. Intern. Conf. Laser'85, Las Vegas* (1985) p. 311-316.

36. F.A. Popp, *DIA-GM* **15** (1988) p. 33.

37. M. Galle, R. Neurohr, G. Altmann, F.A. Popp and W. Nagl, *Experientia* **47** (1991) p. 457.

38. V. Ernst, *Z. Physik* **218** (1969) p. 111-128.

39. K.H. Li and F.A. Popp, Dynamics of DNA Excited States, in *Molecular and Biological Physics of Living Systems*, ed. R.K. Mishra (Kluwer Academic Publishers, 1990) p. 31.

40. M. Gueron, J. Eisinger and A.A. Lamola, Excited States of Nucleic Acids, in *Basic Principles in Nucleic Acid Chemistry* **1**, ed. P.O.P. Tso (Acad. Press. Inc., 1974) p. 311.

41. F.A. Popp, K.H. Li and W.P. Mei, *Experientia* **44** (1988) p. 576.

42. H. Paul, *Rev. Mod. Phys.* **54** (1982) p. 1061.

43. S. Saito, *Ann. Phys.* **187** (1988) p. 249.

44. H.H. Ritze and A. Bandilla, *Opt. Comm.* **28** (1979) p. 241.

45. D.F. Walls, *Nature* **280** (1979) p. 451-454.

46. R. Horak, L. Mista and J. Perina, *Phys. Lett. A* **35** (1971) p. 400.

47. F.A. Popp, in this volume.

48. L. Mandel, Quantum effects in spontaneous parametric down-conversion of light, in *Frontiers in Quantum Optics*, eds. E. Pike and S. Sarkar (Adam Hilger, Bristol and Boston, 1986) p. 318.

49. L. Mandel, *Phys. Rev. Lett.* **49** (1982) p. 136.

50. F. Casagrande and L.A. Lugiato, *Nuovo Cim. B* **55** (1980) p. 173.

51. H.J. Carmichael, Theory of quantum fluctuations in optical bistability, in *Frontiers in Quantum Optics*, eds. E. Pike and S. Sarker (Adam Hilger, Bristol and Boston, 1986) p. 120.

52. H.P. Yuen, *Phys. Rev. A* **13** (1976) p. 2226.

53. H.P. Yuen and J.H. Shapiro, *IEEE Trans. Inf. Theory* **26** (1980) p. 78.

54. R. Loudon, *G. Mod. Optics* **34** (1987) p. 709.

55. A. Jannussis and V. Bartzis, *IL Nuovo Cimento B* **102** (1988) p. 33.

56. F.A. Popp, B. Ruth, W. Bahr, J. Boehm, P. Grass, G. Grolig, M. Rattemeyer, H.G. Schmidt and P. Wulle, *Coll. Phenom.* **3** (1981) p. 187.

Chapter 5

Coherent Radiation from DNA Molecules

Ke-hsueh Li

Abstract

We discuss first the possibility of coherent emission of DNA molecules caused by the cooperative motions of DNA bases on the basis of Dicke's radiation theory which represents a general model of the real properties of all radiation sources. Then we consider the interactions between electromagnetic waves and mechanical base oscillations which can be assigned to polaritons, as in solid state physics, e.g. the joint vibrations between photons and phonons in DNA molecular skeleton.

5.1 Introduction

Although some success in the study of radiation properties of molecules has been achieved by use of spectral analysis similar to that of simple atoms, the difficulties to describe the electronic structure of middle-sized or big molecules (like biopolymers) are so enormous that even the rapid progress in computer application can not help to overcome this situation.

Of course, radiation from matter (atoms, molecules, macromolecules including e.g. DNA) does carry the necessary information allowing to comprehend the motion of electrons and the structure of energy levels, but, we are already facing continuum or quasi-continuum spectra for relatively small molecules in cellular media.

On account of this difficulty which arises evidently by interactions among different components taking part in the processes under investigation, we may certainly not ask for precise knowledge about individual electron orbitals and even relevant energy levels.

It becomes clear that the traditional method of research, which merely approaches some isolated individuality of matter or processes, must be changed. Instead of that we ought more to consider correlations between individualities and collectivity, co-operativity of processes and matter. Thus it is at least methodologically wrong if we try to understand the radiation properties of DNA only through the individual electron structures of its distinct composite parts, the base molecules, just as we cannot expect to understand the property of conductivity of a piece of semiconductor only by studying the individual structures of isolated atoms. Naturally, as an aperiodical "crystal"[1] the DNA molecule is much more complicated than any piece of a semiconductor.

Even for common spontaneous radiation from excited gas molecules we must consider the coherent interaction which causes the radiation law to deviate drastically from the famous Planck's radiation law[2].

The conventional description of the radiation of emitting matter is not satisfactory due to the existence of interference (coherent) patterns of the quantized Maxwell field in which the active atoms or molecules are embedded. On account of the interaction of atoms with this space-time field structure the conventional consideration of statistically independent emission processes is in principle wrong. By using the theory of Weisskopf and Wigner[3] of the natural line width extended to a case of many atoms interacting with one common quantized radiation field, Ernst[4] has shown that subradiance as well as superradiance of the lasing transition in ruby are so strong that the process of natural spontaneous emission of a photon in the sense of the theory of Weisskopf and Wigner practically never occurs.

We know experimentally well that the fluorescence spectra of native and denatured calf thymus DNA, poly dAT and single-stranded DNA from the phage ϕx 174 are ranging around 3500Å, and can be identified as the spectra of excimer- and exciplex-emission of DNA bases[5], where the distance of neighbouring DNA bases is merely ca. 3.4Å. That is to say that every wavelength can cover about 1000 bases. It must be borne in mind that the coherence length is much longer than the wavelength under consideration.

Consequently, the DNA photon emission (or fluorescence) can be by no means treated as an usual spontaneous emission in the sense of Weisskopf and Wigner. Through cooperation of molecular excited states of DNA bases the emission must be coherent. Dicke interpreted the term "cooperative" in the following sense:

> "One interesting and somewhat paradoxical aspect of the correlation and coherence problem being discussed is the fact that the two radiating atoms could be extremely far apart, many, many wavelengths, and still exhibit this correlation effect. One might naively wonder with such a radiating system, initially in a state for which both atoms are excited, how the one atom would ever know about the existence of the other".

Obviously, this second atom must, in some way, get information about the position of the first atom. This isn't as mystical as it may sound, if we consider that the two atoms are in the same coherence volume of photons emitted by themselves. And this coherence volume is identical to the corresponding region of uncertainty space[6]. For the one-dimensional case we have

$$\Delta x \geq \frac{\hbar}{\Delta p} = \frac{1}{2\pi}\frac{\lambda_0^2}{\Delta\lambda} = L,$$

where λ and L is the wave and coherence length respectively, λ_0 denotes the mean value of the wavelengths under consideration. Thus the two atoms are coupled with the same electromagnetic field in such a way that the one atom could know about the existence of the other. In comparison with the usual spontaneous emission the radiation from such a cooperative system has the striking feature that there are special directions for which the emission of photons is strictly forbidden. The calculation provides that the photon field in the coherence volume builds a coherent interference structure along the DNA-helical-axis which can be constructive (superradiance) or destructive (subradiance).

The interaction of radiation with matter is only possible through the distribution of charge whose motion can drive again the mechanical vibrations of the lattice. This lattice vibration excited by electromagnetic waves can be described by lattice equations of motion together with the Maxwell equations as developed by Huang[7] forty years ago for a diatomic cubic crystal, for instance the alkali halides. Electromagnetic and mechanical joint vibrations of this kind may occur in different branches of physics and are hereafter called polaritons[8].

The polariton-like process may occur in DNA lattice driving by photon emission during deexcitation of exciplexes, where a part of the excited energy of the excimer would be transferred to the lattice vibrations and hereafter into increasingly lower frequencies, even into those of chaotic motion (thermal motion) of macromolecules. Such a process of thermalization may be one of the ways for heat generation in cells.

In nonlinear mediums of chaotic systems, the energy transfer from higher frequencies into lower frequencies may find a solid physical basis[9]. We are very inspired by investigations of nonlinear systems following Lorenz[10]. He has modeled chaotic motions of climate only using three coupled nonlinear differential equations of first order with three constants. With similar equations the chaotic phenomena in quantum optics can also be described[9]. Therefore, it may be reasonably assumed that there are some simple principles behind the complexity of fluorescence spectra of DNA molecules observed in experiments. To expose those principles is certainly an actual goal of physics in the study of biological problems.

5.2 Dynamical Consideration for DNA Excited States

The necessity for dynamical considerations of DNA is obvious, as life is a dynamic structure of processes far away from thermal equilibrium displayed in physical space-time.

Since J.D. Watson and F. Crick created their classical model, biologists and biochemists have uncovered an extensive repertoire of DNA architectures. They have blessed researchers with an understanding of how genetic information is encoded. Inspired by such knowledge, and fascinated by the variety of DNA forms, biologists are penetrating deeper into the atomic arrangements. But the structural models tell us only part of the information, because they are static and inert. However, in reality DNA is rich in movement. It vibrates, "breathes" and emits. The generation time of *E.coli* DNA with ca. 2×10^6 nucleotide pairs (molecular weight 10^9) needs only 40 minutes for replication. Actually, it has been known for a long time that DNA molecules are luminescent materials. The discussion of the behaviour of excited states of nucleic acids should serve primarily as an aid for understanding the photochemical events occurring in the whole DNA chain. Any excitation of an individual base must influence the other bases due to the electromagnetic interactions between charge distributions and delocalization of excited energies.

5.2.1 Excited states of DNA bases

As it is true for all molecules, the physical and chemical properties of DNA bases are determined primarily by the electron distribution in the base. Interactions or reactions take place biochemically and photochemically mainly in excited state but actually not thermally in ground states, otherwise the specific character of bases would not be identified.

The electron distributions and energies of a molecule in both the ground state and excited states are in principle available by solution of the Schrödinger equation. In practice a number of approximations are introduced to obtain solutions to the many electron-, many nucleus Schrödinger equation.

On account of the perimeter-free-electron-orbital model of Platt[11] the cata-condensed aromatic hydrocarbons have "closed-shell" $f^4\pi$ electron configurations in the ground state which are similar to the ground-state closed-shell electron configurations of the noble atomic gases. As a consequence, the ground electronic state of an aromatic hydrocarbon molecule or a noble gas atom is a singlet state S_0, and their excited electronic states are either singlet states S_1, S_2, ..., S_p or triplet states T_1, T_2, ..., T_q.

For labeling and characterization of energy levels of pyrimidines we see no difficulty, for pyrimidine it is well-understood as an inductively perturbed benzene, but the same cannot be said for the pyrimidine DNA bases-thymine (uracil) and cytosine. The problem is, most likely, that these are ten π electron systems with eight centers, whereas benzene has six π electrons on six centers[12].

In the spirit of the above discussion on the pyrimidines, both adenine and guanine are expected to show deviations from annulene character because of the extra π electrons and centers contributed by the substituents. However, this problem of energy characterization is not serious for molecules with complex vibrational states, if we compare it with that of noble gases. Birks has shown[13], that

> "the most immediate problem is the identification and assignment of the various singlet and triplet excimer emissions in the noble gases. Monoenergetic photon or electron excitation provides a useful tool to unravel the complexities. Kinetic studies of the atomic and excimer luminescence provide a means of studying the excimer formation and radiative decay. Paradoxically, the physics of simple atomic excimers is complex, while the physics of complex aromatic molecular excimers is relatively simple. In this instance the physicist's preference for simple systems has proved

illusory; atomic excimers are more complicated than aromatic molecular excimers. Kasha's rules explain the paradox: the high density of vibrational states of a complex molecule simplifies its luminescence behaviour. No such simplification occurs in an atomic system."

Kasha's classification of electronic transitions is almost universally used by organic photochemists in practical work. According to that classification the energy levels of molecular electrons may be generally classified in increasing order of energy in terms of σ (bonding, symmetric), π (bonding, antisymmetric), n (nonbonding, lone pair), π^* (anti-bonding, antisymmetric) and σ^* (anti-bonding, symmetric).

This general schema of energy classification has been successfully used in analysis of various DNA bases for understanding the fluorescence of their excited states.

All oligo- and polynucleotides have absorption spectra very similar to those of the equivalent monomer mixture[5], except for hypochromism, which represents really some cooperative phenomenon and will be mentioned below. Even the spectral shape of the low-energy absorption band of dinucleoside phosphates closely resembles an equimolar mixture of its constituent nucleosides, but reduction in the band intensity (hypochromism) has been observed in all sixteen common ribodinucleoside phosphates[14] and in all sixteen common deoxyribodinucleoside phosphates.

Such facts show experimentally that the ground state interactions between nucleosides and nucleotides are really insignificant because they do not differ greatly in their electron affinities as well as their stable electron configurations in the ground state. That could also be indicated by the fact that the bases of the DNA chain can be considered in some cases as individuals, despite of the cooperative state for performing living functions.

As we have mentioned already, a molecule in the excited state has a different distribution of molecular electrons in comparison with that in the ground state. Thus electron affinity and chemical reactivity must become different. In this case it is not surprising that an excited base forms a new molecular complex together with an other one even if this other one is in the ground state, while the complex cannot be built when both are in the ground state. Thereby it is not important, how the excitation is produced. However, the excited molecular complexes by photon excitation have been investigated extensively in last time and termed as photoaddition[15] in chemistry.

The intermolecular chemical bond of such a complex is not ionic for two identical base molecules. Although the ground state is repulsive due to the non-bonding state, the intermolecular binding energy in the excited state may not be very small. Probably, the nature of the intermolecular interaction is the same as that of the ordinary covalent chemical bond. This sort of complex, which exists only in the excited electronic state, is known for many aromatic molecules in solution and in crystal form. It is called excimer (for identical molecules) or exciplex (for different monomers). The question arises, whether also some complexes of higher order by three or more molecules are produced. Theoretically, this cannot be excluded for higher excitations. But the probability is certainly very small due to the high excitation energy. In addition, one must also consider the stability of such an excited complex of higher order.

Here we may point out a very interesting fact namely that the excimer (exciplex) formed by the base molecules of DNA must be classified as a so-called intramolecular excimer. This fact is important because the threshold for generating stimulated coherent emission from intramolecular excimers is much lower than that from intermolecular excimers[16]. The technique to realize the first intramolecular stimulated coherent emission has been devised by Nakashima et al.[17].

5.2.2 Lowest potential energy profiles of excimers (exciplexes)

Like rare gas atoms such as He, Ne, Ar ..., the aromatic hydrocarbons with "closed-shell" electronic structures cannot form stable molecular complexes of higher order in the ground state. Only very weak attractions arise between them due to the Van der Waals dispersion forces at relatively long distances. At short distances mainly the repulsive force due to anti-binding ground state dominates, and the monomers remain in their relative individuality as a result of closed-shell electron structure. This repulsive force will be drastically increase by reducing the interplanar distance.

In the excited state the electron "closed-shell" can be opened by redistribution of the electron cloud and displays some chemical activity. In the building of excimers (exciplexes), e.g. excited state composited systems, the total wave function Φ has to be expressed by the superposition of various electron configurations[18].

For the sake of simplicity and of principal understanding of the building process of exciplexes we consider only the case of two identical monomers, neglecting charge-transfer. In the zero-order approximation the molecular exciton interaction can be expressed by

$$\Phi_{\pm} = \frac{1}{\sqrt{2}}(\Phi_1^a\Phi_2^0 \pm \Phi_1^0\Phi_2^a), \tag{5.1}$$

where Φ_i^a and Φ_j^0 represent molecule i in excited and j in ground state, respectively. The wave functions (5.1) have energies

$$E_+ = E_a - \frac{(|\mu_e| \cdot |\mu_g| + \mu_t^2)}{R^3},$$

$$E_- = E_a - \frac{(|\mu_e| \cdot |\mu_g| - \mu_t^2)}{R^3}, \tag{5.2}$$

in which μ_g is the ground-state dipole moment, μ_e the excited-state dipole moment, and μ_t the transition dipole moment. The strength of the interaction (the splitting) depends strongly on the interplanar distance R [14]

$$E_+ - E_- = -2\frac{\mu_t^2}{R^3}. \tag{5.3}$$

This splitting energy approaches ∞ for $R \to 0$, so that the stable excited state of the composite system requires a repulsive potential to balance it at the equilibrium point R_0 (Fig. 5.1).

This repulsive potential energy can be seen as the same as that of the ground state approximately for many π-electron systems[13]. This approximation does not neccessarily apply to noble gas excimers. From the data of excited and fluorescence spectra we estimate the excitation energy of a monomer base molecule at 2600Å (Fig. 5.1):

$$E_3 - E_0 = 4.77\ eV, \tag{5.4}$$

and the energy of excimer fluorescence :

$$E_2 - E_1 = 3.55\ eV. \tag{5.5}$$

For the interplanar distance of the DNA helix (3.4Å) the magnitude of the exciton splitting has been estimated on the average as 750 (cm^{-1})[14]. The well known covalent bond distance of $C-C$ is 1.54 Å and of $C=C$ is 1.33Å in their ground states, respectively, while we don't have the corresponding data for the excited state and especially for excimer-like complexes.

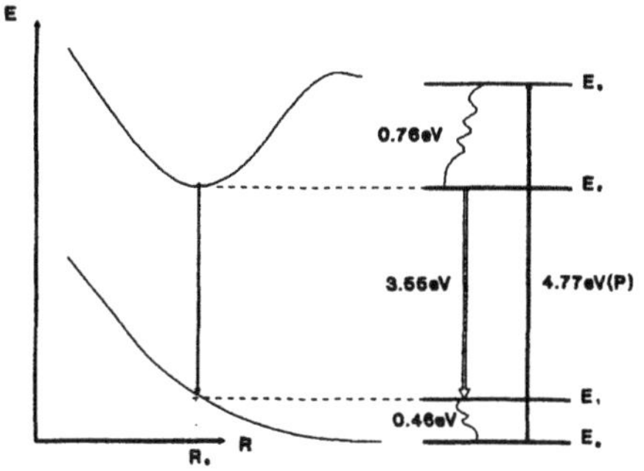

Figure 5.1: *A scheme of a four energy-level system corresponding to the excimer potential profiles. P denotes the pumping energy.*

Let us make a tentative assumption that the interplanar distance between two neighbouring base molecules forming a stable excimer is ca. $2\AA$, then we can, from the exciton splitting, estimate the interaction energy to be about 0.46 eV which must be balanced by the repulsive potential in the excited state. From the above discussion follows the repulsive potential energy of the ground state:

$$E_1 - E_0 \simeq 0.46\ eV, \tag{5.6}$$

which will be assigned to the base monomers to keep them apart, when the excimer would be deexcited at the same stable point R_0 (Fig. 5.1). Then we obtain also the binding energy of the excimer which can be read from the energy scheme (Fig. 5.1) as

$$E_3 - E_2 = (E_3 - E_0) - (E_2 - E_1) - (E_1 - E_0) = 0.76\ eV. \tag{5.7}$$

These rough estimates show that the repulsive energy of the ground state is 0.46 eV and thus significantly exceeds the thermal energy of 0.025 eV at room temperature, while the excimer binding energy equals 0.75 eV, being almost twice that of pyrene excimer in crystal form which bas been analysed by Birks[13].

The existence of excimers (exciplexes) in the organism as a form of excited-state matter is perhaps very general and well known for a long time[19]. Excimers (exciplexes) in connection with photophysical and photochemical processes in DNA have been discussed since a long time[20,21]. Excimer formation in membranes has been used to estimate membrane microviscosity and the rates of lateral diffusion of lipids[22,23].

The theory of the excimer state has been developed primarily for aromatic hydrocarbons[13], while the first experiments on pyrene excimers have been performed by Förster and Kasper[24], where the concentration dependent appearance of the excimer fluorescence spectrum has been established. Actually, neither aromaticity nor heteroaromaticity is necessary for excimer formation by organic molecules. Some saturated organic amines exhibit excimer fluorescence in solution and in the vapour phase. A detailed study of excimer formation by one of these compounds, 1-azabicyclo (2,2,2) octane in the vapour phase has been made by Halpern[13].

Presently, some important fluorescent biomolecules with excimer characteristics have been found already and the coherence features of their radiations have been identified[25–29].

It should be enlightening at this place to cite a passage from Birks:

> "In its extension from physics into chemistry and biology, the excimer has increased in size, complexity and diversity, large aromatic molecules have replaced simple atoms as the excimer constituents. Fluid solutions and liquids, polymers and intramolecular systems, have replaced the vapour phase as the excimer environment. These factors have extended the range of excimer phenomena, but they have not changed their physical nature. The excimer remains a physical entity. The excimer is not a chemical entity, since it exists only while it pocesses electronic excitation energy. On dissipation of this energy, the system reverts to a dissociated pair of molecules."

The existence of excimer emission is normally indicated experimentally if a structureless, red-shifted fluorescence is observed without alteration of the absorption spectrum.

The energy level scheme of an excimer is given in Fig. 5.1 which displays the energy diagram corresponding to a four-level system. Because of energy levels E_3 and E_2 being in the same electron potential the transition rate between them is much quicker than that between E_2 and E_1. A similar case we find also for energy levels E_1 and E_0. For this four-level system we have deduced a perfect laser equation elsewhere[30,31] by using simple rate equations which are satisfying the excimer dynamics, and also discussed some nonlinear behaviour and bifurcation of processes.

5.3 The Dicke Theory

This chapter treats the central issue of this paper. We know that the radiation theory developed by M. Planck at the beginning of this century represents a new era of theory of optics which introduced the quantum of action and exerts a tremendous influence on the development of modern sciences. If we denote the theory of Planck as the first milestone of radiation theory that confines itself to the discussion of independent and random events at thermal equilibrium, then the theory of Dicke can certainly enjoy the prestige of being the second milestone in the history of radiation theory because it breaks through limitations of individuality, randomness and thermal equilibrium.

Dicke[2,32] has convincingly shown that "in the usual treatment of spontaneous radiation by a gas, the radiation process is calculated as though the separate molecules radiate independently of each other. This simplified picture overlooks the fact that all the molecules are interacting with a common radiation field and hence cannot be treated as independent." He first considered the problem under the assumption that the gas distances have dimensions small compared to a radiation wavelength. The result shows that the radiation rate of a coherent spontaneous radiation process is proportional to the square of the molecular concentration rather than linearly to the concentration as the usual radiation theory has predicted. He called this new strange emission "superradiance".

The opposite case, of the states having abnormally low radiation rates, when in the ideal case for an even number of molecules the radiation rate equals zero, be named "subradiance". The molecules in the last case are also highly correlated as discussed by Dicke in his original paper. The subradiance state can also be called the photon trapping state, which is very important in photobiology as predicted by F.A. Popp[33]. The trapping mechanism is a coherent migration of photon (excitation energy) within the system of cooperative molecules. The trapping time can be several orders of magnitude larger than the natural decay time of a single molecule and is mainly a function of the density of the molecules. Theoretically it has, in principle, no upperlimit, i.e., it can also be infinite.

The purpose of this chapter will be to show that

1. the Dicke theory is universal for the discussion of all real systems including biological systems, while the theory of Planck represents a special case for independent radiators in thermal equilibrium.

2. correlation, coherence and cooperation are the important consequences following from the Dicke theory. These are physical concepts on the base of which we can gain insight into a number of modern biological problems, especially, the problem of photon emission which has puzzled many biologists for a long time.

3. DNA molecules are well known to biologists as a fluorescence material. However, the radiation wrongly was considered to be incoherent. The mistake may be seen from the simple fact that the wavelength emitted can cover almost a thousand DNA bases which are the emitters. In the light of the Dicke theory, the conclusion of incoherence is by no means tenable. We will show this by a simple classical model describing how the photon coherently can be delocalized along the DNA chain and can contribute to building a molecular complex between base molecules, the so called excimer and exciplex, and in the near zone approximation of the electromagnetic wave, how bases of the same sort can recognize each other and communicate on long-range by the specific frequency, although many bases of other sorts are embedded in between them.

5.3.1 A universal theory for real radiation systems

The essential presuppositions to derive Planck's radiation law are:

1. All oscillators of a given frequency are independent.

2. The whole system is enclosed in a cavity with walls of 100% reflectivity in thermal equilibrium which could be realized by a tiny black-body introduced into the cavity and having the function only to transfer energy between oscillators of different frequencies but not to contribute energetically to the system.

3. The energy of the oscillators can only be divided into an integral finite number of equal parts (quantum condition) which have to be kept finite.

However, the first two presuppositions can never be satisfied by a real system, therefore Planck's black-body radiation which can be counted as one of the most successful achievements of this century remains still an ideal extreme case in the theory and can not be found in real nature. Two oscillators can only be considered as independent, if the distance between them is longer than the *coherence length* of the waves emitted by them. Within the *coherence length* (which is actually the uncertainty length in the sense of Heisenberg[34]) the photon is completely delocalized. Thus the two oscillators in principle cannot be separated.

On account of thermal perturbations the *coherence length* can be drastically reduced, but the thermal perturbation can only affect the long wavelengths (thermal radiations), but hardly short wavelengths. That is just the reason why the intensity measurement of the sun's infrared and far-infrared spectra at the outer surface of our atmosphere approaches the theoretical values of Planck's black-body radiation while the short wavelength parts of the same measurements deviate evidently from Planck's theoretical values[35]. Hence, different wavelengths have different colour-temperatures, also the sun's radiation is by no means at thermal equilibrium.

Therefore we need a more general theory to describe phenomena of radiation and of interaction of radiation with matter. We find it in the Dicke theory which has shown in its initial stage only that the coherent emission follows laws completely different from those of usual spontaneous emissions. The significance of this theory was later gradually recognized by different authors[36–41]. With convincing arguments Ernst[4] has shown that even if the theory presented by Dicke is applied to the lasing transition in ruby, one finds that subradiance as well as superradiance are so strong that the process of natural spontaneous emission of a photon in the sense of the theory of Weisskopf and Wigner[42] practically never occurs.

Dicke used the usual perturbation theory to calculate the radiation from a collection of identical two-level atoms initially prepared in certain excited states. He found that under certain circumstances the radiation rate from such atoms which are in a volume smaller than the wavelength, is proportional to the square of the number of atoms (superradiance, coherent radiation rate); in an other case when the number of atoms is even, the radiation rate even can take the value of zero (subradiance, coherent photon trapping). On the other hand, the incoherent radiation rate, i.e. the case of emitting independently, is proportional to the number of atoms (a linear phenomenon). This universal principle of physics eventually cannot be evaded for the consideration of biological problems, especially, if interactions of electromagnetic radiation with living matter is concerned.

5.3.2 A preliminary consideration of Dicke's theory

When discussing Dicke's theory, be it quantum-mechanically or classically, an important factor may not be overlooked, that is the uncertainty principle of Heisenberg. The importance of this principle lies in the fundamental rule with which the probability of finding particles or an intensity of waves at a certain space-time point has to be calculated. We have shown in other papers in this book[6,43] that the identification of the coherence volume with the uncertainty volume in ordinary space-time may be traced back to Heisenberg's principle itself.

Heisenberg and also Feynman have clearly deliberated the fundamental difference in evaluation the probability of independent particles and that of particles within the same uncertainty volume. Here it may be unnecessary to be repeated. But, we want to show some practical results by the following example :

We consider all excited atoms within the uncertainty volume of electromagnetic waves emitted by them as tiny dipoles. The total amplitude of all dipole radiations is the sum of all amplitudes of individual radiators (dipoles):

$$E = E_1 + E_2 + E_3 + ... = \sum_i E_i, \qquad (5.8)$$

in accordance with Dicke's original supposition.

To find the total intensity at every spatial point we have to build the square of this result at the point of observation:

$$E^*E = \sum_{i=1}^{N} \mid E_i \mid^2 + \sum_{i,j=1}^{N} E_i^* E_j \qquad (i \neq j). \qquad (5.9)$$

On account of the fact that every measurement can only be made in a finite time interval, the evaluation of results carries certainly a statistical character. Thus the second term of Eq. (5.9) represents a series of correlation functions between different dipoles:

$$< E_i^* E_j > \simeq \mid E_i \mid^2 e^{-\frac{r_{ij}}{L}}, \qquad (5.10)$$

where $r_{ij} = (r_j - r_i)$, L is the *coherence length.*

For an ideal extreme case, when $r_{ij} \ll L$ and $e^{-\frac{r_{ij}}{L}} \simeq 1$, we get from equation (5.9)

$$< E^*E >= N^2 < E_i^2 > . \qquad (5.11)$$

This result has been denoted by Dicke as superradiance, a case of the best coherence and cooperation among dipoles where the constructive intercerence dominates the field. For the case of r_{ij} is increasing and $e^{-\frac{r_{ij}}{L}} \leq 1$ we have

$$< E^*E > \leq N^2 < E_i^2 >, \tag{5.12}$$

and finally, for $r_{ij} \gg L$ the particles are separated by distances longer than the dimension of the *coherence volume*, thus becoming individual, the correlation functions will approach the value zero:

$$< E_i^* E_j > \rightarrow 0. \tag{5.13}$$

Therefore, the Eq. (5.9) gets the usual form for individual and independent dipoles:

$$< E^*E > = N < E_i^2 >, \tag{5.14}$$

which gives the usual linear dependence of the total intensity measured on the number of independent dipoles.

Now we have yet to discuss another case, that is if all the correlation functions (5.10) take the negative sign instead of the positive one as discussed above. In such a case we can even obtain a vanishing total intensity:

$$< E^*E > = 0, \tag{5.15}$$

where the field of every pair of dipoles would exactly be compensated to zero. That is the case of destructive interference and was denoted by different authors as a trapping of excited energy (photons) in the system under consideration. The lifetime of photon trapping can be infinite in principle[34]. This phenomenon has been named subradiance. The ideal subradiance state appears only if the total dipole number N is even and the population occupation of two energy-levels is completely equal in amount, i.e. $\frac{1}{2}N$ in the ground state and $\frac{1}{2}N$ in the excited state. As a consequence, one photon released by a molecule must be absorbed immediately by another one and vice versa. This perfect cooperation amongst molecules must also presuppose the existence of some favourable space-time order of the considered system. Evidently, that order is nothing else than the interference pattern of waves from different radiation sources (molecules or dipoles) and contradictory to the famous statement of Dirac on the optical interference[44].

Such cooperative effect in spontaneous emission from many molecules was actually predicted as early as 1954 by Dicke in Eq. (92) of his paper[2]. Paul[44] has specialized Dicke's result to the case of two initially excited atoms ($N = 2$) located at definite positions r_1 and r_2, respectively, and arrived at the following interesting relation for the stationary state:

$$p(\vec{k_2}) = p_0(\vec{k_2})\{1 + \cos[\Delta\vec{k}(r_2 - r_1)]\}, \tag{5.16}$$

where $\Delta\vec{k} = \vec{k_2} - \vec{k_1}$.

The physical meaning of this expression is the following: It is assumed that the emission of a photon in the direction $\vec{k_1}$ has been observed. The quantity $p(\vec{k_2})$ is the probability (per unit time and unit solid angle) that the second photon will be emitted in the direction $\vec{k_2}$, and $p_0(\vec{k_2})$ is the corresponding radiation probability for a single isolated excited atom. The factor

$$1 + \cos[\Delta\vec{k}(r_2 - r_1)] \tag{5.17}$$

describes the interference between two dipoles. It demonstrates the strong correlation between the two emitted photons and shows also evidently cooperation between the two dipoles (atoms). All these orderliness appears only within the *coherence volume* (uncertainty volume). Beyond this volume the two dipoles cannot be dependent on each other, they are individual.

The high temporal-spatial order of interference patterns naturally determines the regular behaviour of emissions of coherent and cooperative photons, the photons emitted from cooperative systems, as opposed to ordinary spontaneous processes displayed in chaotic systems (thermal emission from systems at equilibrium state). Especially for destructive interference (subradiance state), the cooperative emission has the best chance to produce the phenomenon of nonclassical photon emissions, the so called antibunching effect and the squeezed state photons, because

1. the intensity of the photon field can be reduced to arbitrary weakness, and

2. all distances separating photons may be almost equal and arbitrarily extended.

This regularity removes any possible fluctuation of the photon field. We have shown elsewhere in this book[34] that the process of cooperation and coherent collectivation is at the same time also a process of reduction of fluctuations. The typical cooperative system in quantum mechanics has the minimum uncertainty, while the opposite may not always be true.

5.3.3 Dicke's theory for an aperiodical one-dimensional molecular chain

(a)

Using the Schrödinger equation, Stephen investigated the interaction energy between two atoms, one of which is excited[45]. If $\frac{R}{\lambda} < 1$, where R represents the distance between two atoms and λ the wavelength of radiation emitted, the photon may be trapped between the two atoms leading to the well-known dipole interaction

$$\frac{e^2}{m}(\vec{p} - 3\vec{n}(\vec{n} \cdot \vec{p}))\frac{e^{i\vec{k}\cdot\vec{R}}}{R^3}, \tag{5.18}$$

where $\vec{n}$ is an unit vector defined by $\vec{n} = \frac{\vec{R}}{|R|}$.

The mode of state of the two interacting atoms can be damped very slowly due to the photon (excited energy) trapping. The life-time τ of such a mode has been estimated by Stephen:

$$\tau \simeq (\frac{R}{\lambda})^2 \gamma_0, \tag{5.19}$$

where γ_0 is the spontaneous emission probability of an idividual atom.

Now we want to consider the dynamics of a one-dimensional molecular chain. Firstly, we assume that it is composed of two parallel dipoles perpendicular to their binding line. The two dipoles are linked by the dipole field (5.18). Here we take the simple case of two coupled vibrational equations:

$$\ddot{x}_1 + \omega_0^2 x_1 = g x_2,$$

$$\ddot{x}_2 + \omega_0^2 x_2 = g x_1, \tag{5.20}$$

where the coupling constant g can be either of mechanical or of electromagnetic character, for example, the dipole field (5.18). The classical description of super- and subradiance by equation system of oscillators (5.20)[46] can exhibit the salient features of Dicke's theory. This model is very convenient for us, in order to discuss the molecular chain composed by more than two charged harmonic oscillators. We can disregard the external driving and damping terms, as it can be understood as if the two terms were just compensated in the sense of a stationary state at equilibrium of an open system[47]. Because the magnetic interaction is by a factor $(\vec{k} \cdot \vec{R})$ smaller than the electric part in the region $\vec{k}\cdot\vec{R} < 1$, the field in the near-zone approximation is dominantly electric in nature. The magnetic induction vanishes, of course, in the static limit $k \to 0$. Consequently the near-zone stretches right to infinity.

For two parallel dipoles perpendicular to $\vec{R}$ we have from (5.18)

$$g = \frac{e^2}{m}\frac{1}{R^3}. \tag{5.21}$$

The exponential factor $e^{i\vec{k}\cdot\vec{R}}$ has been set approximately equal 1.

For the solution and discussion of the system of equations (5.20) we introduce the following collective coordinates:

$$x_\pm = x_1 \pm x_2, \tag{5.22}$$

where + denotes the symmetric and − the antisymmetric case. We then have

$$\ddot{x}_+ + (\omega_0^2 - g)x_+ = 0 \tag{5.23}$$

and

$$\ddot{x}_- + (\omega_0^2 + g)x_- = 0 \tag{5.24}$$

instead of (5.20).

The frequencies of the symmetric and antisymmetric vibration are

$$\omega_\pm = \sqrt{\omega_0^2 \mp g}, \tag{5.25}$$

respectively. The solutions of (5.23) and (5.24) take the simple form :

$$x_\pm = A_\pm e^{i\omega_\pm t}. \tag{5.26}$$

The general solutions of the original equations (5.20) can be written as

$$x_1 = \frac{1}{2}A_+ e^{i\omega_+ t} + \frac{1}{2}A_- e^{i\omega_- t}, \tag{5.27}$$

$$x_2 = \frac{1}{2}A_+ e^{i\omega_+ t} - \frac{1}{2}A_- e^{i\omega_- t}. \tag{5.28}$$

For simulating the simultoneous excitation of both molecules we impose on equations (5.27), (5.28) the following initial conditions

$$x_1(t=0) = x_2(t=0) = 0,$$
$$\dot{x_1}(t=0) = \dot{x_2}(t=0) = v. \tag{5.29}$$

This means that at $t = 0$ both oscillators are at a position of equilibrium, but simultaneously subjected to a push with the velocity of v.

Under this condition the real part of the amplitudes must satisfy

$$Re(A_+ + A_-) = Re(A_+ - A_-) = 0$$

or

$$Re A_\pm = 0, \tag{5.30}$$

and the imaginary part reads

$$Im\ A_\pm = \frac{i}{2} a_\pm, \tag{5.31}$$

where $a_\pm$ are real.

On account of (5.29) we have from (5.27) and (5.28)

$$a_+\omega_+ + a_-\omega_- = a_+\omega_+ - a_-\omega_- = -2v,$$

resulting in

$$a_-\omega_- = 0$$

and

$$a_- = 0, \tag{5.32}$$

due to $\omega_- \neq 0$.

Hence, the two solutions of Eq. (5.27), (5.28) take the same form[48]

$$x_1 = x_2 = -\frac{1}{4} a_+ \sin \omega_+ t. \tag{5.33}$$

Now we understand the process described above as follows: both oscillators under the initial conditions (5.29) are simultaneously excited. In the symmetric case they perform a collective, coherent vibration. The frequency shifts from ω_0 slightly to the red side. The antisymmetric vibration does not occur, due to the constraint (5.32). According to classical electrodynamics[54] the field intensity of dipole radiation is proportional to

$$|x_+|^2 = x_+^* x_+.$$

Hence we have

$$I \simeq < x_+^* x_+ > = < x_1^* x_1 + x_2^* x_2 + x_1^* x_2 + x_2^* x_1 >, \tag{5.34}$$

where $< \cdots >$ denotes the statistical average of the physical quantity under consideration. The statistical consideration of real measurements is still necessary and does not need further explanation.

The last two terms of (5.34) represent a correlation function which can be evaluated for any concrete problem. For the ideal case of complete coherence we have

$$I \simeq 4I_1, \tag{5.35}$$

where I_1 is the intensity of the radiation field emitted by an individual dipole. Here we see that the total intensity (5.34) for complete coherence is proportional to the square of the number of the radiators of the source, instead of the linear proportion for the usual chaotic light source. That is just the superradiance case predicted by Dicke in his paper "Coherence in Spontaneous Radiation Processes"[2].

If the correlation terms in Eq. (5.34) vanish in case when the statistical average has been performed for chaotic processes at thermal equilibrium state, then the total intensity (5.34) is linearly proportional to the number of radiators of the light source as usually predicted by Planck's theory.

(b)

What happens, however, when only one of both oscillators is excited initially? For such a case we must modify the initial condition of velocity imposed to (5.29) as follows:

$$\dot{x}_1(0) = v, \qquad \dot{x}_2(0) = 0, \tag{5.36}$$

while the initial condition of positions remains unchanged. The condition (5.36) tells us that the oscillator x_1 at $t = 0$ has been excited by a push with the velocity v and the other remained unchanged.

Now we get instead of Eq. (5.29)

$$\dot{x}_1(0) = -\frac{1}{2}(a_+\omega_+ + a_-\omega_-) = v, \tag{5.37}$$

$$\dot{x}_2(0) = -\frac{1}{2}(a_+\omega_+ - a_-\omega_-) = 0. \tag{5.38}$$

For simplicity we consider here a special case for $a_+ = a_- = a$ without loss of generality. Hence we have

$$v = -\frac{a}{2}(\omega_+ + \omega_-) = -a\bar{\omega}, \tag{5.39}$$

where we define

$$\bar{\omega} = \frac{1}{2}(\omega_+ + \omega_-) \tag{5.40}$$

as the average frequency. Hence one can also define

$$\omega_\pm = \bar{\omega} \mp \frac{1}{2}\Delta\omega, \tag{5.41}$$

which may be reformulated according to Eq.(5.25)

$$\bar{\omega} = \omega_0\sqrt{\frac{1}{2} + \frac{1}{2}\sqrt{1 - \frac{g^2}{\omega_0^4}}} \tag{5.42}$$

$$\frac{\Delta\omega}{2} = \omega_0\sqrt{\frac{1}{2} - \frac{1}{2}\sqrt{1 - \frac{g^2}{\omega_0^4}}} \tag{5.43}$$

and

$$\mid \dot{x}_2(0) \mid = \frac{a\Delta\omega}{2} \ll v, \tag{5.44}$$

instead of Eq. (5.38)[48].

Thus we can rewrite the equations (5.27) and (5.28) in the following form

$$x_1 = \frac{1}{2}e^{i\bar{\omega}t}(A_+e^{-\frac{i}{2}\Delta\omega t} + A_-e^{\frac{i}{2}\Delta\omega t}) \tag{5.45}$$

$$x_2 = \frac{1}{2}e^{i\bar{\omega}t}(A_+e^{-\frac{i}{2}\Delta\omega t} - A_-e^{\frac{i}{2}\Delta\omega t}), \tag{5.46}$$

on account of the assumption above $a_+ = a_- = a$ we have the simple form

$$x_1 = ae^{i\bar{\omega}t}\cos\frac{\Delta\omega}{2}t, \tag{5.47}$$

$$x_2 = -iae^{i\bar{\omega}t}\sin\frac{\Delta\omega}{2}t. \tag{5.48}$$

If we want to consider the energy changes of two oscillators with time we ought to form the expression $E_i = \frac{1}{2}\omega_i^2 \mid x_i \mid^2$ for $i = 1, 2$. Then we have

$$E_1 = \frac{1}{2}v^2(1 + \cos\Delta\omega t) \tag{5.49}$$

$$E_2 = \frac{1}{2}v^2(1 - \cos\Delta\omega t). \tag{5.50}$$

We see here that the excitation energy which initially has been imposed to the oscillator x_1 (5.37) by a push transfers periodically up and down between both oscillators with a frequency of $\Delta\omega$. This energy is trapped in the system without loss due to the total energy $E = E_1 + E_2$ remaining a constant. It represents an ideal case of subradiance according to Dicke's theory.

(c)

So far, we have discussed only a special case where the two dipoles must be identical. If we remove this restriction we have to take the following coupled differential equations instead of Eqs. (5.20):

$$\ddot{x}_1 + \omega_1^2 x_1 = g_2 x_2, \tag{5.51}$$

$$\ddot{x}_2 + \omega_2^2 x_2 = g_1 x_1. \tag{5.52}$$

This system of equations reduces to that of Eqs. (5.20) by assuming

$$\omega_1^2 = \omega_2^2 = \omega_0^2$$

and

$$g_1 = g_2 = g.$$

From the secular equation

$$(\omega^2 - \omega_1^2)(\omega^2 - \omega_2^2) - g_1 g_2 = 0 \tag{5.53}$$

of the equation system (5.51) and (5.52) there follows the general collective frequency:

$$\omega_\pm = \sqrt{\frac{(\omega_1^2 + \omega_2^2)}{2} \mp \sqrt{g_1 g_2 + (\frac{\omega_1^2 - \omega_2^2}{2})^2}} \tag{5.54}$$

instead of Eq. (5.25) for the special case.

From Eq.(5.54) we have two other relations

$$\omega_+^2 + \omega_-^2 = \omega_1^2 + \omega_2^2, \tag{5.55}$$

$$\omega_+^2 \omega_-^2 = \omega_1^2 \omega_2^2 - g_1 g_2 < \omega_1^2 \omega_2^2, \tag{5.56}$$

and on account of these relations we have the following inequality

$$\omega_- \geq \omega_1 \geq \omega_2 \geq \omega_+, \tag{5.57}$$

which follows from the fact that the sum of two collective frequencies in square ω_+^2 and ω_-^2 is exactly equal to the sum of individual oscillator eigenfrequencies in square ω_1^2 and ω_2^2, but the product of ω_+^2 and ω_-^2 is always smaller than the product of ω_1^2 amd ω_2^2. Hence , the values of ω_+ and ω_- must be distributed broader than the values of ω_1 and ω_2. Fröhlich[49] has discussed also some interesting problems concerning these relations for his selective long range dispersion forces between giant molecules.

The general solutions of vibration equations (5.51) and (5.52) can be represented by[48]

$$x_1(t) = A_+e^{i\omega_+t} + A_-e^{i\omega_-t}, \tag{5.58}$$

$$x_2(t) = \frac{(\omega_1^2 - \omega_+^2)}{g_1}A_1e^{i\omega_+t} - \frac{(\omega_1^2 - \omega_-^2)}{g_1}A_-e^{i\omega_-t}, \tag{5.59}$$

and

$$\bar{\omega} = \frac{1}{2}\sqrt{(\omega_1 + \omega_2)^2 - 2\omega_1\omega_2(1 - \sqrt{1 - \frac{g_1g_2}{\omega_1^2\omega_2^2}})}, \tag{5.60}$$

$$\Delta\omega = \sqrt{(\omega_1 - \omega_2)^2 + 2\omega_1\omega_2(1 - \sqrt{1 - \frac{g_1g_2}{\omega_1^2\omega_2^2}})}. \tag{5.61}$$

As before, one finds also in this case the beat of oscillations for non-stationary solutions, where $x_1(t)$ takes the same form of Eq.(5.27), (the factor $\frac{1}{2}$ is merged into the amplitudes $A_\pm$), but, the expression of $x_2(t)$ deviates not unessentially from that of Eq. (5.28). The amplitude $x_2(t)$ fluctuates with frequency $\Delta\omega$, but does not go anymore to zero. This means that only a part of the excitation energy can be exchanged between the oscillators. The amount of this part of the energy depends of the distinguishability of the two oscillators under consideration.

(d)

Now we come to discuss vibrations of a molecular chain composed of more than two coupled dipoles which are not all identical. Therefore we will ask how different dipoles take part in collective vibrations and how they exchange excitation energies represented by photons.

For simplicity we consider three coupled atoms as illustrated in Fig. 5.2, where the atoms 1 and 3 are identical, but different from the atom 2.

Consequently, we have the following coupled differential equations:

Figure 5.2: *A section of an aperiodic molecular chain with two identical particles at the terminals.*

$$\ddot{x}_1 + \omega_1^2 x_1 = g_2 x_2 \tag{5.62}$$

$$\ddot{x}_2 + \omega_2^2 x_2 = g_1 x_1 + g_3 x_3 \tag{5.63}$$

$$\ddot{x}_3 + \omega_3^2 x_3 = g_2 x_2, \tag{5.64}$$

where we have only considered the nearest neighboured interactions, since the ratio of this interaction to the next one is already 8:1. On account of the shielding which is always existent, the above consideration is sufficient.

Due to the identity of atoms 1 and 3 the relations $\omega_1 = \omega_3$ and $g_1 = g_3$ are valid, hence we have

$$\ddot{x} + \omega^2 x = 2 g_2 x_2 \tag{5.65}$$

$$\ddot{x}_2 + \omega_2^2 x_2 = g_1 x, \tag{5.66}$$

with a collective coordinate

$$x = x_1 + x_3.$$

With these equations we come back again to the case of (**c**) where we discussed the energy transfer between two different oscillators as described by equations (5.51) and (5.52). The result was that the energy exchange between two systems (here we denote them by $x = x_1 + x_3$ and x_2) can also occur, but, not very well due to non-resonance $\omega \neq \omega_2$. On the other hand, the energy exchange between x_1 and x_3 within the system denoted by the collective coordinate x remains still perfect. In order to show this we substitute a trial solution

$$x_2 = A_2 e^{i\omega t} \tag{5.67}$$

into equation (5.63), obtaining

$$x_2 = \frac{1}{\omega_2^2 - \omega^2}(g_1 x_1 + g_3 x_3). \tag{5.68}$$

Inserting (5.68) into (5.62) and (5.64), respectively, we get

$$\ddot{x}_1 + (\omega_1^2 - \frac{g_1 g_2}{\omega_2^2 - \omega^2})x_1 = \frac{g_2 g_3 x_3}{\omega_2^2 - \omega^2}, \tag{5.69}$$

$$\ddot{x}_3 + (\omega_3^2 - \frac{g_1 g_2}{\omega_2^2 - \omega^2})x_3 = \frac{g_2 g_1 x_1}{\omega_2^2 - \omega^2}. \tag{5.70}$$

On account of $\omega_1 = \omega_3$ amd $g_1 = g_3$ we obtain

$$\ddot{x}_1 + \bar{\omega}^2 x_1 = \bar{g} x_3 \tag{5.71}$$

$$\ddot{x}_3 + \bar{\omega}^2 x_3 = \bar{g} x_1, \tag{5.72}$$

with

$$\bar{\omega}^2 = (\omega_1^2 - \frac{g_1 g_2}{\omega_2^2 - \omega^2})$$

and

$$\bar{g} = \frac{g_1 g_2}{\omega_2^2 - \omega^2}.$$

Hence we return again to the problem of two identical dipoles which we discussed at the beginning of this subsection (5.20). If the central particle does not have a dipole field, i.e. $g_2 = 0$, the electromagnetic link between particles 1 and 3 will break down. In this case a collective motion is impossible due to $\bar{g} = 0$. We see that the central particle plays an essential role for holding the three particles together forming a collective state.

Figure 5.3: *A section of an aperiodic molecular chain with two identical particles at the terminals and with two different ones in between.*

(e)

Now let us further consider the dynamics of four particles forming a one-dimensional chain, with particle 1 and 4 being identical (Fig. 5.3)

For this segment of aperiodic chain we have the following equations

$$\ddot{x}_1 + \omega_1^2 x_1 = g_2 x_2, \tag{5.73}$$

$$\ddot{x}_2 + \omega_2^2 x_2 = g_1 x_1 + g_3 x_3, \tag{5.74}$$

$$\ddot{x}_3 + \omega_3^2 x_3 = g_2 x_2 + g_4 x_4, \tag{5.75}$$

$$\ddot{x}_4 + \omega_4^2 x_4 = g_3 x_3. \tag{5.76}$$

In order to ascertain how the two identical dipoles x_1 and x_4 collectively vibrate, we use the same method as before to insert trial solutions similar to Eq. (5.67) into (5.74) and (5.75). We then get

$$x_2 = \frac{1}{\Delta_2^2 \beta}(g_1 x_1 + \frac{g_{34}}{\Delta_3^2} x_4) \tag{5.77}$$

$$x_3 = \frac{1}{\Delta_3^2 \beta}(g_4 x_4 + \frac{g_{12}}{\Delta_2^2} x_1) \tag{5.78}$$

with $\Delta_i^2 = \omega_i^2 - \omega^2$, $g_{ij} = g_i g_j$, $\beta = 1 - \frac{g_{23}}{\Delta_2 \Delta_3}$.

Inserting these equations into (5.73) and (5.76), respectively, we get:

$$\ddot{x}_1 + (\omega_1^2 - \Omega_1^2) x_1 = \alpha_1 x_4 \tag{5.79}$$

$$\ddot{x}_4 + (\omega_4^2 - \Omega_4^2)x_4 = \alpha_4 x_1, \tag{5.80}$$

where

$$\Omega_1^2 = \frac{g_{12}}{\beta\Delta_2^2}, \quad \Omega_4^2 = \frac{g_{34}}{\beta\Delta_3^2}, \quad \alpha_1 = \frac{g_{234}}{\beta\Delta_2^2\Delta_3^2}, \quad \alpha_4 = \frac{g_{123}}{\beta\Delta_2^2\Delta_3^2}$$

with $g_{ijk} = g_i g_j g_k$.

Here we obtain again an equation system of the same kind as we have discussed in the case (**c**) Eqs. (5.51) and (5.52). That is to say, in this case the two identical dipoles cannot anymore exchange energy as well as in the case of three dipoles. For example, from strong couplings $\frac{g_2}{\Delta_2^2} \gg 1$ and $\frac{g_3}{\Delta_3^2} \gg 1$ there result

$$\Omega_1^2 \simeq -\frac{g_1\Delta_3^2}{g_3}, \quad \Omega_4^2 \simeq -\frac{g_4\Delta_2^2}{g_2}, \quad \alpha_1 = -g_4, \quad \alpha_4 = -g_1$$

hence:

$$\ddot{x}_1 + (\omega_1^2 + \frac{g_1\Delta_3^2}{g_3})x_1 = -g_4 x_4, \tag{5.81}$$

$$\ddot{x}_4 + (\omega_4^2 + \frac{g_4\Delta_2^2}{g_2})x_4 = -g_1 x_1. \tag{5.82}$$

However, if we take account of the equalities

$$\omega_1 = \omega_4, \qquad g_1 = g_4$$

and of $\frac{g_1\Delta_3^2}{g_3}$, $\frac{g_4\Delta_2^2}{g_2}$ as small perturbations, then the energy exchange between the two identical dipoles 1 and 4 may be very good in comparison to that between different dipoles. This is also valid especially in the case of weak couplings, where we have $\beta \simeq 1$ and the following equations instead of (5.81) and (5.82):

$$\ddot{x}_1 + (\omega_1^2 - \frac{g_{12}}{\Delta_2^2})x_1 = \frac{g_{234}}{\Delta_2^2\Delta_3^2}x_4, \tag{5.83}$$

$$\ddot{x}_4 + (\omega_4^2 - \frac{g_{34}}{\Delta_2^2})x_4 = \frac{g_{123}}{\Delta_3^2\Delta_2^2}x_1, \tag{5.84}$$

with $g_1 = g_4$, $\omega_1 = \omega_4$ and $g_{234} = g_{123}$ and

$$\frac{g_{12}}{\Delta_2^2} \simeq \frac{g_{34}}{\Delta_3^2} \ll \omega_1^2.$$

From this we see that the original identical two dipoles become different ones in the molecular chain because of the different neighbouring influences:

$$g_2 \neq g_3, \qquad \Delta_2 \neq \Delta_3,$$

but, the coupling constants remain equal $g_1 = g_4$, and hence also $\alpha_1 = \alpha_4$. For weak coupling cases we have $\frac{g_2}{\Delta_2^2} \ll 1$, $\frac{g_3}{\Delta_3^2} \ll 1$.
On account of the weakness of couplings, the eigenfrequencies ω_1 and ω_4 are actually influenced only slightly, the energy exchange between them stays almost intact, in spite of the dipoles of different kinds in between. Of course, if there are more intermediate members of different kinds, the energy exchange between two identical dipoles is more difficult to ascertain. Fortunately, we have only four bases of different kinds in DNA molecules. Therefore, the above mathematical description is actually sufficient for understanding the principles of energy exchange between identical bases along the DNA skeleton. Such an energy transfer is also a consequence of energy delocalization along the molecular chain, or coherent photon trapping, which can be understood only in terms of Dicke's theory.

So far we have always discussed only the case of two identical dipoles in the chain performing coherent cooperative motions, in order to keep mathematical simplicity and exactness. For such cases we can very well use Dicke's model to examine classically constructive and destructive interference between them as we have shown at the beginning of this subsection. Without any doubt, the method presented here can be extended to the case of more than two dipoles of identical kind, of course, with more mathematical complexity.

5.4 DNA-Polaritons: Interactions between Photons and Phonons

From the discussions up to this point we know that the photon emission from DNA bases is closely connected to the mechanical vibration of base dipoles. Therefore it is also necessary to consider the coupled states of photons (electromagnetic waves) and phonons (base vibrations or mechanical lattice vibralions) within the DNA molecule skeleton. In solid state physics, such joint vibration modes were firstly discussed in 1951 for the simple NaCl crystal by K. Huang[7,8], and later they were called polaritons. A number of papers have been published on different aspects of lattice vibrations in DNA molecules. For instance, Maret et al.[50] have observed the longitudinal acoustic mode in oriented DNA fibres and films in a Brillouin scattering experiment. Up to now these observations[51] and the possible soliton excitations in DNA double helices[52] have been the bases of all discussions on the lattice-dynamical model of homopolymer DNA.

However, the important process of base excimer (exciplex) formation, which has been identified by fluorescence spectra of excited bases[5,14,53] was overlooked in all the mentioned discussions. The importance of the effect of excimer emission on the base vibrations, and also on DNA lattice dynamics, can be seen in our own model in the energy difference between the absorption spectrum of monomers and the fluorescence spectrum of excimers (Fig. 5.1). Just this part of the energy must be responsible for the excitation of base vibrations due to the changes in the charge cloud during emission.

5.4.1 Mathematical formulation

The process from the excitation of monomers to the formation of an excimer, and that from the emission of an excimer to the excitation of base vibration modes, are very complicated. We know, however, from Dicke's theory that the cooperative problem of N-dipoles can be approximately understood by any two of them as a model. This approximation is even very accurate, especially for the discussion of subradiance in which we have got increasing interest.

In order to examine the joint vibrations of an electromagnetic wave and the base lattice, we begin with the assumption that the excimer dipole d representing the motion of the electron cloud may oscillate with eigenfrequency ω_0 and is coupled with the mechanical vibration mode of base molecules Q with frequency Ω. The dipole oscillation may obey the differential equation

$$\ddot{\vec{d}} + \omega_0^2 \vec{d} = -g\vec{d} \cdot \vec{D}, \tag{5.85}$$

where g represents the coupling constant. Without loss of general validity we suppress the vector characters of $\vec{d}$ and $\vec{D}$ afterwards. The negative sign represents the fact that the energy will be used for the excitation of base vibration Q.

Two monomers building one excimer may roughly be considered as a two-atomic molecule, which can be reduced to the motion of a single particle of reduced mass M, whose displacement from its equilibrium position can be represented by a generalized coordinate Q which obeys the equation of motion:

$$\ddot{Q} + \Omega^2 Q = gQd \tag{5.86}$$

with the positive driving term gQd.

In classical electrodynamics[54], the radiation field as represented by the electric vector $\vec{E}$ is proportional to the dipole vector $\vec{d}$, so that $\vec{d}$ must satisfy the Maxwell equations just as $\vec{E}$. The nonlinearly coupled equations (5.85) and (5.86) describe correctly the joint vibrations of radiation field and mechanical base waves, which can be briefly designated as polaritons. By solving these coupled equations we must consider the direction of energy dissipation as part of the physical constraint. The question of how the dipole of excimer (exciplex) is been excited has to left open first, but, the excitation of base vibrations always must be performed by the coupling term gQd, and the energy dissipates from the dipole system to the system of base vibrations. Of course, the energy can dissipate , with longer wavelengths further from the base system to the DNA skeleton, where it ultimately may be thermalized. It builds an unidirectional channel of energy dissipation.

5.4.2 Solutions and some physical interpretations

The mechanical vibrations of Q have the frequency $\Omega \simeq 600$ to $800\ cm^{-1}$ for DNA base ring-breathing modes[55]. For base oscillations as a mass cluster, the frequency can be set far lower than the breathing frequency given above. On the contrary, the frequency of the excimer dipole oscillation lies in the spectral range from violet to green. Thus, in Eq. (5.85), Q is a slowly varying time-dependent driving factor which can be considered as a constant by first approximation. On account of this fact we get the solution of (5.85) as

$$d = d_0 \cos \omega t, \qquad d = d_0 \sin \omega t \tag{5.87}$$

with

$$\omega = \sqrt{\omega_0^2 + gQ}. \tag{5.88}$$

We require $\omega \to 0$, and $d \to d_0$, to be a constant permanent dipole, so that we must take $d = d_0 \cos \omega t$ as a purely physical solution of our problem and set it consequently into Eq. (5.86) to solve it . We have it in the form

$$\ddot{Q} + [\Omega^2 - g d_0 \cos \omega t]\Omega = 0, \tag{5.89}$$

or writing it in a standard form:

$$\frac{d^2}{d\tau^2}Q + [\lambda - 2h\cos 2\tau] = 0, \tag{5.90}$$

where

$$\tau = \frac{\omega}{2}t, \quad \lambda = \frac{4\Omega^2}{\omega^2}, \quad h = \frac{2gd_o}{\omega^2}. \tag{5.91}$$

The equation (5.90) is so-called Mathieu equation which can be solved exactly by the Floquet-Theorem. The solution reads:

$$Q_1(\tau) = e^{\mu\tau} \sum_{K=-\infty}^{\infty} C_K e^{i2K\tau},$$

$$Q_2(\tau) = e^{-\mu\tau} \sum_{K=-\infty}^{\infty} C_K e^{-i2K\tau} \tag{5.92}$$

and the parameter μ satifies relation

$$\cosh \pi\mu = Q. \tag{5.93}$$

According to the theory of the Mathieu equation, for the case $\frac{2h}{\lambda} \ll 1$ (which is valid for our problems), the Floquet-exponent μ takes real values (which renders the solutions of (5.92) unstable) only, if

$$\lambda = \frac{4\Omega^2}{\omega^2} = n^2 \qquad (n = 0, 1, 2, \cdots). \tag{5.94}$$

When this condition is not satisfied, μ becomes imaginary (then the solutions (5.92) are stable). In our case λ can surely not take even values, so that the excimer vibration mode Q is stable. The decay of an excimer into monomers means, however, unstable modes, corresponding to a photochemical reaction, whose mechanism has not been introduced into the premise of our mathematical formulations. Therefore, the unstable solution cannot explain the decay process of excimers. Putting the solution (5.92) (which is composed of stable oscillations with different frequencies and imaginary μ) back again into the dipole equation of motion (5.85) for the next approximation, we get:

$$\ddot{d} + [\omega_0^2 + gQ(\tau)]d = 0, \tag{5.95}$$

where $Q(\tau) = Q_1(\tau) + Q_2(\tau)$, λ, h are real parameters, and $Q(\tau)$ a periodic function which can be represented by a Fourier series.

This kind of equation was first investigated by Hill in 1886 and named later on as Hill's differential equation. The solutions of Hill's equation have a form similar to those of the Mathieu equation (5.90), but, with more parameters depending on the actual concrete problem.

Here we will show some important properties of biological interest of this equation which are worth to be investigated for the next time:

1. Due to the feedback from the mechanical vibration of the base molecules the motion of dipoles (i.e. the distribution of electrons) takes the much more complicated form of Eq. (5.95) than that of Eq. (5.87) without perturbations. Evidence for complexity is not only due to the fact that the frequency spreads drastically from a single defined line to an almost continuous spectrum described by a Fourier series composed of all the possible combinations of eigenfrequencies ω and Ω and their higher harmonics, but also due to the change from a two parametric solution to a many-parametric one which implicates a greater number of possible instable solutions. This complexity clearly originates from the nonlinear couplings, and arises in spite of the simplicity of the physical back-ground (Eqs. (5.85) and (5.86). This leads to the conclusion that some simple principle, which we do not know so far, may be hidden in the very complicated continuous spectra of photon emission from biological tissues which are observed every day.

2. According to the Floquet-Theorem the frequencies emitted from the coupled system remain very stable due to the mathematical stability of Hill's equation, even if the amplitudes of the driving terms are subject to some perturbations. This means that the external influence can evoke changes of intensity, but, without alteration of frequency.

3. The Hill equation actually describes the frequency-modulated processes, since, for instance, the frequency-modulated equation:

$$(1 + a\cos 2x)y'' + \lambda y = 0$$

can be transformed into Hill's equation by the Liouville transformation. The frequency modulation is caused by the nonlinear coupling between the two systems under consideration.

4. Hill's equation belongs to the type of equation of parametric excitation, which physically represent open systems[47]. The driving parameters, which are responsible for the modulation of frequency, can be smaller than the eigenfrequency of the system by many orders of magnitude. But, the effect will be significant for a long time.

5. On account of the instability of the solution and of the direction of energy dissipation from short wavelengths to longer wavelengths[9] we see that the excimer formation (excited electronic states of DNA bases) constitutes the energy source for the excitation of DNA lattice vibrations, and therefore also, at least, it represents one part if not the only one of the heat source of the cells when the excited energy ultimately has been thermalized.

5.5 Some Conclusions

We have presented evidence that the Dicke theory of radiation is universally applicable for phenomena involving electromagnetic emission, while the Planck theory represents the limited case of radiating particles which are independent from each other in the state of thermal equilibrium.

Excimer (exciplex) formation is the consequence of DNA base excitation[53] due to the stacking effect between base molecules[56]. According to Dicke's theory the fluorescent emission from DNA base molecules must be coherent due to the cooperation of base molecules contrary to the assumption that bioluminescence may be incoherent. We have shown in the context that the identical base molecules of the well known four kinds A, C, T(U) and G can cooperate very well in spite of other bases in between. This is surely a solid physical basis for coherent emission and cooperation of molecules with aperiodic structures, as in the case of the usual lasing materials.

By discussion of the information variety of aperiodic solids, Schrödinger[1] has shown in comparison with periodic arrangement of atoms that "a well-ordered association of atoms, endowed with sufficient resistivity to keep its order permanently, appears to be the only conceivable material structure, that offers a variety of possible ('isomeric') arrangements, sufficiently large to embody a complicated system of 'determinations' within a small spatial boundary. Indeed, the number of atoms in such a structure needs not be very large to produce an almost unlimited number of possible arragements. For illustration, think of the Morse code, ..."

It may be suitable here to provide an additional argument for excimer (exciplex) formation from excited states of DNA bases: In the usual treatment of spontaneous radiation, the radiation process is calculated as though the separate molecules radiate independently of each other. In such a case the radiation rate is linearly proportional to the molecular concentration. When interaction (cooperation) is taken into account, then the linear law is broken. Experimental evidence for this was presented by Goldschmidt et al.[57] on an excimer-excimer interaction in pyrene solutions and crystals containing a high concentration of pyrene molecules. They have got the rate equation

$$\frac{dn}{dt} = -\alpha n - \beta n^2,$$

whose solution will furnish a hyperbolic law

$$n = \frac{1}{(\beta + \frac{\alpha}{I_0})t + \frac{1}{I_0} + O(t^2)}$$

as every coherent emission source does[47,58].

In addition, the converted quantum yield of excimer fluorescence from the intensity measurements changes downwards from a small concentration of pyrene molecules to a higher concentration up to crystals due to cooperation of molecules. The similar behaviour of quantum yields has been observed by different groups of DNA fluorescence researchers[5,14,20]. Their results were:

"The fluorescence quantum yield of DNA is only about one-tenth that of an equimolar mixture of the four constituent nucleotides".

"... its fluorescence yield is very low compared to the yield of the mononucleotides, making its study difficult..."

" The monomer emission is undetectable in DNA fluorescence, which shows that its yield is below 10^{-3}, or at least 100 times lower than the equivalent monomer mixture".

5.6 References

1. E. Schrödinger, *What is Life?* (Cambridge University Press, 1955).

2. R.H. Dicke, *Phys. Rev.* **93** (1954) p. 99

3. V. Weisskopf and E. Wigner, *Z. Physik* **63** (1930) p. 54

4. V. Ernst, *Z. Physik* **218** (1969) p. 111-125

5. M. Gueron, J. Eisinger and A.A. Lamola, in *The Basic Principles of Nucleic Acid Chemistry* **Vol. 1**, ed. P.O.P. Tso (Acad. Press, Inc., 1974) p. 329.

6. K.H. Li, in this volume.

7. K. Huang, *Proc. Roy. Soc. A* **208** (1951) p. 352.

8. R. Claus, L. Merten and J. Brandmüller, in *Springer Tracts in Modern Physics* **75** (Springer, Berlin, 1975).

9. F. Mitschke, *Naturwissenschaften* **78** (1991) p.97.

10. E. Lorenz, *J. Atmos. Science* **20** (1963) p.130.

11. J.R. Platt, *J. Chem. Phys.* **17** (1949) p. 484-495.

12. P.R. Callis, *Ann. Rev. Phys. Chem.* **34** (1983) p. 329-357.

13. J.B. Birks, *Rep. Prog. Phys.* **38** (1975) p. 903-974.

14. W.W. Hauswirth and M. Daniels, Excited states of the nucleic acids, in *Photochemistry and Photobiology of Nucleic Acids*, ed. S. Y. Wang, **Vol. 1** (Academic Press, New York San Francisco London, 1976) p. 109-167.

15. J.J. McCullough, *Chem. Rev.* **87** (1987) p. 811-860.

16. W. Klöpffer, in *Organic Molecular Photophysics* **Vol.1, chapt. 7**, ed. J.B. Birks (John Wiley & Sons, 1973).

17. N. Nakashima, C. Yamanaka, R. Ide and S. Misumi, *Chem. Phys. Lett.* **18** (1973) p. 386.

18. N. Mataga and T. Kubota, *Molecular Interactions and Electronic Spectra* (Marcel Dekker, Inc., New York, 1970) p. 414-482.

19. K. D. Gundermann, *Chemilumineszenz organischer Verbindungen* (Springer, Berlin, 1968).

20. Paul O.P. T'so, *Basic Principles in Nucleic Acid Chemistry* **Vol. 1, 2**, ed. P.O.P. T'so (Acad. Press, Inc., New York and London, 1974).

21. S.Y. Wang, *Photochemistry and Photobiology of Nucleic Acids* **Vol.1, 2**, ed. S.Y. Wang (Acad. Press, New York, San Francisco, London, 1976).

22. M.J. Vanderkooi and J.B. Callis, *Biochem.* **13** (1974) p. 4000.

23. H.J. Galla and W. Hartman, *Chem. Phys. Lipids* **27** (1980) p. 199.

24. T. Förster and K. Kasper, *Z. Phys. Chem.* (Frankfurt am Main) **1** (1954) p. 275.

25. M.J. Yuen, G.L. Closs, J.J. Katz, J.A. Roper, M.R. Wasielewski and J.C. Hindman, *Proc. Natl. Acad. Sci. USA* **77** (1980) p. 5589.

26. U.A. Kahn and M. Kasha, *Proc. Natl. Acad. Sci. USA* **80** (1983) p. 1767.

27. F.A. Popp, B. Ruth, W.B. Bahr, J. Böhm, P. Grass, G. Grolig, M. Rattemeyer, H.G. Schmidt and P. Wulle, *Collective Phenomena* **3** (1981) p. 187.

28. K.H. Li, *Laser + Elektro-Optik* **No. 3** (1981) p. 32.

29. F.A. Popp, in this book.

30. K.H. Li and F. A. Popp, Dynamics of DNA Excited States, in *Molecular and Biological Physics of Living Systems*, ed. R.K. Mishra (Kluwer Academic Publishers, 1990) p.31-52.

31. K.H. Li, *Phys. Lett. A* **114** (1986) p. 405.

32. R.H. Dicke, Quantum Electronics, eds. N. Bloembergen and P. Grivet (Dunod et Cie., Paris, 1964) p. 35.

33. F.A. Popp, *Electromagnetic Bio-Information*, eds. F.A. Popp, et. al. (Urban & Schwarzenberg München-Wien-Baltimore, 1979, 2nd ed. 1989), see also the paper in the present book.

34. K.H. Li, Coherence in Physics and Biology, in this volume.

35. A.C. Giese, *Living with Our Sun's Ultra Violet Rays*, Figure 2.1 (Plenum Press, New York, London, 1976) p. 18.

36. J.H. Eberly, Superradiance Revisited, *AJP* **40** (1972) p. 1374.

37. M.J. Stephen, First-order Dispersion Forces. *J. Chem. Phys.* **40** (1964) p. 669.

38. R.H. Lehmberg, Radiation from an N-Atom System. *Phys. Rev. A* **2** (1970) p. 883.

39. A. Crubellier, S. Liberman, D. Pavolini and P. Pillet, *J. Phys. B* **18** (1985) p. 3811.

40. D.A. Hutchinson and H.F. Hameka, Interaction Effects on Lifetimes of Atomic Excitations. *J. Chem. Phys.* **41** (1964) p. 2006.

41. M. Gross and S. Haroche, Superradiance: An Essay on the Theory of Collective Spontaneous Emission. *Phys. Rep.* **93** (1982) p. 301-396.

42. V. Ernst and P. Stehle, *Phys. Rev.* **176** (1968) p. 176 and reference 4.

43. K.H. Li and F.A. Popp, Coherence and some Quantum Paradoxes, in this volume.

44. H. Paul, Interference between independent photons. *Rev. Mod. Phys.* **58** (1986) p. 209-231.

45. C. Mavroyannis and M.J. Stephen, *Mol. Phys.* **5** (1962) p. 629 and reference 37

46. N.E. Rehler and J.H. Eberly, Superradiance. *Phys. Rev. A* **3** (1971) p. 1735.

47. K.H. Li, Physics of Open Systems. *Phys. Rep.* **134** No. 1 (1986).

48. W. Macke, *Wellen – Ein Lehrbuch der Theoretischen Physik* (Akad. Verlag, Leipzig, 1958).

49. H. Fröhlich, *Phys. Lett.* **39A** (1972) p. 153.

50. G. Maret, R. Oldenbourg, G. Winterling, K. Dransfeld and A. Rupprecht, *Colloid and Polymer Science* **257** (1979) p. 339.

51. J.W. Powell, G.S. Edwards, L. Genzel, F. Kremer and A. Wittlin, *Phys. Rev. A* **35** (1987) p. 3929.

52. C.T. Zhang, *Phys. Rev. A* **35** (1987) p. 886.

53. G. Ge and S. Georghiou, *Photochem. Photobiol.* **54** (1991) p. 301-305.

54. J.D. Jackson, *Classical Electrodynamics* (John Wiley & Sons, Inc., 1975).

55. E. Koglin and J.-M. Sequaris, in *Topics in Current Chemistry* **134** (1986) p. 1.

56. P. Wu, T.M. Nordlund, P. Gildea and L.W. McLaughlin, Base Stacking and Unstacking As Determined from DNA Decamer Containing a Fluorescent Base, *Biochemistry* **29** (1990) p. 6509.

57. C.R. Goldschmidt, Y. Tomkiewicz and I.B. Berlman, An Intermolecular Interaction of Excited States in Crystals and Liquids, *Chem. Phys, Lett.* **2** (1986) p. 520-522, p. 536-538.

58. K.H. Li and F.A. Popp, *Phys. Lett. A* **93** (1983), and F.A. Popp in this book.

Chapter 6

Technical Notes to Biophoton Emission

C. Mieg, W.P. Mei and F.A. Popp

Abstract

Some basic problems of biophoton measurements are discussed. One of the equipments which has been used for most of the measurements of this volume is described in more detail. Some results of signal/noise-ratio, spectral quantum efficiency and long-time stability are presented.

6.1 Introduction

A review on detector systems for biophoton emission has been presented not long ago by Inaba[1]. We therefore confine ourselves to a discussion of some basic problems and to the description of an equipment which has been used for most of the measurements which are topics of this book.

Due to the low biophoton intensity of a few up to some hundred photons/$(s \cdot cm^2)$ in the range of at least 200 to 800 nm, the sensitivity of a detector system of, say, a light-sensitive area of 1 cm^2 and high enough quantum efficiency of the order 1, has to reach about 10^{-18} W.

By reducing the area of the light-sensitive material, the spatial resolution of the image improves. However, since the sensitivity decreases, a higher amplification of the original signal is necessary such that a compromise between the length of the measurement time interval for detecting at least a signal and the effective area of the detector material has to be found. This means that both spatial and time-resolution cannot be optimized at the same time. Once one has chosen a rather large area of the photosensitive detector, one renounces a high spatial resolution in favour of a quite high time-resolution which, in view of the non-stationarity of living systems, enables one to investigate dynamical processes.

One way of overcoming the antagonism between spatial and temporal resolution to some extent is the use of light fibers which condense low intensities at the surface area of the biological sample to rather high intensities at the end of the fiber by narrowing the cross-section and focussing the beam onto an adequately small area of the light-sensitive detector. This effect increases the detectability of biophoton emission considerably[2]. However, obviously it falsifies the spatial distribution as well as the spectral and the statistical properties of the light emission.

As methods like charge-coupled device (CCD) do not cover the whole optical range from 200 to 800 nm and, in addition, display too low sensitivity, the only available detectors for measuring biophotons are at present photomultipliers.

One may divide the available techniques into two categories, one for achieving a high spatial resolution by use of rather small cathodes working at room temperature, and the other for getting a high time-resolution by use of large cathode areas whose working temperature is rather low (about -30 °C). The material of the cathode has to be selected in regard to the spectral quantum efficiency.

Once one has decided for the one or the other method, the number of dynodes, the cooling temperature and a compromise between the working points of the voltage and the discriminator threshold after amplification have to be found out. The ideal working point enables one, at highest possible signal/noise-ratio, to vary the working parameters by a small amount without changing considerably the detected signal. Table 6.1 qualitatively displays the different influences of the parameters on the signal.

The optimization procedure has to be carried out rather carefully in order to get a reliable detector system for biophoton emission.

Since our group is mainly interested in the dynamics of "low level luminescence" from living systems, we designed three different types of equipment with always rather large cathode areas. One of these devices serving for "standard" measurements will be explained now in more detail.

Table 6.1: Influence of the working parameters on the measurements.

Parameter	Qualitative effects
Number of Dynodes	Increasing number increases sensitivity and decreases signal/noise-ratio
Material of Cathode	Material determines spectral quantum efficiency
Area of Cathode	Increasing area increases sensitivity and decreases signal/noise-ratio
Temperature of Photomultiplier	Increasing temperature increases noise
Voltage	Increasing voltage increases sensitivity and noise
Discriminator Threshold	Increasing threshold decreases sensitivity and increases signal/noise-ratio

6.2 Materials and Methods

Fig. 6.1 sketches the design of the detector system, Fig. 6.2 displays it in full detail[3].

The biological sample, always inserted in a quartz cuvette (inside dimensions, $2.2\times2.2\times3.8\ cm^3$, thickness 0.15 cm), is put into the dark chamber, where the photons from the specimen hit the photocathode of the multiplier. As one chooses, a filter wheel with interference filters can be rotated in front of the multiplier in such a way that different wavelengths with an uncertainty of about ±10 nm are selected out. In addition, a chopper disk can be switched on, in order to distinguish between real count rate and dark count rate of the equipment alone.

The cathode (EMI 9558 QA, selected type) has a diameter of 44 mm and is sensitive within the range of 200 to 800 nm. The impinging photons induce electric pulses in the photomultiplier, which are fed then by means of an electronic network into an amplifier and then into a discriminator. The device provides a logical pulse whenever the incoming pulse exceeds an adjustable threshold. This means that the equipment works as a photon counter with a quantum efficiency determined by the cathode of

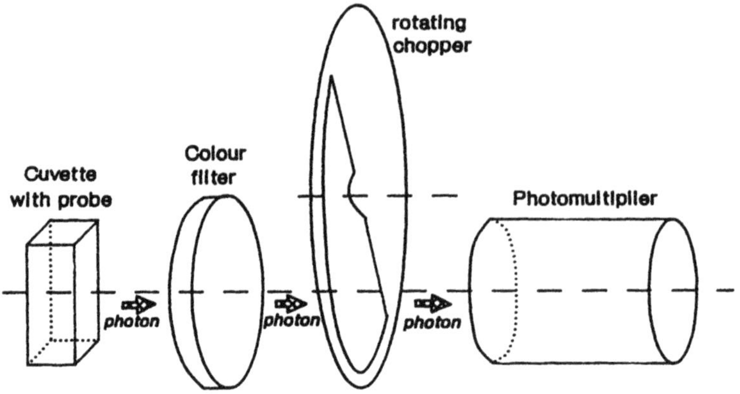

Figure 6.1: *Principle of photon counting system. The quartz cuvette in a dark chamber contains the biological sample. A colour filter allows to select certain modes of biophoton emission. A chopper serves for separating dark count rate (originating almost completely from the multiplier) and real count rate. The photomultiplier registers the impinging photons.*

the photomultiplier (Fig. 6.3).

The background (noise) originates mainly from pulses of the multiplier itself. It shows the temperature dependence (Fig. 6.4).

In order to reduce the background, the photomultiplier is cooled by a cooling jacket KM. Copper wool (KW) provides the thermal contact, and a ground metal cylinder MM protects the multiplier from electric and magnetic fields. In order to prevent the photomultiplier from freezing, the whole tube is kept in vacuum together with the cooling jacket. The quartz glass (QP) in front of the cathode has, therefore, no thermal contact with the cooled cathode and cannot become covered by moisture.

By the aid of a chopper, the equipment is able to detect a real signal current density of 2 photons/$(s \cdot cm^2)$ at a significance level of 99.9% within 6 hours. The disc of the chopper has an open sector, providing that photons from the sample reach the cathode. In this position, the detector measures the signal plus background. If the disc covers the cathode, the background alone is registered. The disc is driven by a geared motor (SM, UG) at one cycle per second. This means that both values of the background and signal plus background are measured within the time-interval of

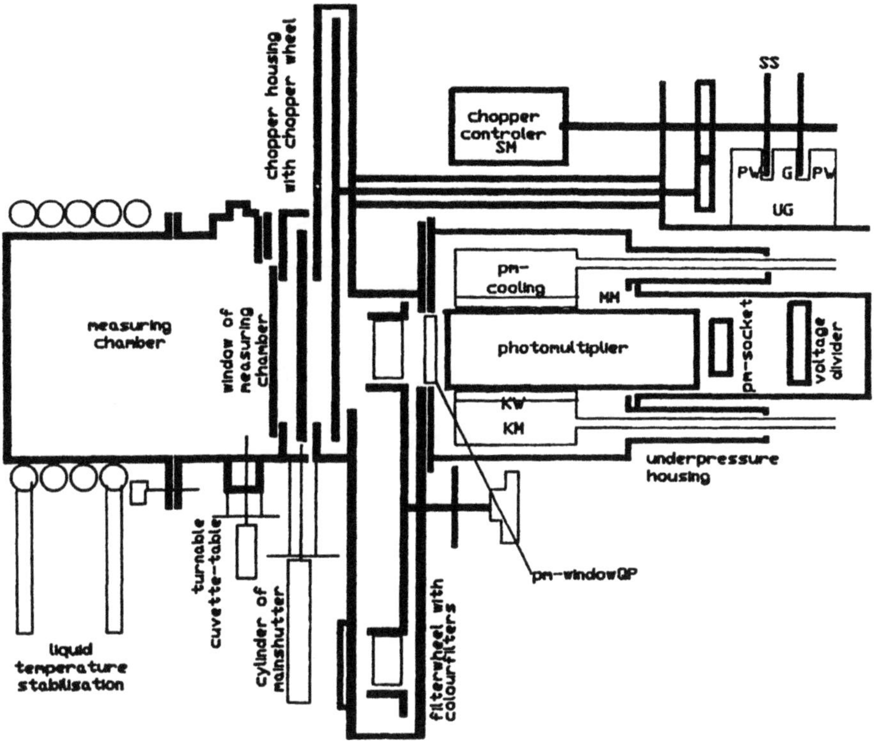

Figure 6.2: *One of the equipments which has been used for the results of this book.*

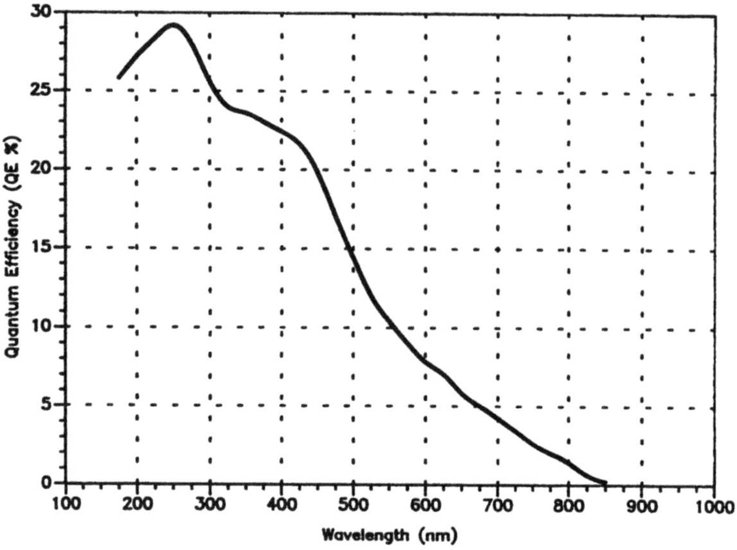

Figure 6.3: *The quantum efficiency of the S20-cathode of the EMI-955QA photomultiplier.*

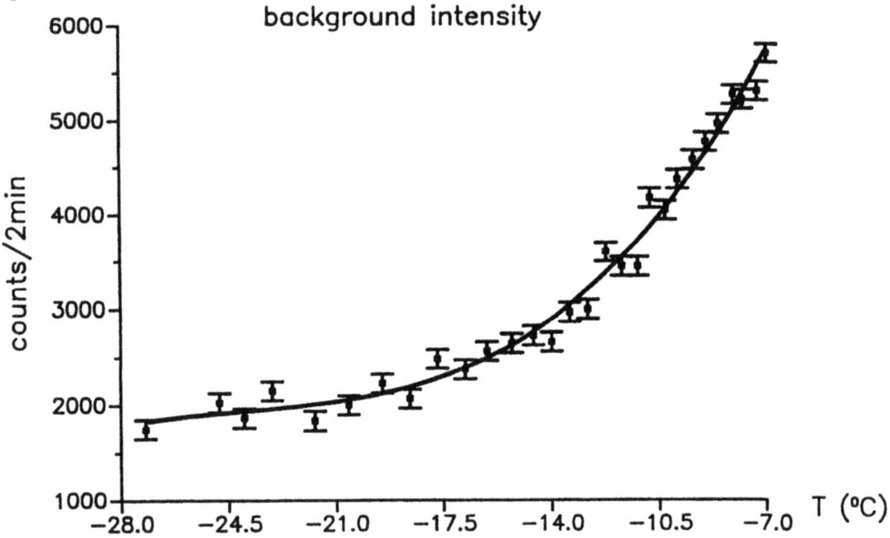

Figure 6.4: *The dark count rate in dependence on the cooling temperature of the photomultiplier.*

one second. For this goal the additional sector-discs SS, by means of a lamp G and photosensitive resistors PW, provide control pulses to indicate whether the cathode is open or closed. The logical pulses of the discriminator are fed into two scalers. One is active when the control pulse indicates an open cathode, the other, when the cathode is closed. Both scalers sum up the counts during the measurement time-interval which can be adjusted with running chopper to between 2 s and 2^{11} s, while with resting open chopper the measurement time-interval can be chosen as low as 10 ms. With rotating chopper-disc the difference between both scalers constitutes the signal amplitude, which is proportional to the photon intensity from the sample during the measurement time.

Both background and signal plus background are subject to errors, but the errors diminish with increasing measurement time t according to $\frac{1}{\sqrt{t}}$, provided that both, signal and background, remain stable.

For optimization and calibration, attenuated light of about 10^3 photons/s has been directed onto the cathode. The working voltage and temperature of the photomultiplier and the threshold of the discriminator are adjusted to the values at which the test-intensity is optimally detected.

The total number of photons emitted per units of wavelength and time interval can be calculated by taking account of the solid angle, transmission and bandwidth of the filters and the quantum efficiency of the cathode. An integration of the spectral distribution yields the total number of emitted photons. For a more detailed description see reference[3].

In order to measure "delayed luminescence", a monochromator can be connected to the dark chamber in such a way that the sample is illuminated perpendiculary to the axis between cuvette and photomultiplier. Immediately after light-exposure, the connection is disrupted and the cathode is opened with a delay time of about 250 ms. A mechanical shutter avoids the exposure of the sample to the afterglow of the lamp.

Fig. 6.5 shows the variation of the real count rate of a light-indicator (left side) and the dark count rate (right side) in dependence on the voltage and the discriminator level.

The signal/noise-ratio takes then the values of Fig. 6.6, where on the left side the noise has not been substracted from the real count rate, while on the right side this (small) correction has been performed.

Fig. 6.6 demonstrates that an optimum as well for the working voltage as the discriminator level can be obtained.

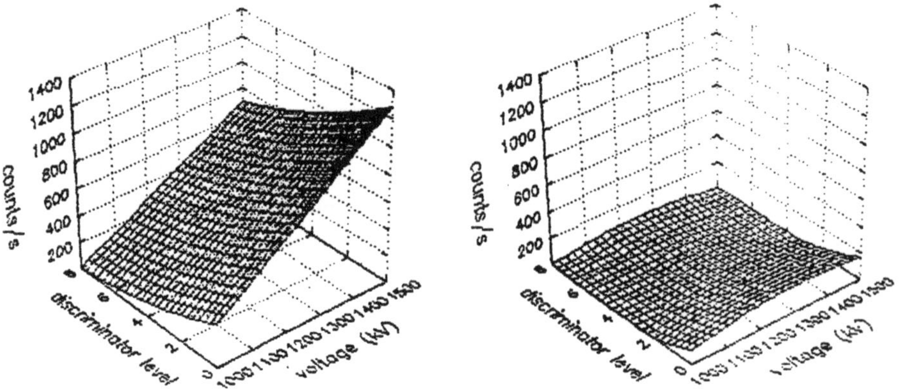

Figure 6.5: *The count rate of a light indicator of fixed intensity (left side) and the dark count rate (right side) in dependence on the working voltage of the multiplier tube and the discriminator level.*

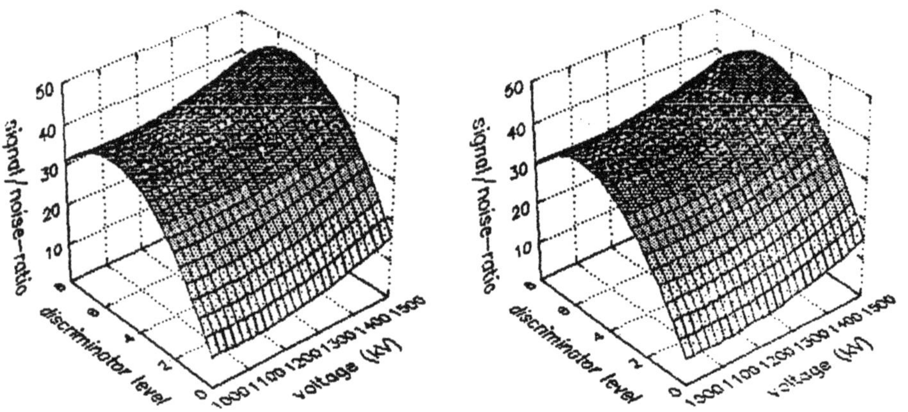

Figure 6.6: *The signal/noise-ratio of the light indicator without correction (left) and after subtraction of the background (right) in dependence on the working voltage of the multiplier tube and the discriminator level.*

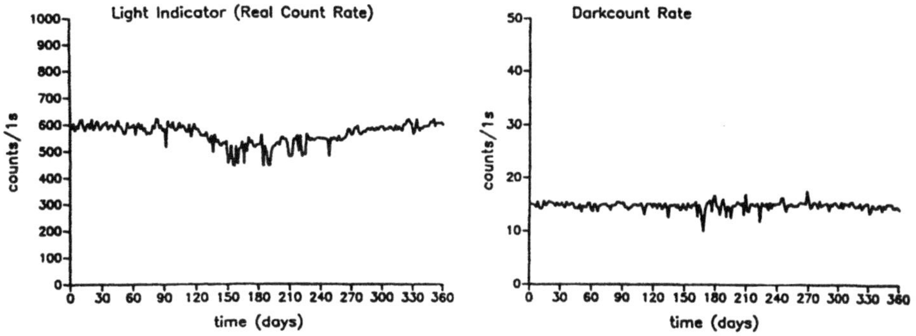

Figure 6.7: *Long-time stability of the equipment: The count rate of the light indicator (left) and the dark count rate (right) over the period of always one year (1991).*

Fig. 6.7 displays an example of the long-time stability of the equipment, where the left side represents the photon intensity of the light-indicator, the right side the dark-count rate.

6.3 References

1 . H. Inaba, Super-high sensitivity systems for detection and spectral analysis of ultraweak photon emission from biological cells and tissues. *Experientia* **44** (1988) p. 550-559.

2. Y.N. Liu, *Personal Communication.*

3. B. Ruth, Experimenteller Nachweis ultraschwacher Photonenemission aus biologischen Systemen. *Dissertation* (Experimental Physics, Marburg 1977).

Chapter 7

Spontaneous and Light-Induced Photon Emission by Rat Hepatocytes and by Hepatoma Cells

Roeland van Wijk and Hans van Aken

Abstract

We have investigated spontaneous and light-induced photon emission by suspensions of rat hepatocytes and of hepatoma cell lines. Rat hepatocytes exhibit spontaneous biophoton emission but that of hepatoma cell lines was not detectable. After irradiation with white light, the reemission intensity was also found to be different for the various cell types studied with hepatocytes yielding the lowest value and the least differentiated tumor cell line the highest value.

The decay kinetics and wavelength dependency of light-induced photon emission of HTC hepatoma cells was investigated after illumination with white light and light passing through different coloured filters. The mechanisms of spontaneous and light-induced photon emission by mammalian cells are completely different.

Induced photon emission was neither influenced by anaerobiosis nor by the intactness of the cells. Cell-fractionation studies demonstrate that the induced photon emission was caused by the nuclear fraction and by isolated chromatin. Phenol-extracted DNA, however, has lost this capacity. Our data suggest that differences in the chromatin structure may explain the cell-specific induced photon emission.

7.1 Introduction

Emission of weak radiation in the visible range of the electromagnetic spectrum by the liver has been described by several authors[1,2,3]. From liver in situ and isolated liver a spontaneous emission of approximately 7-12 photons/$(s \cdot cm^2)$ was registered, depending on the spectral sensitivity of the photomultiplier[2,3]. The phenomenon of light emission was still present after isolation of individual hepatocytes, amounting to approximately 30 photons/s per 10^6 cells[4].

Furthermore, photon emission has been readily detected for other organs or tissues, like heart[5,6], lung[7], nerve[8] and muscle[9,10].

In the past decades increasing emphasis has been placed on comparative biochemical and biophysical studies of parental and tumor cells. However, hardly any attention has been paid to the phenomenon of weak photon emission. Such studies are complicated because from the very beginning of cellular pathology a high degree of diversity and morphological heterogenity of neoplasms, even when derived from the same tissue or cell type, has been recognized. In recent years several tumor model systems, originating from various cell types, have been developed for experimental oncology, the prime example of such a model system is a spectrum of rat hepatomas of different growth rate[11]. In addition to a different growth rate and a morphological heterogeneity this series of hepatomas exhibits heterogeneity at the biochemical level, expressed as quantitative differences in enzyme levels and in regulation of these levels[12].

Differences in gene expression between various hepatomas are not random and coordinated alternations have been recognized. For example, coordinated alternations have been observed in key glycolytic and gluconeogenic enzymes[13]. It is important to realize that the individual tumors show a considerable biological stability. This stability is maintained even when the various hepatomas are maintained for many years in the animals by transplantation.

A variety of cell lines were derived from primary induced or transplantable rat hepatomas. As is the case for hepatomas, the rat hepatoma cell lines express the original liver characteristics to a different degree. Reuber H35 and MH_1C_1 are considered to be well differentiated cell lines in contrast to HTC and RLC. The ultrastructure of the cytoplasm, the mitochondrial volume and a number of adult hepatic functions and iso-enzyme activities appear to have persisted in the well-differentiated tumor cell lines[14–18]. Furthermore, these cells have retained certain features of regulatory circuits, which control these processes as well as the sensitivities for hormones[14,15,19–21].

In general, the well differentiated hepatoma cell lines (Reuber H35 and MH_1C_1) are characterized by a high growth and division potential but with relatively small alterations of gene expression, whereas the poorly differentiated hepatoma cell lines (HTC and RLC) combine high division potential with large alterations in gene expression as compared to hepatocytes.

We have explored the photon emission in rat hepatoma cell lines in the hope of finding variants which could be of use in elucidating the mechanisms involved in regulation of this emission.

Besides spontaneous photon emission (P.E.) we investigated light-induced photon emission (I.P.E.) of these cell suspensions. A series of experiments on I.P.E. has been described for plant-systems[22,23]. The decay curves of photon emission observed after pre-illumination of cucumber seedlings and leaves of Bryophyllum were not found to be easy to interpret. Pre-illumination with quasi-monochromatic light showed that the decay kinetics of I.P.E. were independent of the excitation wavelength and followed a hyperbolic rather than an exponential law[22].

Using the microsporogenesis of larch microsporocytes as a model system, it was found that both the intensity of I.P.E. and the characteristics of decay varied considerably during the successive stages of development. At 8 out of 13 developmental stages decay of I.P.E. could be described as being essentially hyperbolic[23]. Futhermore, as in other plant systems the process of decay of I.P.E. continues for nearly an hour or even longer. Recently, several mammalian systems have been carefully studied for their decay curves after exposure to light. The first mammalian system consisted of isolated rat hepatocytes and the liver-derived hepatoma cell lines Reuber H35 and HTC[24,25,26]. The second mammalian system compared Wish tumor cells and their normal counterparts, human amnion cells[27].

In this context we present an overview of our studies on several features of spontaneous and light-induced photon emission of the hepatocyte-hepatoma model system. A variety of experiments are described which indicate that

1. rat hepatocytes exhibit spontaneous biophoton emission but that of hepatoma

cells was not detectable.

2. after irradiation with white light, the emission intensity is lowest for hepatocytes and highest for the least differentiated tumor cell line.

3. spontaneous and induced photon emission are different with respect to their dependency on cell density, suggesting that cell suspensions have a high transparency for I.P.E.

4. I.P.E. is traced to the chromatin structure.

7.2 Materials and Methods

7.2.1 Cell cultures

Rat hepatocytes were isolated from liver as decribed before[28] and suspended in Dulbecco's Minimal Essential Medium (DMEM) without phenol red. The rat hepatoma cell lines Reuber H35[29] and HTC[30] were grown as monolayer cultures in DMEM supplemented with 10% fetal calf serum. Cell suspensions were prepared by trypsinization[28]. After trypsinization the cells were suspended in DMEM without phenol red.

7.2.2 Preparation of nuclei, chromatin and DNA

Nuclei of hepatocytes and hepatoma cells were prepared as described before[31]. In short, cells were centrifuged for 15 min at 500g at 4°C and resuspended in a sucrose buffer (0.27M sucrose, 0.1 mM EDTA, 5 mM Tricine-KOH; pH7.4) and homogenized by 10-50 strokes of a tight-pestle Dounce homogenizer, depending on the cell type. After disruption of the cells, KCl and $MgCl_2$ were added to the final concentrations of 7.5 and 1mM. The homogenate was centrifuged for 15 min at 500g. The sediment, which contained the nuclei, was washed twice with the sucrose buffer containing 7.5 mM KCl and 1mM $MgCl_2$ (pH 7.0).

Chromatin was isolated by suspending the nuclei in 76 mM NaCl, 24 mM EDTA, pH 7.0, centrifuging the swollen nuclei for 15 min at 500g, and resuspending them in 0.85M sucrose followed by centrifugation for 30 min at 10,000g. The sediment, which contained the chromatin, was resuspended in sucrose buffer containing 7.5 mM KCl and 1 mM $MgCl_2$ (pH 7.0).

For preparation of the DNA the procedure was essentially as described by Perucho et al.[32], with some modifications. The chromatin was centrifugated for 30 min at 10,000g and resuspended in 150 mM NaCl, 10 mM EDTA, and 10 mM Tris-HCl (pH 7.5). Proteinase K and SDS were added to bring the final concentrations to 200 μg/ml and 0.4%, respectively. Chromatin was heated to 65°C for 30 min, and then incubated overnight at 37°C. Equal volumes of 650 mM NaCl, 10 mM EDTA, and 10 mM Tris-HCl (pH 8.0) were added, and the lysates were gently extracted two to three times with an equal volume of buffer-equilibrated phenol.

The aqueous phase was extracted once more with an equal volume of a mixture of phenol, chloroform, and isoamylalcohol (25:24:1) and once more with chloroform and isoamylalcohol (25:1) before precipitation with 2 vol of pure ethanol. The DNA precipitates were removed with a glass rod and washed successively with two changes of 70% ethanol and two changes of 100% ethanol. The precipitates were then air-dried, after which they were dissolved in sucrose buffer containing 7.5 mM KCl and 1 mM MgCl2, pH 7.0.

7.2.3 Photon emission (P.E.) measurements

For detection and registration of spontaneous ultraweak photon emission we used the single photon counting described by Ruth[33] and Popp et al.[22]. This device was equipped with a cooled EMI 9558QB photomultiplier tube.

It must be emphasized that the number of photons registered per cell depends on a variety of factors. Among them are instrumental characteristics like the size and shape of the sample cuvette, the use of a mirror, the distance between cuvette and photomultiplier and the spectral sensitivity of the photomultiplier.

In this respect it can be mentioned that the cathode has a diameter of 44 mm and is sensitive in the range between 200 and 800 nm. The average quantum efficiency in this range was approximately 10%. The photomultiplier was cooled to -30°C, which reduced the dark count rate to about 35 cps. It was further equipped with an optical filter unit and a shutter to block the optical path to the sample compartment. A 2 cm quartz cuvette containing 11 ml of cell suspension or medium was placed in the darkened sample compartment at a distance of 125 mm from the photomultiplier. The sample compartment was kept at 37°C. Prior to the commencement of measurements, the samples were kept in complete darkness for 5 min at 37°C and continuously stirred. The photon-emission intensity was measured and integrated over intervals of 5 or 50ms, by computer.

7.2.4 Light-induced photon emission (I.P.E.)

For stimulation of photon emission a 150W-tungsten halogen lamp (nr 7158 Philips, Eindhoven, The Netherlands) was used. It was situated at right angles to the photomultiplier and was equipped with a second filter unit.

The distance between lamp and cuvette was 19 cm. The light passed through a heat-reflection filter which cuts off wavelengths above 720 nm and below 310 nm. No temperature changes were observed during the illumination period.

During the illumination period the shutter was closed. Each measurement cycle started by irradiating the sample for 10 s. The shutter was opened 300 ms after turning off the lamp: the emission was recorded from 0.3 to 5 or 15 s after the end of illumination.

In some experiments additional filtration was performed using glass optical-absorption filters (Schott, Mainz, Germany).

7.3 Results

7.3.1 Spontaneous photon emission (P.E.)

Figure 7.1 is an illustration of P.E. by hepatocytes and the two hepatoma cell lines, Reuber H35 and HTC. Rat hepatocytes in suspension exhibit a spontaneous photon emission which can be clearly distinguished from the "background emission" that consisted of electronic or other sources of noise as well as possible filter phosphorescence. The latter was estimated by measuring the dark count rate of a quartz cuvette filled with medium (Fig. 7.1). This signal was low compared to the signal from rat hepatocytes; 15% or less depending on the density of the suspension. The hepatocyte-specific P.E. was measured at cell densities of 0.5 - 2.2 $\times 10^6$ cells/ml.

The hepatocyte-specific signal increased and then decreased with increasing cell densities (Fig. 7.1 and Tab. 7.1). The number of cells relate to the measured P.E. since it influences the absorption by cells other than the emitter cell. The most simple equation for calculation of P.E. intensity is

$$I'_{PE} = I_{PE}^{meas.} \cdot 10^{A.d.},$$

where $I_{PE}^{meas.}$: measured value of PE, d: mean path length of PE (half diameter of cuvette, in cm), and A: absorption (optical density) at measured wavelength of emission (through 1 cm path length).

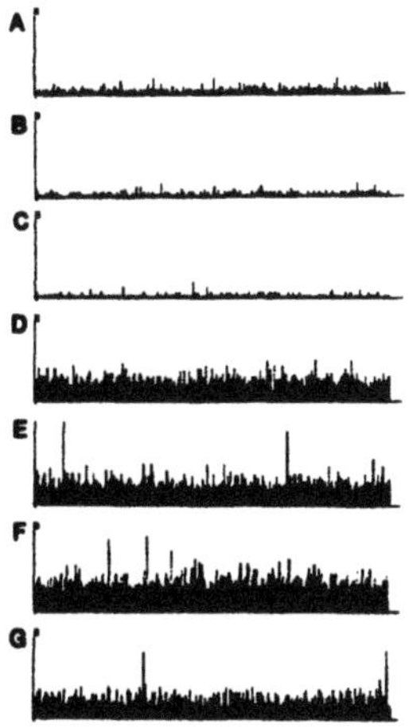

Figure 7.1: ***Photon emission of isolated hepatocytes and hepatoma cell lines. Suspensions of Reuber H35 hepatoma cells at a density of 2 $\times 10^6$ (B), or HTC hepatoma cells at a density of 2 $\times 10^6$ (C), and isolated hepatocytes at densities of 0.5 $\times 10^6$ (D), 1 $\times 10^6$ (E), 1.6 $\times 10^6$ (F), 2.2 $\times 10^6$ (G) were incubated in DMEM medium as described in materials and methods under constant stirring. Photon counts per 5 ms were registered for a period of 5 s. The quartz cuvette without any medium was no different from that containing DMEM medium alone (A) or hepatoma cells (B, C).***

Usually P.E. is measured over a broad range of wavelengths. P.E. absorbance is a function of the wavelength. Figure 7.2 shows the wavelength - dependent absorbance for hepatocytes.

At high cell densities the absorbance of light by hepatocytes increases more strongly than at shorter wavelengths. The ratio of transmission at 460 nm and transmission at 700 nm decreases from 0.53 to 0.11 at densities increasing from 0.5 $\times 10^6$ to 2.2 $\times 10^6$ cells/ml. Furthermore, the measured emission is also a function of the wavelength - dependent photomultiplier sensitivity (Fig. 7.2B). According to Cadenas et al.[4] P.E. by isolated hepatocytes occurs in the range of 460 – 750 nm, with major peaks at 560 nm and 640 nm. This range is within the spectral response region of the currently used photocathode; however, effective sensitivity decreased above 500 nm and quantum efficiencies decreased from 20 to 5% at wavelengths increasing from 460 to 700 nm. The combination of preferential decrease of transmission at shorter wavelengths plus the high quantum efficiency at this shorter wavelength requires

Table 7.1: Photon emission of hepatocyte suspensions with different cell concentrations prior to and after correction for transmission, and photocathode sensitivity[a].

Concentration cells/ml	Measured emission (average cps)	Corrected photon emmision/(s · cell) for S.A. 460/700 ratio of	
		S.A. = 1	S.A. = 0.5
2.2×10^6	880	0.028	0.024
1.6×10^6	1346	0.031	0.029
1.0×10^6	1248	0.019	0.019
0.5×10^6	1014	0.016	0.018
None	148	–	–

[a]Photon emission of a 11-ml suspension of hepatocytes was registered for 5 s by measuring 1000 consecutive 5-ms intervals. Transmission at 460 and 700 nm were detemined spectrophotometrically. Corrections were made by 460 and 700 nm only, assuming a simple spectral analysis (S.A.) ratio of 1 and 0.5.

a complex correction. Based on these corrections we estimated P.E. at about 0.02 photon/s per cell, independent of cell density (Tab. 7.1).

Several attempts have been made to measure P.E. of suspensions of the two cell lines Reuber H35 and HTC. These attempts were unsuccessful, although

1. the absorbance studies have shown that absorbance of light by the HTC and Reuber H35 cells was less than that by hepatocytes (Fig. 7.2) and
2. the cell densities of the tested suspensions ranged from 10^5 to 10^7 per ml.

An explanation of the loss of P.E. in the two hepatoma cell lines may be due to alterations of biochemical processes. The underlying mechanism of P.E. has been studied extensively. Cell fractionation studies showed emission from mitochondria[34] and submitochondrial particles[35]. Furthermore, the microsomal fraction of liver cells was recognized as a major contributor of emission[34,36–39]. P.E. of these organelles requires a membrane-bound electron transfer system and can be optimized in the presence of oxygen and hydrogenperoxide. Spectral analysis has been extensively applied in order to identify more precisely the source of the P.E.. In general, various emission bands in the range of 400 - 700 nm were found[36,38] and the data suggest the free radical process accompanying lipid peroxidation as a major source of P.E..

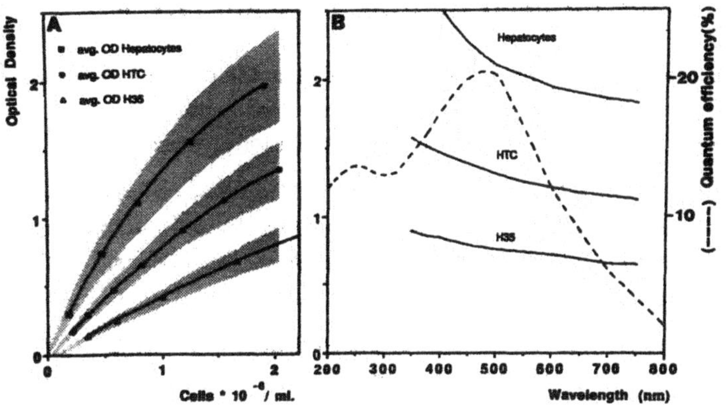

Figure 7.2: *Optical density of isolated hepatocytes and hepatoma cell lines. Suspensions of different densities (ranging from 0.1-2 $\times 10^6$ cells/ml) of isolated hepatocytes, Reuber H35 and HTC were incubated in DMEM media as described in materials and methods. Optical densities were determined for various cell densities (panel A) and for wavelength ranging from 350-700 nm for a cell density of 2 $\times 10^6$ cells/ml (panel B). The spectral response of the used photocathode is included in panel B by dotted line. In panel A the upper and lower limits of the marked areas represent optical densities at 400 and 750 nm respectively.*

This leads to the question whether Reuber H35 and HTC cells have an alteration in lipid peroxidation. Galeoti et al.[40] compared the generation of superoxide and lipid peroxidation of the microsomal fraction of various hepatomas with different degrees of differentiation. Their results showed that the production of superoxide radicals is reduced with increasing growth rate of tumors, the latter being linked to a decreasing degree of differentiation[13].

It will be clear that the alteration in lipid peroxidation in Reuber H35 and HTC cells as compared to that process in hepatocytes as an explanation for loss of P.E. in the two hepatoma cells requires further biochemical study.

With respect to our findings on P.E. of hepatocytes it can be summarized that

1. emission intensity is low (0.02 photon/s per cell) and

2. emission has transparency characteristics that are similar to that of artificial light.

7.3.2 Light-induced photon emission (I.P.E.)

From the preceding section it was concluded that the photons escape from liver cells at low frequency, but none could be detected from hepatoma cells. In addition it is of interest to consider the events after illumination of these cells.

Photon emission of HTC cells after prior illumination of various durations was studied at several cell densities. Durations of 1 second up to 5 minutes produced similar I.E.P. curves from cell populations with densities ranging from 0.27 $\times 10^6$ to 7.27 $\times 10^6$ cells per ml.

Figure 7.3 shows that the number of photons detected during a 50 ms period immediately after the start of the measurement was higher at increased cell densities and ranged from approximately 9.0×10^2 at the lowest up to 1.1×10^4 at the highest density tested in this experiment. I.P.E. from 2.7 $\times 10^5$ cells per ml could not be distinguished from the luminescence after illumination of a cuvette with medium but without cells. The delayed luminescence of an empty quartz cuvette was similar to that of a cuvette filled with water or with DMEM (data not shown). In contrast, illumination of an empty cuvette chamber showed hardly any delayed luminescence. These data demonstrate that quartz glass is responsible for a background pattern of I.P.E. which must be taken into consideration in order to calculate cell-specific I.P.E.-values.

Figure 7.4 shows a comparison between I.P.E. of HTC and Reuber H35 cells as determined in the same experiments. At the start of measurement I.P.E. of suspensions of HTC and Reuber H35 cells at a cell density of 2 $\times 10^6$ cells per ml were 5 $\times 10^3$ and 2 $\times 10^3$ counts per 50 ms, respectively, compared to 0.7 $\times 10^3$ for the illuminated cuvette containing DMEM only.

The decay curves for I.P.E. of cell-specific emission resulted from the subtraction of the cuvette plus medium value from those including the HTC or Reuber H35 cell population and showed a resemblance with the decay kinetics of the cuvette plus medium. The decay curves for I.P.E. decreased continuously, a systematic deviation being apparent from exponential and hyperbolical decay and without any sign of a transition point. After 5 s the value is only 1 SD above average background value, but after 30 s the average values were similar to those obtained without prior illumination.

I.P.E. from hepatocyte suspensions of 10^6 and 2 $\times 10^6$ cells per ml were approximately equal to or less than that of the cuvette and cell-specific emission decay curves could not be obtained. It is concluded from these data that the highest I.P.E.

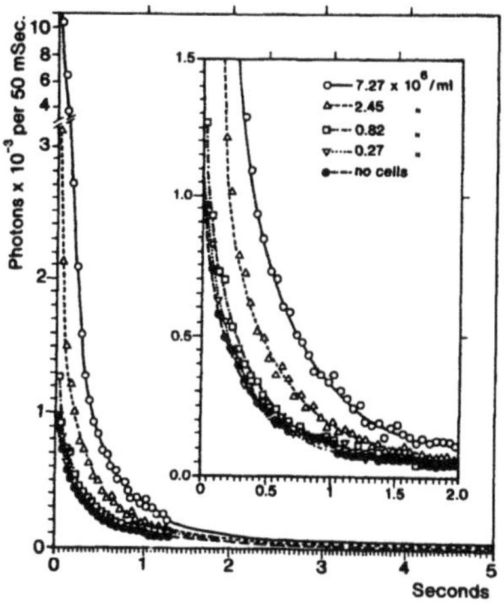

Figure 7.3: *Photon emission of HTC cell populations with different densities (cells/ml). Photons per 50 ms were detemined at various times after the start of the measurement, which was 300ms after the end of a 10 s illumination period.*

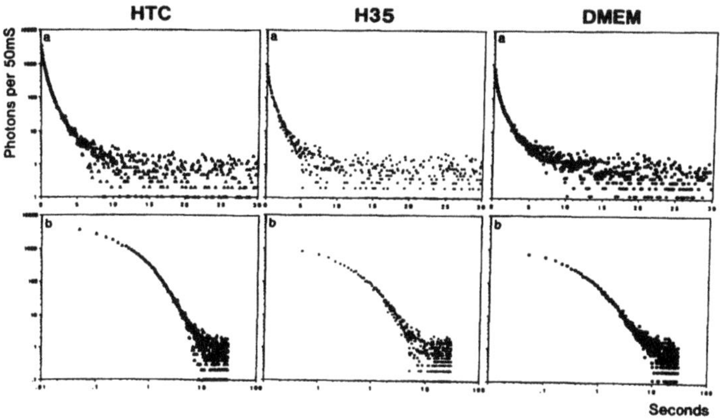

Figure 7.4: *Decay kinetics of photon emission and suspensions (2 $\times 10^6$ cells/ml) of HTC and Reuber H35 hepatoma cells and of a quartz cuvette with medium following irradiation with white light approximated by exponential (a) or hyperbolic decay law (b). Decay of photon emission of cell suspensions was obtained after subtraction of the value of the quartz cuvette with medium.*

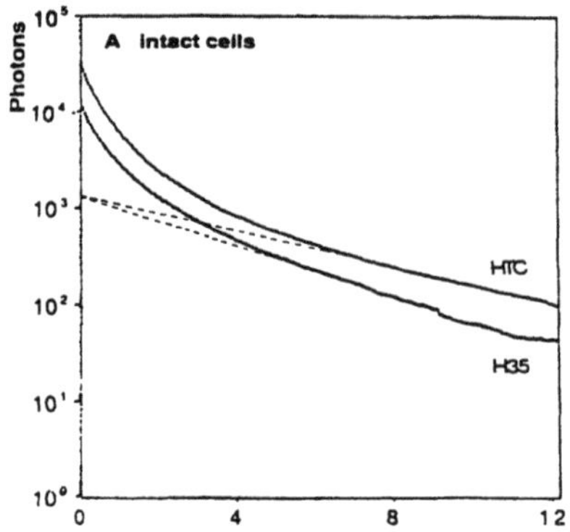

Figure 7.5: *Exponential plot of the number of photons which were not yet emitted by intact HTC- and Reuber H35 cells at the times of measurement specified on the X-axis.*

is obtained from the poorly differentiated HTC cell line and that I.P.E. of hepatocytes is nearly absent.

For a careful evaluation of the extensive decay curve we performed measurements in ten-fold in order to eliminate disturbances caused by irregularities in the dark count rate. By this analysis we could extend the kinetics into the last part ("tail") of the I.P.E. decay curve.

Subsequently we calculated

1. at each time point the amount of I.P.E. that had been produced up to that time and

2. the maximal I.P.E. which could be produced.

In that way cumulative data curves were constructed for a period of about 30 s, and finally we analyzed these cumulative data curves by calculating the I.P.E. fraction that was still left, i.e. not yet emitted at each time point. Figure 7.5 presents the logarithmic plot of fraction vs. time, showing a straight line for the tail of the curve. This might indicate an amount of I.P.E. as a consequence of a decay principle with a constant probability and with a half time of approximately 3-4 s. This (exponential) decay rate occurs in the period of 5 to 25 s after the start of the measurement. At earlier time points the kinetics are more complex.

In order to study the earlier kinetics we eliminated the contribution of the 3-4 s half time decay principle from the earlier data points. Again cumulative data curves were

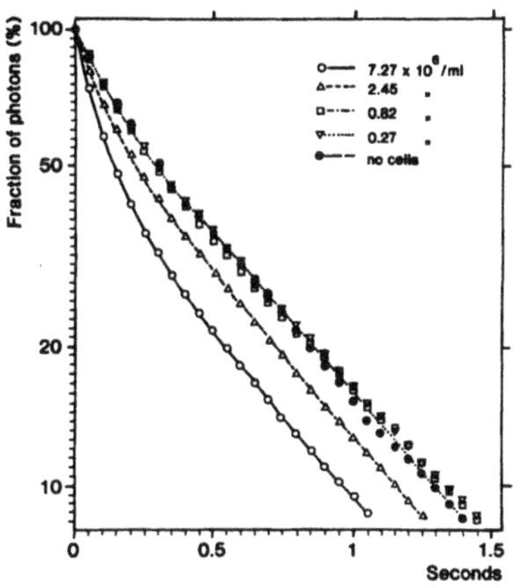

Figure 7.6: *Detailed exponential plot of the fraction of photons which were not yet emitted by HTC cell populations of different densities, at the times of measurement as specified on the X-axis.*

constructed for a period of about 1.5 s and we analyzed these cumulative curves by calculating at each point the I.P.E. fraction that was not yet emitted. The decay had a value that is variable and cell density dependent. The decay is more rapid at higher cell densities as shown for HTC cells (Fig. 7.6).

7.3.3 Transparency for I.P.E.: cell density dependency

I.P.E. from populations of rat hepatocytes and the hepatoma cell lines Reuber H35 and HTC were studied after prior illumination of populations of different cell densities (Fig. 7.7).

Cell densities ranged from 0.05 - 8 $\times 10^6$ cells/ml of medium and were obtained by dilution. By expressing the total number of photon counts emitted in 5 s as a function of cell density HTC cells displayed an increasing emission with increasing cell density. In the case of Reuber H35 cells the photon counts were lower and increased only slightly with increasing cell density. In contrast, hepatocytes exhibited a significantly lower emission of photons as compared to the emission by the other cell types and the medium at cell densities above approximately 3 $\times 10^6$ cell/ml.

The data presented in Fig. 7.7 were not corrected for decreased transmission at high cell densities. The transmittance at cell densities of 2.3 $\times 10^6$ cells per ml is

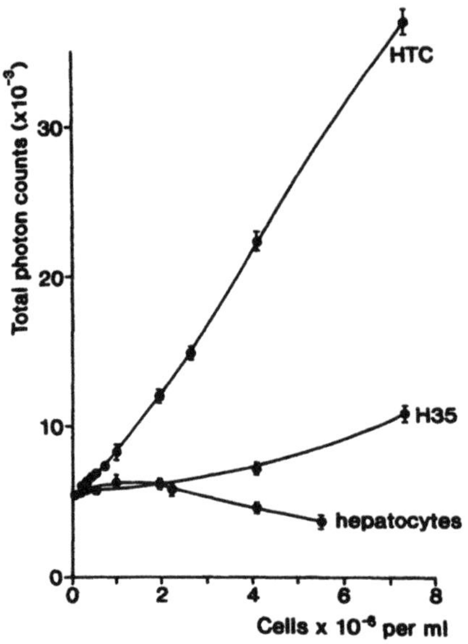

Figure 7.7: *Cell density dependence of total emitted photons of suspensions of isolated hepatocytes, Reuber H35 and HTC in a quartz cuvette, after illumination with white light. The total number of photons was calculated by adding up the number of photons registered in a 5 s period after illumination.*

reduced to 10% and less, depending on the cell type (Fig. 7.2). Any correction will cause a dramatic, non-linear, increase at the higher cell densities for HTC as well as, although to a lesser extent, for Reuber H35 cells. Only the data for hepatocytes can be explained by the assumptions that

1. cell specific I.P.E. is very low and

2. cuvette specific I.P.E. is (from one side) reduced by low transmittance at higher cell densities.

These results led to the conclusion that there can only be a linear relationship between HTC cell density and photon emission if suspensions of hepatoma cells have a far greater transparency for I.P.E. than for artificial light.

7.3.4 I.P.E. counts at various colour filter combinations

The decay of I.P.E. from an illuminated cuvette containing a HTC cell suspension was studied using various combinations of coloured filters (Fig. 7.8). In these studies the decay of I.P.E. was registered for 5 s by measuring photon emission during one thousand consecutive 5 ms intervals.

The induced photon emission was analysed by interposing glass absorption filters between the cuvette and the photocathode. Transmission is reduced to 7% by a BG 3 filter. This filter has maximum (100%) transmission at 300 nm, and 1% transmission at 260 nm and 510 nm. In contrast transmission by a BG 39 filter is 44% of the value without any filter. The latter filter has maximum (100%) transmission at 500 nm, and 1% transmission at 320 and 660 nm. Furthermore a cut-off filter RG610 which usually has no transmission below 590 nm, shows 16% transmission. These data suggest that I.P.E. occurs mainly at wavelengths ranging from 500-650 nm.

In some experiments cell suspensions were irradiated with light that had passed through glass absorption filters. Illumination with a cut-off filter OG 550 and G 610 did not cause any I.P.E.. The original light is filtered through these cut-off filters such that virtually no transmission occurs below 530 and 590 nm. The use of the cut-off filter CG 400 resulted in 35% induction of I.P.E. and this value is even increased by using a blue filter (BG 25) which displays maximum transmission (95%) at 390 nm and 1% transmission value at 310 and 510 nm.

Each coloured-filter condition of illumination was ultimately analysed with respect to the decay of emitted light after passing various filters (Figure 7.8). It is concluded from these experiments that induction as well as emission was dependent on the wavelength in the visible range.

Table 7.2 shows the cell-specific I.P.E., i.e. total number of photons registered in the 5 s period after illumination of HTC cells; analysed with different filters. This cell-specific emission was obtained after subtraction of the values of the quartz cuvette with medium, obtained under similar conditions. It is concluded that maximal induction of photon emission was caused by light with a wavelength smaller than 500 nm and emission occurred mainly in the wavelength range of more than 500 nm.

7.3.5 Possible sources of I.P.E. and P.E.

To study the oxygen dependency of I.P.E. we flushed the cuvette containing the cell suspension with water-saturated argon at a flow rate of 150 ml/min for at least

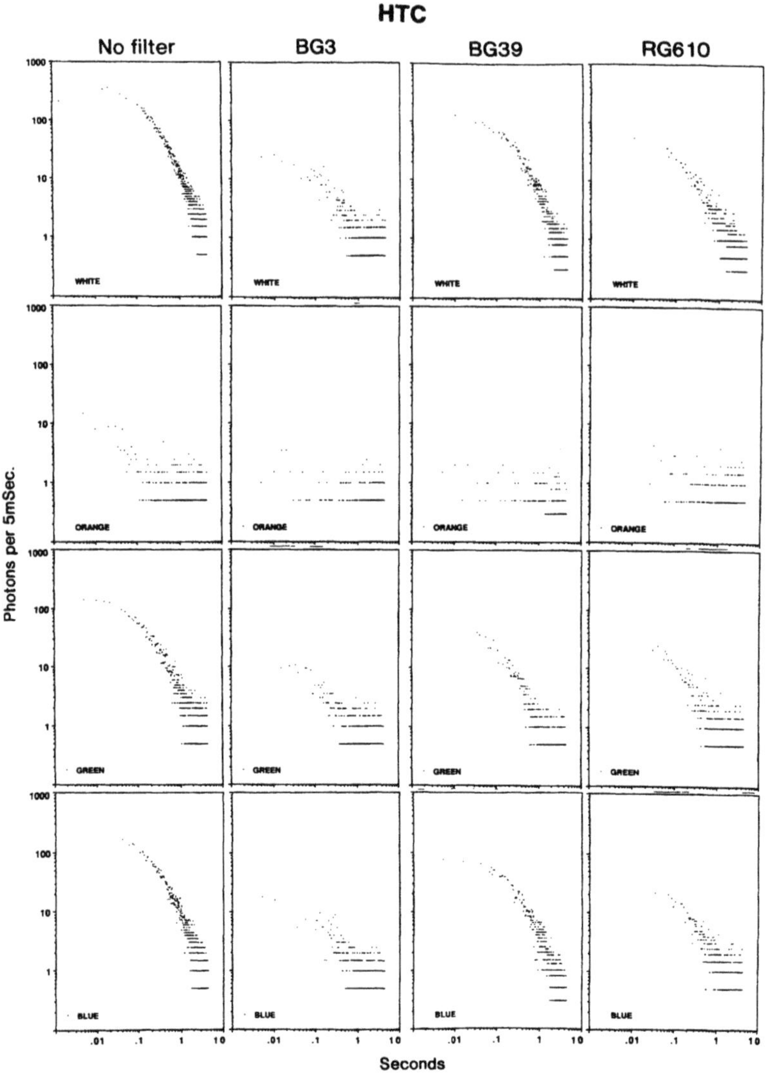

Figure 7.8: *Decay kinetics of photon emission of a suspension of HTC cells after illumination and analysis without or with different filters. For the illumination the following filters were used: no (1st horizontal row), OG 550 (2nd horizontal row), GG 400 (3rd horizontal row), BG 25 (4th horizontal row). For the analysis were used: no filter (1st vertical row), BG 3 (2nd vertical row), BG 39 (3rd vertical row), RG 610 (4th vertical row).*

Table 7.2: Light-induced photon emission of HTC cells after illumination and analysis with different filters[a]

	Filters for measurement during postillumination period			
Filters during illumination	None	BG3 λ_{max}=360nm	BG39 λ_{max}=500nm	RG610 $\lambda > 610$nm
None	11788±122	593±14	4282±84	1479±24
BG25(λ=380nm)	6830±33	418±50	2615±74	964±53
GG400($\lambda > 400$nm)	3246±39	340±23	970±30	680±30
OG550($\lambda > 550$nm)	14±11	11±3	6±4	58±12
RG610($\lambda > 610$nm)	20±9	31±2	5±3	26±14

[a]For the illumination of HTC cell suspensions the following filters were placed between lamp and cuvette (vertical row): none, BG25 (blue, λ_{max} = 380 nm), GG400 (green, $\lambda > 400$ nm), OG550 (orange, $\lambda > 550$), RG610 (red, $\lambda > 610$ nm). For the analysis during the post-illumination the following filters were placed between cuvette and photocathode (horizontal row): none, BG3 (blue, λ_{max} = 360 nm). BG39 (green, λ_{max} = 500 nm), RG610 (red, $\lambda > 610$ nm). The total number of photons was calculated by adding up the number of photons registrated in 5 s period after illumination. Measurements were performed in triplicate and cell-specific emission was obtained after subtraction of the value of the quartz cuvette with medium obtained under similar conditions.

15 min prior to the start of I.P.E. measurements. The suspension was agitated by stirring throughout the gassing procedure. Even keeping the culture under these conditions for 1 hour did not decrease I.P.E. values.

Reoxygenation of the hypoxic cells was performed similarly except that the cell suspension was flushed with water-saturated 95% oxygen/5% CO_2 mixture.

This reoxygenation did not increase the I.P.E. values, indicating that I.P.E. does not require any oxygen-dependent metabolism.

In this respect I.P.E. differs from P.E.. Extensive studies considering the underlying mechanism of P.E. have revealed that P.E. from liver appeared to be oxygen-dependent. Furthermore P.E. was derived from submitochondrial particles[35,41], mitochondria[34] and the microsomal fraction[34,36,38,39,42]. It requires a membrane

Table 7.3: Light-induced photon emission of HTC cells and cellular fractions[a]

	Total photon counts in 0.5 s corresponding with 10^6 cells	Transmission at 500 nm, %
Whole cells	4053±770	18
Cytoplasmic fraction	1736±581	87
Nuclear fraction	5670±1463	36
Chromatin fraction	13230±3206	44
DNA (phenol-extracted)	388±72	96

[a]Cell fractions were measured at amounts that correspond with the original cell number. For details of the fractionation procedure, see Methods. Values (total photon counts in 0.5 s per 10^6 cells) were obtained in 5 different experiments and are given for the mean ± SD after subtracting the background values of cuvette with buffer. Measured IPE was not corrected for absorption of light by the cell suspension. Transmission at 500 nm was determined spectrophotometrically.

bound electron transfer system and can be optimized in the presence of oxygen.

7.3.6 Chromatin as source of I.P.E.

The process or principle leading to I.P.E. was studied with disrupted and fractionated HTC cells. Several methods for cell disruption were applied. We found remarkably similar I.P.E. intensity of broken cells compared to intact cells, indicating that the process leading to I.P.E. requires no intact cells. This is not a characteristic of freshly broken cells, but it was observed in broken cells, which were incubated at 37°C for several hours without (extra-) energy source (data not shown).

The I.P.E. of cell fractions were tested, starting with a preparation that was obtained by Potter-Elvejem homogenization (table 7.3). Centrifugation of the homogenate at 600 rpm results in the separation of cell nuclei.This fraction retains most of the activity. Several fractions of the cytoplasmic fraction were tested in more detail. Mitochondria and subribosomal particles demonstrate a very low activity, less than 5% of that of the nuclear fraction (not shown).

The further fractionation of the nuclei shows that intactness of nuclei is not required for I.P.E.. The isolated chromatin has I.P.E. values that are even higher than I.P.E.

values of nuclei. In contrast, phenol-extracted DNA did not produce any I.P.E.. A comparison of the I.P.E. of different cell fractions requires a correction for absorption of light by the sample. However, transparency for I.P.E., at least that from cells, is higher than that for artificial light, and the correction factor is essentially unknown. Nevertheless the transmission values (table 7.3) cannot explain the differences in I.P.E. of chromatin and phenol-extracted DNA, indicating that the DNA-protein-complex is mainly responsible for I.P.E..

7.4 Discussion and Conclusion

Four important conclusions were derived from our studies on light-induced photon emission so far.

1. Under similar conditions of measurement I.P.E. is lowest for hepatocytes and highest for the least differentiated hepatoma cell line.

2. The chromatin structure, i.e. the DNA-protein-complex, is responsible for I.P.E.

3. Maximum induction of I.P.E. was caused by wavelengths below 500 nm, and emission occurred above 500 nm.

4. Transparency of the cell suspension for I.P.E. is much higher than for artificial light.

The first, second and fourth conclusion are different for I.P.E. and P.E.. P.E. has been shown for hepatocytes, but it was not detectable from the hepatoma cell lines. In general P.E. characteristics suggest the free radical process accompanying lipid peroxidation as a major source of P.E.. Finally our data showed that the application of the extinction coefficient of cells, as assessed spectrophotometrically, to the measured P.E. value lead to a linear relationship between P.E. and cell density. This means that spontaneous photon emission has a comparable transparency to artificial light.

The characteristics of I.P.E. of hepatoma cells, especially those of HTC, may be explained by a model including chromatin dependent emission of light with a coherent character. According to Li and Popp[43] the hyperbolic instead of exponential decay kinetics is in favour of coherence. After exposure of living cells to weak light decay functions have been observed whose agreement with an exponential law can be clearly rejected, and it corresponds better to a hyperbolic relaxation[22,44,45]. On the basis of these experimental results a model of "low-level luminescence" was developed which suggests that exciplex formation in DNA linked to photons in its lattice system (thus forming polaritons) is responsible for photon emission[43,46].

Nevertheless for an unequivocal interpretation of I.P.E. further clarification of the mechanism is necessary. However, this system and the one previously described[27] show a distinction between normal and malignant cells. They indicate that alterations of I.P.E. characteristics are indicative of alterations in the gene structure and expression in differentiation and malignancy. Therefore this phenomenon deserves further attention.

7.5 Acknowledgements

The authors wish to thank F.A. Popp for his critical advice, and H. Ovelgonne for critically reading the manuscript.

7.6 References

1. L. Shlyakhtina and A. A. Gurvich, *Biophysics* **17** (1972) p. 1146.

2. A.Boveris, E. Cadenas, R. Reiter, M. Filipkowski, Y. Nakase and B. Chance, *Proc. Natl. Acad. Sci. USA* **77** (1980) p. 347.

3. A. Boveris, E. Cadenas and B. Chance, *Fed. Proc.* **40** (1981) p. 195.

4. E. Cadenas, H. Wefers and H. Sies, *Eur. J. Biochem.* **119** (1981) p. 531.

5. V. V. Perelygin and B. N. Tarusov, *Biophysics* **11** (1966) p. 616.

6. R. Barsacchi, P. Camici, U. Bottigli, P. A. Salvadori, G. Pelosi, M. Maiorino and F. Ursini, *Biochim. Biophys. Acta* **762** (1983) p. 241.

7. E. Cadenas, I. D. Arad, A. Boveris, A. B. Fisher and B. Chance, *FEBS Letters* **111** (1980) p. 413.

8. V. V. Artemey, A. S. Goldobin and L. N. Guskov, *Biophysics* **12** (1967) p. 1278.

9. V. V. Blokha, G. V. Kossova, A. D. Sizov, V. A. Fedin, Y. P. Kozlov, O. R. Kols and B. N. Tarusov, *Biophysics* **13** (1968) p. 1084.

10. I. G. Shtrankfeld, L. L. Klimensko and N. N. Komarov, *Biophysics* **13** (1968) p. 1082.

11. H. P. Morris, H. M. Dyer, B. P. Wagner, H. Miyali and M. Rechcigl, *Advan. Enzyme Regul. 2* (1964) p. 321.

12. V. R. Potter, M. Watanabe, H. C. Pitot and H. P. Morris, *Cancer Res.* **29** (1969) p. 55.

13. G. Weber and M. A. Lea, in *Methods in Cancer Research*, **Vol. 2**, ed. H. Busch, (Academic Press, New York, 1967) p. 523.

14. W. D. Wicks, R. Van Wijk, K. Clay, C. Bearg, M. M. Bevers and J. Van Rijn, in *The Role of Cyclic Nucleotides in Carcinogenesis*, eds. J. Schultz and H.G. Gratzner (Academic Press, New York, 1973) p. 103.

15. W. D. Wicks, R. Van Wijk and J. B. McKibbin, *Advan. Enzyme Regul.* **11** (1973) p. 117.

16. H. Van Rijn, M. M. Bevers, R. Van Wijk and W. D. Wicks, *J. Cell Biol.* **60** (1974) p. 181.

17. H. Volman, *Virchows Arch. B.: Zellpathologie* **26** (1978) p. 249.

18. D. H. J. Schamhart, K. W. Van De Poll and R. Van Wijk, *Cancer Res.* **39** (1979) p. 1051.

19. R. Van Wijk, W. D. Wicks and K. Clay, *Cancer Res.* **32** (1972) p. 1905.

20. R. Van Wijk, W. D. Wicks, M. M. Bevers and J. Van Rijn, *Cancer Res.* **33** (1973) p. 1331.

21. R. Van Wijk, *Intern. Rev. Cytol.* **85** (1983) p. 63.

22. F. A. Popp, B. Ruth, R. W. Bahr, J. Bohm, P. Grass, G. Grolig, M. Rattemeyer, H. G. Schmidt and P. Wulle, *Coll. Phenomena* **3** (1981) p. 187.

23. W. D. Chwirot, R. S. Dygala and S. Chwirot, *Cytobios* **47** (1986) p. 137.

24. D. H. J. Schamhart and R. Van Wijk, in *Photon Emission from Biological Systems*, eds. B. Jezowska-Trebiatowska, B. Kochel, Slawinski and W. Stek (World Scientific, Singapore, 1987) p. 137.

25. R. Van Wijk and D. H. J. Schamhart, *Experientia* **44** (1988) p. 586.

26. R. Van Wijk, H. Van Aken and D. H. J. Schamhart, in *Biological Luminescence*, eds. B. Jezowska-Trebiatowska, B. Kochel, J. Slawinski and W. Strek (World Scientific, Singapore, 1990) p. 460.

27. W. Scholz, U. Staszkiewics, F. A. Popp and W. Nagl, *Cell Biophys.* **13** (1988) p. 55.

28. D. H. J. Schamhart, W. Berendsen, J. Van Rijn and R. Van Wijk, *Cancer Res.* **44** (1984) p. 4507.

29. H. C. Pitot, C. Peraino, P. A. Morse and V. R. Potter, *Natl. Cancer Inst. Monogr.* **13** (1964) p. 229.

30. E. B. Thompson, G. M. Tomkins and J. F. Curran, *Proc. Natl. Acad. Sci. USA* **56** (1966) p. 296.

31. B. Van Dongen, W. L. M. Geilenkirchen, J. Van Rijn and R. Van Wijk, *Exp. Cell Res.* **166** (1986) p. 427.

32. M. Perucho, M. Goldfarb, K. Shimizu, C. Lama, J. Fogh and M. Wigler, *Cell* **27** (1981) p. 467.

33. B. Ruth, in *Electromagnetic Bio-Information, 2nd edition*, eds. F. A. Popp, U. Warnke, H. L. Koenig, W. Peschka (Urban and Schwarzenberg, Munich/Baltimore, 1989) p. 128.

34. A. Boveris, B. Chance, M. Filipkowski, Y. Nakase and K.G. Paul, in *Frontiers of Biological Energetics, Electron to Tissue* **Vol. 2**, eds. A. Scarpa, P.L. Duutton and J.S. Leigh (Academic Press, New York, 1978) p. 975.

35. E. Cadenas, A. Boveris and B. Chance, *Biochem. J.* **186** (1980) p. 659.

36. E. Cadenas and H. Sies, *Eur. J. Biochem.* **124** (1982) p. 349.

37. J.P. Hamman, D.R. Gorby and H.H. Seliger, *Biochem. Biophys. Res. Comm.* **75** (1977) p. 793.

38. M. Nakano, T. Noguchi, K. Sugioka, H. Fukuyama, M. Sato, Y. Shimizu, Y. Tsuji and H. Inaba, *J. Biol. Chem.* **250** (1975) p. 2404.

39. J. R. Wright, R. C. Rumbaugh, H. D. Colby and P. R. Miles, *Arch. Biochem. Biophys.* **192** (1979) p. 344.

40. T. Galeoti, G. M. Bartoli, S. Bartoli and E. Bertoli, in *Development in Biochemistry*, Vol. 11b, eds. W. H. Bannister and J. V. Bannister (Elsevier, North Holland, New York, 1980) p. 106.

41. E. Cadenas, A. Boveris and B. Chance, in *Developments in Biochemistry* Vol. 11A, eds. J. V. Bannister and H. A. O. Hill (Elsevier, North Holland, New York, 1980) p. 92.

42. K. Sugioka and M. Nakano. *Biochim. Biophys. Acta* **423** (1976) p. 203.

43. K. H. Li and F. A. Popp, *Phys. Lett. A* **93** (1983) p. 262.

44. W. B. Chwirot, R. S. Dygdala and S. Chwirot, *Cytobios* **44** (1985) p. 95.

45. W. B. Chwirot, R. S. Dygdala and S. Chwirot, *Cytobios* **44** (1985) p. 239.

46. F. A. Popp, in *Electromagnetic Bio-Information*, eds. F. A. Popp, G. Becker, H. L. Koenig and W. Peschka (Urban and Schwarzenberg, Muenich/Baltimore, 1979) p. 107.

Chapter 8

Biophoton Re-emission Studies in Carcinogenic Mouse Melanoma Cells

Hugo J. Niggli

Abstract

Practically all cells emit light at a steady rate from a few photons per cell per day to several photons per cell per second, which is readily detected with a sensitive photomultiplier placed with the cells in a dark chamber. It is widely accepted that at least a part of the emitted light results from radical reactions, but chromatin has also been suggested to be the source of this ultra-weak radiation in living cells.

Although the emission of this light is very low in mammalian cells, recent studies have demonstrated that the nature of this radiation can also be studied as rescattered emission after a brief illumination with an ordinary light source. Using this experimental design, we have investigated the contribution of nuclei to the ultra-weak photon re-emission in Cloudman S 91 mouse melanoma cells. In all experiments reported here, it turns out that the nuclei make the greatest, if not the entire contribution to this radiation indicating that chromatin acts as a photon store emitting partially coherent light which may be involved in the control of gene activity, cell metabolism and cell communication.

8.1 Introduction

Photons necessarily participate in all atomic and molecular interactions and changes in the physical universe. At the beginning of this century, Carl Huter (1861-1912) ingeniously suggested that radiation processes are involved in cell metabolism and cell communication[1,2]. It was Alexander G. Gurwitsch in the early twenties who started biological experiments in order to show that ultra-weak photons transmit information in living systems[3]. Some authors have proved the existence of this very weak radiation emanating from biological tissues, whereas others have been unable to find such an effect, although working with measuring equipment of a similiar detectivity[3]. In the meantime it is commonly agreed that plant, animal and human cells emit ultra-weak photons also called biophotons[3-7].

Most of the investigators think that this radiation results from radical reactions as for instance lipid peroxidation[3,8]. However, there exists also a highly interesting model published in 1983 by Nagl and Popp[9] suggesting that there is a negative feedback-loop in living cells which couples together states of a coherent ultra-weak photon or biophoton field and the conformational state of the cellular DNA. The authors assume photon transfer or radiationless chemical pumping from the cytoplasmic metabolism which results in changes of the DNA conformation via exciplex/excimer formation.

Their hypothesis is based on experimental data reviewed by Birks[10] who also suggested these excimers as precursors of the pyrimidine photodimers which play a key role in the radiation damage of DNA. Since the conformation of the DNA molecule is believed to be of importance for the regulation of the nuclear information transfer, such processes in turn influence the metabolic activity of a cell, thus closing the feedback-loop. The experimental material collected up to date does not allow to answer the question of the source of the ultra-weak photon emission definitely. Therefore in this work ultra-weak photon re-emission studies[11,12] in mouse melanoma cells and their isolated nuclei are performed to find further proof that the cell-nucleus contribute to this radiation.

8.2 Materials and Methods

8.2.1 Cell culturing

Cloudman S91 mouse melanoma cells of clone M-3 (CCL53.1), a melanin-producing cell-line, were purchased from the American Type culture Collection (Rockville, Maryland). The cells were grown as monolayers in Dulbecco's modified Eagle's medium (DMEM) with 10% fetal calf serum and 100U/ml penicillin/100 μg/ml streptomycin as described by Friedman and Gilchrest[13]. They were routinely passaged 1:4 each 3-4 days as described previously[14]. Melanoma cells to be irradiated were grown in 75 cm^2 tissue culture flasks in 15 ml DMEM.

8.2.2 Tyrosinase assay

For the determination of tyrosinase activity, 3×10^4 cells were seeded into 35 mm tissue culture dishes and grown for several days. For the experiments with melanin synthesis induction, the cells were stimulated for 48-72h with 10^{-11}M cholera toxin (Sigma, St. Louis, MO) or 10^{-5} M forskolin (Sigma). The modified version of the Pomerantz charcoal adsorption method[15] described in detail for cell extracts[14,16] was performed to measure tyrosinase activity.

8.2.3 Preparation of cells, nuclei and DNA

Medium was removed from the tissue culture flasks prior to irradiation, the cells were trypsinized and the action of trypsin was stopped by the addition of 50μl of a 5% aqueous solution of soybean trypsin inhibitor to each flask. The content of 10-20 flasks were pooled ($100\text{-}200\times10^6$) cells, centrifuged at approximately 1'000 rpm for 3 min and resuspended in 10ml Earle's balanced salt solution (EBSS).

After biophoton measurements in whole cells described below, the same cells were used for isolation of nuclei at room temperature as described by Niggli and Cerutti[17]. In short, the nuclei were prepared by homogenization in 0.25 M sucrose, 0.01 M Tris (pH 7.5), 0.01 M $CaCl_2$, and 0.5 mM PMSF with a dounce homogenizer (Kontes, Vineland, NJ) using the "B" pestle at a concentration of 6×10^6 cells/ml. At this stage and later on, the extent of purification of nuclei free of cytoplasmic membranes was followed microscopically, after colouring the nuclei in a small aliquot of the homogenized cell suspension with a drop of Coomassie brillant blue.

After centrifugation at 2'500 rpm for 10 min, the pellet was resuspended in 0.25 M sucrose, 0.01 M Tris (pH 8.0), 0.01 M $CaCl_2$, 0.5 mM PMSF, and 0.01 M Triton x-100. Triton is added to facilitate the release of membranes from nuclei. The sample is then recentrifuged as specified above. The nuclear pellets were suspended finally in 10 ml of 0.25 M sucrose, 0.01 M Tris (pH 7.5), 0.0001 M $CaCl_2$, and 0.5 mM PMSF and biophoton detection was performed as described below. DNA was isolated by a filter method described previously[18].

8.2.4 Irradiation facilities

For photon emission measurements, the cell suspension in a volume of 10 ml was transferred into a quartz sample glass (2.2×2.2×3.8 cm^3). For detection and registration of re-emitted photons in various cell concentrations the test sample was kept in a dark chamber in front of a single photon counting device equipped with a EMI 9558 QA photomultiplier tube (diameter of the cathode 48mm, cooled down to -25°C). This high-sensitivity photon counting device is described in detail by Popp and co-workers[3,19,20] and measures photon intensities as low as 10^{-17} W in the range between 200 and 800 nm.

The integral intensity values within given time intervals (4–40ms) were stored and processed by an interfaced computer. For re-emission experiments, irradiation of the test sample was performed perpendicular to the detecting direction with focused white light from a 150 watt tungsten lamp (Type 58.8297; Osram, München, G). Each measuring cycle started by irradiating the sample for 30s. The re-emission was measured by two-shutter techniques; one to stop the irradiation and the other to open a light path to the detector as described before[21]. The interval (dead time) between the end of an excitation and the start of re-emission was 150 ms.

8.3 Results and Discussion

Figure 1 shows white light-induced photon re-emission dynamics in whole untreated mouse melanoma cells (Fig. 8.1A) and their isolated nuclei (Fig. 8.1B). Suspension densities ranged from 0.5-2.0×10^7 cells/ml and 1.2-7.5×10^6 nuclei/ml, respectively. It is noteworthy that nuclei reveal a significant higher re-emission intensity than whole cells. The finding that nuclei emit efficiently biophotons confirms the most recent results of Inaba and co-workers[7] who dedected an extremly weak native light emission from rat liver nuclei.

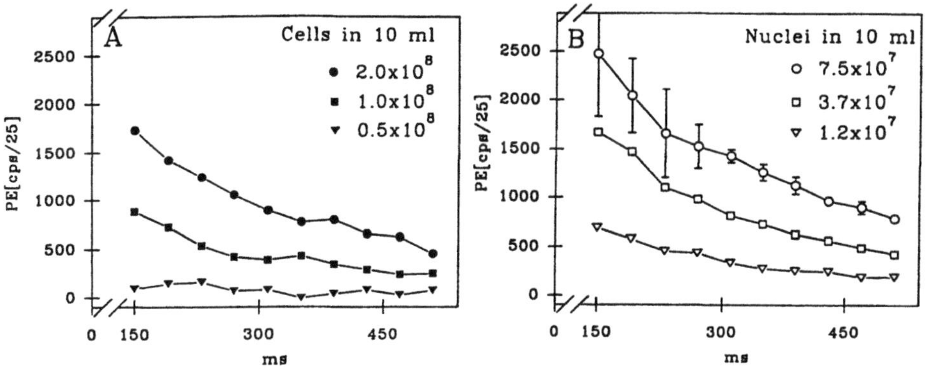

Figure 8.1: *Biophoton re-emission of Cloudman S91 mouse melanoma cells after 30s pre-illumination with white light. Points represent the average of three determinations in the same cells. Standard deviations are less than 10%. Background values obtained with medium are subtracted. Maximal error obtained in two fully distinct experiments with different cells is 25% as shown by error bars in Fig. 8.1B. A: Whole mouse melanoma cells. B: Nuclei of mouse melanoma cells.*

The presented re-emission relaxation dynamics of whole untreated mouse melanoma cells (Fig. 8.1A) and their isolated nuclei (Fig. 8.1B) all follow without exception a hyperbolic decay function. This especially significant observation indicates coherent re-scattering[22]. It should be noted that this relaxation cannot be assigned to the usual optical transition of isolated- or triplet-states, but are actually due to collective excitation within the matter under study[3].

This follows from the rather long decay time, lasting at least a few seconds, and the relaxation dynamics[23,24]. Therefore, the source of this radiation can no longer be regarded as a single molecule exhibiting spontaneous chemiluminescence from some kind of metabolic aberration[8]. The finding of efficient re-emission of biophotons in nuclei of mouse melanoma cells strongly supports the notion that chromatin emits this radiation. The observation that pure DNA has no re-emission capacity (data not shown) points to an essential importance of whole chromatin - possibly for conformational reasons.

These interpretations are also consistent with earlier observations of Popp and co-workers[3,19,20]. They found that conformational changes of DNA induced in vivo with the intercalating agent ethidium bromide are clearly reflected by changes of the ultra-weak photon emission of cells. In these experiments the highest biophoton emission was measured in fully uncondensed chromatin.

They also reported that human erythrocytes which do not contain nuclei lack ultra-weak photon emission even if they have been irradiated with white light. In addition, Chwirot[3,5] found a close positive correlation between changes in DNase activity and in the intensity of white light induced photon re-emission of microsporocytes from larch trees in successive stages of meiosis. Two further observations contribute to the hypothesis of chromatin as source of radiation:

1. Eisinger et al.[25] reported that intramolecular excimer fluorescence, due to the interaction of two identical aromatic groups in the same molecule, also occurs in the dinucleotides of thymine and cytosine as well as in polycytosine.

2. Vigny and Duquesne[26] showed that such excited complexes of nucleic acids represent an excellent excitation energy trap.

Thus there is further substantial experimental evidence that chromatin may act as coherent radiator and photon store, respectively. Although the number of measured photons which escape from the cells and nuclei is very low, the number of photons actually generated within the system may be far higher. In that case a photon has a low probability of escaping from the emitters after generation, which means that there is an effective intracellular mechanism of photon trapping. It is conceivable that this trapping can influence metabolic and cellular events by triggering amplification mechanisms and promote photochemical processes even in the dark as proposed by Cilento[3].

It is noteworthy that our dilution experiment shown in Fig. 8.2 exhibits a non-linear increase of the induced re-emission intensity with higher cell density for the carcinogenic Cloudman S91 mouse melanoma cells. This behaviour seems typical for cancer cells and confirms previously reported results with rat hepatoma and human tumorigenic amnion cells[3,11]. In fact, these phenomena can be interpreted in terms of Dicke's theory of photon-emission from an ensemble of emitters[27]. He showed that photon emission tends to bifurcate into two branches of superradiance and subradiance as soon as the wavelength of the emitted light is large compared to the distances between the emitters which are also absorbers.

Superradiance is the increase of emission intensity concomitant with a shortening of the relaxation time. The opposite branch describes the regimen of subradiance

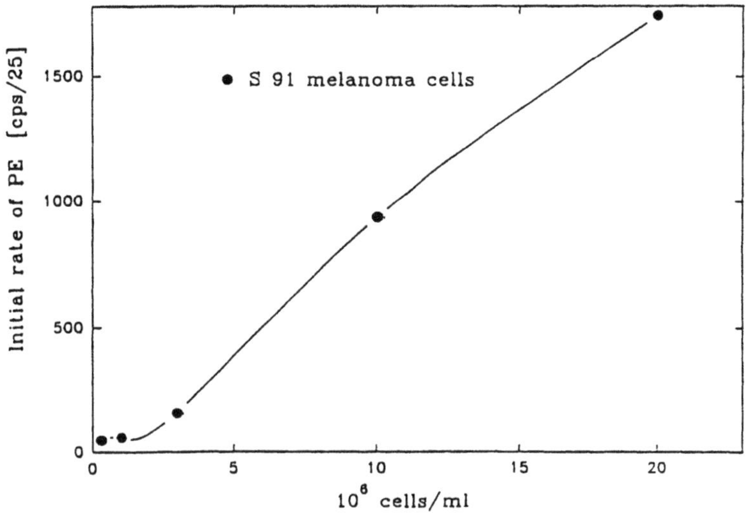

Figure 8.2: *Initial rates of biophoton re-emission of Cloudman S91 mouse melanoma cells after 30s pre-illumination with white light at suspension densities ranging from 0-2×10⁷ cells/ml. Points represent the average of 2-13 determinations with standard deviation in two fully distinct experiments. Background values obtained with medium are subtracted.*

where emission intensity decreases with a more and more prolonged decay time, corresponding to photon storage within the system. Thus, in terms of Dicke's theory, cancer cells have a lower capacity for subradiance the closer they are together leading therefore to increasing biophoton re-emission values as depicted in Fig. 8.2.

Most interestingly in this concept, the intermolecular distances d of the original photon absorbers have to satisfy Dicke's condition $d \ll \lambda$. Again, the DNA molecule in chromatin is the most likely absorber candidate, because the distances of adjacent base pairs is three orders of magnitude smaller than the wavelengths of the re-emitted light. Note that theoretically the highly elongate, threadlike DNA molecule may act as much better photon store than the significantly shorter histone and non-histone proteins.

The most impressive proof of an at least partial contribution of the nucleus in ultra-weak photon re-emission can be deduced from our re-emission relaxation dynamics for cholera toxin-treated cells and their nuclei shown in Fig 8.3A and 8.3B, respectively. Suspension densities ranged from $0.3\text{-}1.9\times10^7$ cells/ml and $1\text{-}5\times10^6$ nu-

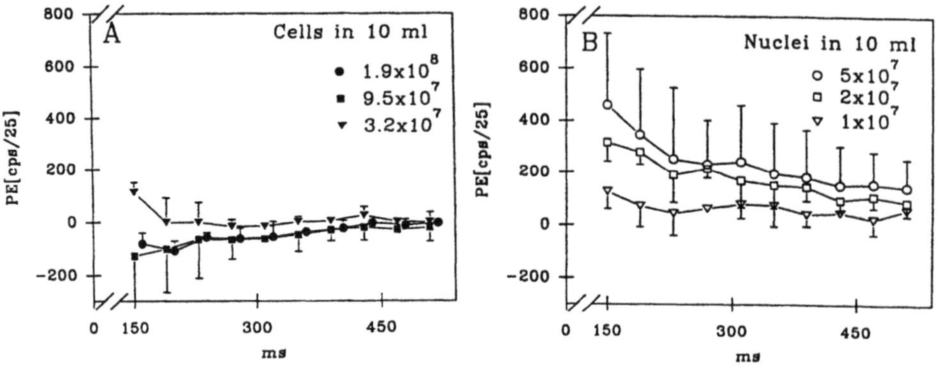

Figure 8.3: *Biophoton re-emission of Cloudman S91 melanoma cells after 30s pre-illumination with white light. The cells were treated for 48-72h with 10^{-11}M cholera toxin. Points represent the average of three determinations in the same cells. Standard deviations are less than 10%. Background values obtained with medium are subtracted. Results of two fully independent experiments are shown. Maximal error obtained in two fully distinct experiments with different cells is 40%. A: Whole cholera toxin-treated cells. B: Nuclei from cholera toxin-treated cells.*

clei/ml, respectively. Qualitatively similar results of biophoton re-emission were obtained with forskolin-treated cells (data not shown). Both substances induce melanin synthesis via the c-AMP pathway.

Most interestingly, in such treated melanoma cells no biophoton re-emission is measured as depicted in Fig. 8.3A for cholera toxin-treated cells. Most probably the re-emitted light is absorbed by melanin. The important finding, however, that the isolated nuclei of these melanin-rich cells represent hyperbolic re-emission dynamics (Fig. 8.3B) with an approximately 3 times higher intensity than whole, untreated cells (Table 8.1) tends indeed to identify chromatin as a main source of biophoton emission. It is noteworthy that there exists a physical association between the centriole and the interphase nucleus as shown by Bornens[28]. This author even suggested the centriole as a gyroscopic oscillator and consequently, a pace maker for cell metabolism[29]. Therefore, it can be assumed that the ultra-weak photon radiation emitted by the nucleus may be controlled by the centriole.

Table 8.1: Tyrosinase activity as an indicator of melanin content in melanoma cells and their isolated nuclei compared with initial rates of biophoton emission.

Type	CT-treatment* [M]	DOPA produced per 10^6 cells or nuclei/hr [pMol]	initial rate of biophotons per 10^6 cells or nuclei/ml [cps/25]
S91 nuclei	none	0.1	683
S91 nuclei	10^{-11}	0.3	158
S91 cells	none	5.1	59
S91 cells	10^{-11}	21.1	30

*CT = cholera toxin

We have previously shown that melanin at higher concentrations in Cloudman S91 mouse melanoma cells seems to confer little protection against UV-induced pyrimidine photodimer formation[14]. However, as depicted in Table 8.1 increasing re-emission intensities are observed in mouse melanoma cells and their nuclei with decreasing tyrosinase activity which is a sensitive marker for melanin content[13]. On a per nucleus or per cell basis the treated nuclei emit 5-fold more photons than the treated cells and the untreated nuclei emit 10-times more than the untreated cells.

Since each cell contains one nucleus this would imply that the cells do not emit as much light as their component parts, or more likely that there is something in the cells (e.g. melanin) that absorbs 80-90% of the re-emitted photons. As most recently reported by Grasso et al.[30], normal cells emit significantly less biophotons than cells from cancer tissue. Thus, returning to Dicke's theory[27], normal cells have a greater capacity for subradiance the closer they are together, while tumour cells show the opposite behaviour, that is, the loss of subradiance.

According to this model Popp and Li[31] suggested, that long range interaction between cells is based on the coherence of the subradiance regimen and therefore, tumour cells seem to be unable to communicate. Based on this assumption, melanin in melanoma cells may act, in a kind of emergency, as an organizational trigger[32] in order to minimize this loss of subradiance and maintain therefore cellular communication. This interpretation is further sustained by the recent observation that photoprotection afforded by melanin in the skin of blacks may delay diagnosis of Xeroderma Pigmentosum until severe ocular damage brings the patient to the attention of an ophthalmologist[33].

8.4 Acknowledgements

I am deeply indebted to Dr. F.A. Popp, Technologiezentrum Kaiserslautern, for kind permission to use the single photon counting equipment and many helpful discussions. I would like to thank Elena Alborghetti, Corina Benkel, Anne-Lise Etter, Beate Hemmer and Dr. Wei-Ping Mei for excellent technical assistance. I am grateful to Urs Pfefferli for help with the graphics. Especially, I am indebted to Drs. Thomas C. Brown, Richard B. Setlow and Elizabeth T. Snow for critical reading of the manuscript.

8.5 References

1. C. Huter, Menschenkenntnis (Carl-Huter Verlag, Basel, 1991).

2. D. von Zerssen, Die Psychologie des 20. Jahrhunderts, **Band V**, ed. G. Strube (Kindler Verlag, Zürich, 1977).

3. W.B. Chwirot, G. Cilento, A.A. Gurwitsch, H. Inaba, W. Nagl, F.A. Popp, K.H. Li, W.P. Mei, M. Galle, R. Neurohr, J. Slawinski, R.V. Van Wijk and D.H.J. Schamhart (Multi-author review), *Experientia* **44** (1988) p. 543.

4. E. Hideg and H. Inaba, *Photochem. Photobiol.* **53** (1991) p. 137.

5. B. Chwirot and F.A. Popp, *Photochem. Photobiol.*, in press.

6. T. Ichimura, M. Hiramatsu, N. Hirai and T. Hayakawa, *Photochem. Photobiol.* **50** (1989) p. 283.

7. B. Devaraj, R.Q. Scott, P. Roschger and H. Inaba, *Photochem. Photobiol.* **54** (1991) p. 289.

8. D. Slawinska and J. Slawinski, *Photochem. Photobiol.* **37** (1983) p. 709.

9. W. Nagl and F.A. Popp, *Cytobios* **37** (1983) p. 45.

10. J.B. Birks, *Rep. Progr. Phys.* **38** (1975) p. 903.

11. F.A. Popp, in *Disequilibrium and Self-Organisation*, ed. C.W. Kilmister (D. Reidel Publishing Company, Boston, 1986) p. 207.

12. W. Scholz, U. Staszkiewicz, F.A. Popp and W. Nagl, *Cell Biophys.* **13** (1988) p. 55.

13. P.S. Friedmann and B.A. Gilchrest, *J. Cell Physiol.* **133** (1987) p. 88.

14. H.J. Niggli, *Photochem. Photobiol.* **52** (1990) p. 519.

15. S. Pomerantz, *Biochem. Biophys. Res. Commun.* **16** (1964) p. 188.

16. J. Auböck, D. Köfler, M. Sifter and P. Fritsch, *Br. J. Dermatol.* **109** (1983) p. 413.

17. H.J. Niggli and P.A. Cerutti, *Biochem. Biophys. Res. Commun.* **105** (1982) p. 1215.

18. S.A. Leadon and P.A. Cerutti, *Anal. Biochem.* **120** (1982) p. 282.

19. F.A. Popp, U. Warnke, H.L. König and W. Peschka, eds., Electromagnetic Bio-Information (Urban & Schwarzenberg, München, Baltimore, 1989).

20. M. Rattemeyer, F.A. Popp and W. Nagl, *Naturwissenschaften* **68** (1981) p. 572.

21. C. Strzalkowska, D. Slawinska, A. Slawinski and J. Slawinski, *J. Photochem. Photobiol. B* **3** (1989) p. 259.

22. K.H. Li, and F.A. Popp, *Phys. Lett.* **93** (1983) p. 262.

23. B. Fain, *Phys. Rev. A* **24** (1981) p. 2685.

24. K.L. Ngai, A.K. Rajagopal, R.W. Rendell and S. Teitler, *Phys. Rev. B* **28** (1983) p. 6073.

25. J. Eisinger, M. Gueron, R.G. Shulman and T. Yamane, *Proc. Natl. Acad. Sci. USA* **55** (1966) p. 1015.

26. P. Vigny and M. Duquesne, in *Excited States of Biological Macromolecules*, ed. Birks J. (Wiley, London, 1976) p. 167.

27. R.H. Dicke, *Phys. Rev.* **93** (1954) p. 99.

28. M. Bornens, *Nature* **270** (1977) p. 80.

29. M. Bornens, *Biol. Cellulaire* **35** (1979) p. 115.

30. F. Grasso, C. Grillo, F. Musumeci, A. Triglia, G. Rodolico, F. Cammisuli, C. Rinzivillo, G. Fragati, A. Santuccio and M. Rodolico, Photon Emission from Normal and Tumour Tissues. *Experientia* **48** (1992) p. 10-13.

31. F.A. Popp and K.-H. Li, this volume.

32. F.E. Barr, *Med. Hypotheses* **11** (1983) p. 1.

33. J.A. Johnson, *J. Invest. Dermatol.* **96** (1991) p. 156.

Chapter 9

Ultraweak Photon Emission from Synchronized Yeast (Saccharomyces cerevisiae) as a Function of the Cell Division Cycle

Wei-Ping Mei

Abstract

Ultraweak photon emission as well as biochemical and other parameters, DNA synthesis, and cell number were measured as a function of the cell division cycle in yeast cultures, *Saccharomyces cerevisiae*, synchronized with the α factor.

The investigations show that the self-emission of yeast follows a cell division cycle pattern with increases in the late S phase to the G2 phase. The spectrum of the emitted photons is continuous with a maximum in the UV- and blue-region.

This characteristic pattern of photon emission is obliterated only in the presence of the uncouplers of the electrontransport DNP, or FCCP (above a concentration of 10^{-5}M), and the protein synthesis inhibitor cycloheximide.

There is no influence on the photon emission pattern after addition of:

- the inhibitors of the respiratory chain, oligomycine(B),
- the mtDNA synthesis inhibitor of yeast, ethidium bromide.

All the results suggest that the ultraweak photon emission is associated with a biological function in which the cell nucleus may play a fundamental role. They support the electromagnetic hypothesis of cell differentiation[38,39] which states that the DNA acts as a photon store, emitting at least partially coherent light.

9.1 Introduction

Ultraweak photon emission from biological systems is now well investigated. The recent experimental and theoretical results are extensively reviewed by several investigators in this volume.

In this paper the results of investigations on synchronized yeast cultures *Saccharomyces cerevisiae* are present. In many ways yeast is an ideal organism for studies of cell growth and division. This organism progresses through a number of well defined morphological and genetic steps during cell division, giving a precise delineation[1–5].

In the pioneering work of Konev et al. (1966)[6] it was first shown that *Torula utilis*, synchronized with nutrient limitation, emit photons with two peaks in intensity during one generation time. Almost at the same time Stauff and Esser[7,8] found that with *Saccharomyces cerevisiae* ultraweak photon emission was detectable only in the presence of O_2. Quickenden et al. (1974, 1976, 1983)[9–11] investigated the dependence of cell growth on aerobic and anaerobic conditions in cultures of synchronized yeast *Saccharomyces cerevisiae*. Watanabe et al. (1980)[12] investigated the *bioluminescence* and cell growth of *Photobacterium phosphoreum* and found that the bioluminescence intensity per cell mass increased linearly at a rate of 1.5 times per doubling time.

However, up to now, nobody has examined the ultraweak photon emission as a function of the cell division cycle with sufficient accuracy. For this it is necessary to use a high synchronous culture. We achieve this with the pheromone α factor, which gives highly synchronization by blocking the cell cycle at the G1 phase[13–15].

9.2 Material and Methods

9.2.1 Cell culturing

The standard wild-type haploid strains of *Saccharomyces cerevisiae*, XMW-18A mating type **a** (lys 1, his 6, ssl 2-1) were obtained from *Dr. Duntze* at the University of Düsseldorf, F.R. Germany. The cells were grown aerobically in a 500ml flask, which was continuously agitated at 150rpm in the Junior orbit shaker (LAB-LINE Instruments, INC), at 30°C. The culture medium was MV-Medium, containing per Litre: 13.4g yeast nitrogen base w/o amino acids (Difco), 20g glucose, 20mg histidine, lysine and arginine, 10mg uracil, 5mg adenine, 0.1% yeast extract and 0.1% bactopeptone (Difco). Under these conditions the generation time is 95 to 100 min.

9.2.2 Synchronization of mating type cells

Before any measurement is carried out, the cells are inoculated from a colony taken from an agar plate. After 16 hours incubation of the culture, the stationary phase is reached at 10^8 cells/ml. Three ml from the stationary culture are transferred into 20ml new medium in another flask. After incubation for 2 hours, the α factor is added at a concentration of 4nM, and the culture is further incubated for 2 hours in order to synchronize the **a** mating type cells. The synchronized cells are washed with new medium (30°C) twice and recovered by centrifugation (500g). For the measurement of photon emission a cell concentration of about 5×10^7/ml is used.

9.2.3 The measurement of ultraweak photon emission

The equipment for the measurement of ultraweak photon emission is described in Chapter 6.

The measurements are always carried out with 15ml culture ($\sim$ 5×10^7/ml) in a $20\times20\times40$ mm^3 quartz cuvette under aerobic conditions. All additions of chemicals were performed before the measurements. The interval between measurements is 10 seconds. One experiment takes between two to three hrs. Data acquisition, and data processing are performed with PC. The whole software was developed by the Institute of Biophysical Cell Research in Kaiserslautern (FRG).

9.2.4 Determination of DNA synthesis

Two different methods were used to determine the DNA synthesis:

DNA content

After removal of the α factor, 50μl formalin (25%) were added to 200ml yeast suspension. For each point of measurement, 5ml cell culture were centrifuged in the cold at 4500rpm (Heraeus Sepatech) for 20 sec, washed once with 3ml distilled cold water, and hydrolysed in 4ml 1M $NaOH$ for 16h at room temperature. The hydrolysates were added to 4ml 1.5M $HClO_4$ at +4°C for 30min and centrifuged. The pellets were added to 0.1 – 0.2ml 0.5M NaOH for several minutes and again added to 2.5ml 1.5M $HClO_4$ at 70°C for 15min. Then 1.5ml colour reagent (4% diphenylamine, 0.01% paraldehyd in 100% vinegar) were added and left overnight. After centrifugation the supernatants were measured with Zeiss photometer at 600nm against the controls.

Incubation of ^{14}C-adenine

The methods are similar to those of *Bücking-Throm* et. al. (1974)[13].

9.2.5 Chemicals

The α factor, DNP*, FCCP†, oligomycin, ethidium bromide, cycloheximide and other chemicals were obtained from Sigma, Serva and Aldrich Co. F.R. Germany. ^{14}C-adenine was obtained from The Radiochemical Centre, Amersham, F.R. Germany.

9.3 Results

9.3.1 Self-emission and spectra of the yeast

Konev et al. (1966)[6] have shown that the ultraweak photon emission of yeast is actually correlated with the cell division cycle. Their investigation did not indicate at which cell division phase the two peaks are located. Our results show that there

*2,4-dinitrophenol

†carbony cyanide *p*-trifluoromethoxyphenylhydrazone

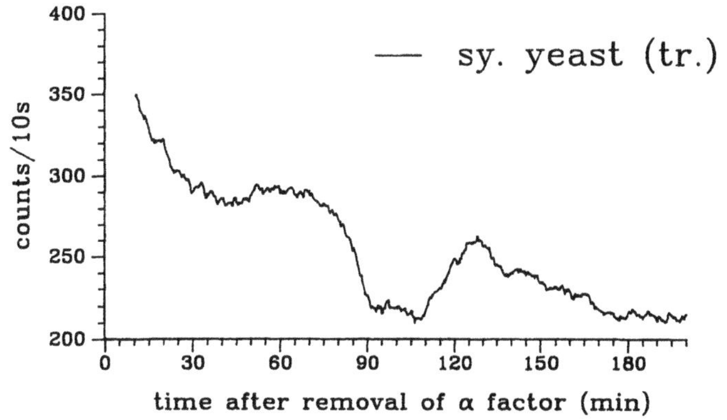

Figure 9.1: *The ultraweak photon emission of synchronized yeast* Saccharomyces cerevisiae *at 30°C in relation to the time after removal of the α factor. There are two division cycles and each peak corresponds to the late S phase to G2 phase.*

is only one peak in a generation, i.e., from late S phase to G2 phase (Fig. 9.1). The second peak is the second S phase and G2 phase of the second generation. Fig. 9.2 presents the average picture of change in the intensity of photon emission with time of the synchronized cell culture for 22 independent experiments. A qualitatively similar pattern was observed in all the individual experiments.

Because it is not possible to perform all procedures in the dark, it is inevitable that both the medium and cells were subject to a small amount of light irradiation, which leads to a decay dynamics (Fig. 9.2).

In order to avoid this disturbance and to obtain better results, the curve of measurement with cells is redrawn with the measurement of the medium alone subtracted at every point. This procedure significantly improves the pattern of photon emission in the cell division cycle (Fig. 9.3).

As is well recognized[16–18], it is impossible to make spectral measurements of selfemission with an interference filter or monochromator, due to the low intensity of the emission. But it is possible to make broad spectral measurements with a glass bandpassfilter (Schott, Mainz, F.R. Germany). With five different filters ranging from the UV to near infrared we find that photon emission has a continuous spectrum mainly in the UV- to blue-region (Fig 9.4). These results agree with the results of Quickenden et al. (1976)[10] and Konev et al. (1966)[6].

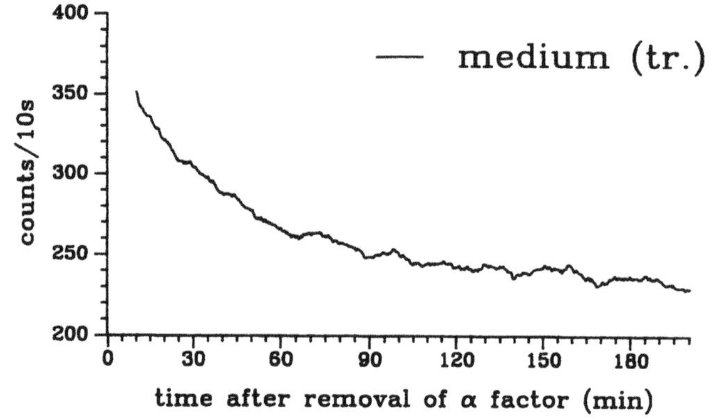

Figure 9.2: *The decay curve of photon emission from the MV-medium after the medium is placed in the measuring chamber. The other conditions are the same as the standard.*

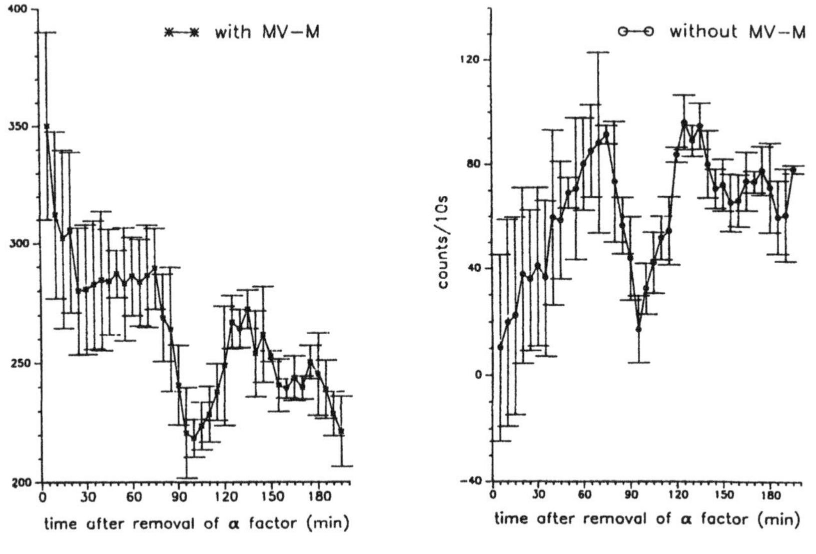

Figure 9.3: *The ultraweak photon emission of synchronized yeast* Saccharomyces cerevisiae *at 30°C in relation to the time after removal of the α factor.*

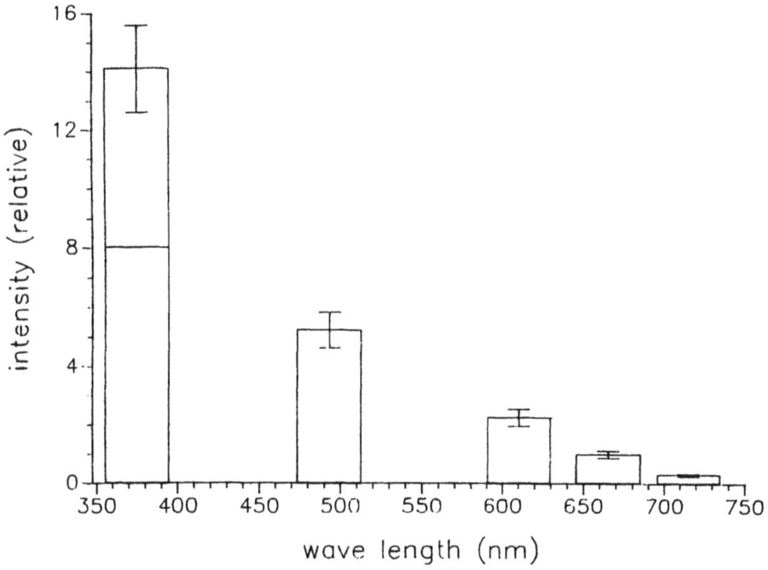

Figure 9.4: *The spectra of the photon emission from synchronized* Saccharomyces cerevisiae *in one generation time after removal of the α factor.*

9.3.2 DNA synthesis

The results confirmed that the first peak of photon emission is in the late S phase and decreases in the G2 phase. With two different methods, i.e. the measurement of DNA content and ^{14}C-adenine incorporation the almost identical results are obtained. That is, the peak of the DNA synthesis is located at 45 min after removal of the α factor. (Fig. 9.5).

9.3.3 Photon emission after the addition of uncouplers

It is well known that DNP and FCCP above the concentration of 10^{-5}M can uncouple the ATP electron transport chain in the cells[19–21]. The measurements with the uncouplers show clearly that they are capable of interrupting photon emission from yeast cells and obliterating the pattern of cell division cycle (Fig. 9.6, 9.7).

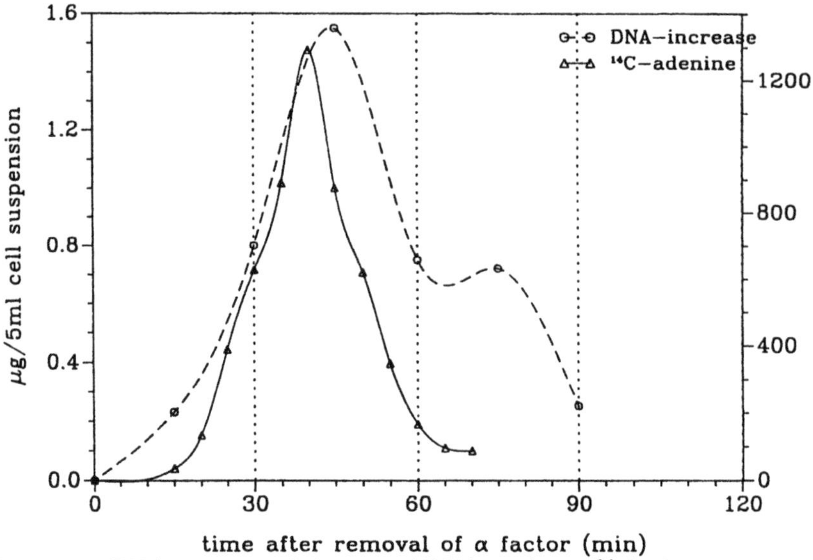

Figure 9.5: *DNA content and the rate of DNA synthesis (^{14}C-adenine incorporation) after removal of the α-factor as a function of the cell division cycle.*

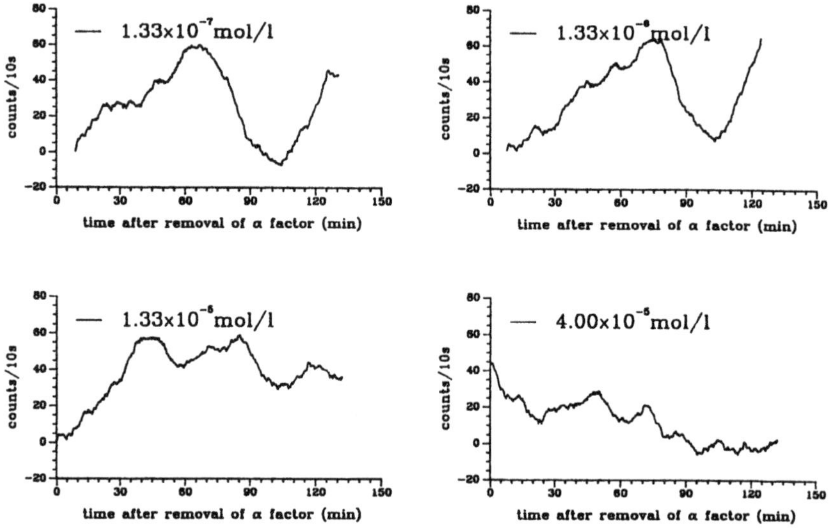

Figure 9.6: *The effect of DNP at different concentrations on the photon emission as a function of the cell division cycle of* Saccharomyces cerevisiae.

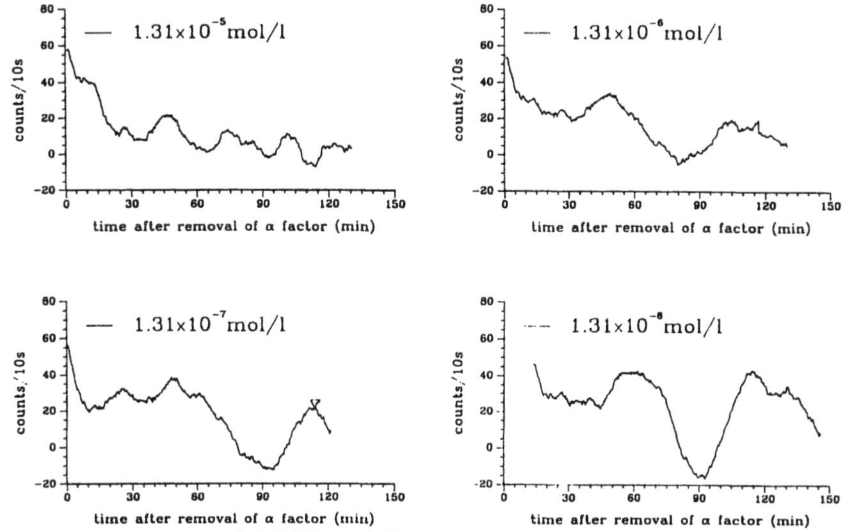

Figure 9.7: *The effect of FCCP at different concentrations on the photon emission as a function of the cell division cycle of* Saccharomyces cerevisiae.

9.3.4 Photon emission after the addition of respiratory chain inhibitors

Similarly, there are no changes in the pattern of photon emission in synchronized yeast cells after addition of respiratory inhibitors, oligomycin(B) at different concentrations (Fig. 9.8).

9.3.5 Photon emission after the addition of cycloheximide

It is known that if cytoplasmic protein synthesis is prevented by cycloheximide, the net synthesis of mitochondrial DNA and nuclear DNA synthesis on the other hand, continues for about one generation afterwards[2,3,22,23]. Our measurements (Fig. 9.9) show that the first peak (that is the first generation after the synchronization) still appeared, but not the second (that is the second generation after the synchronization).

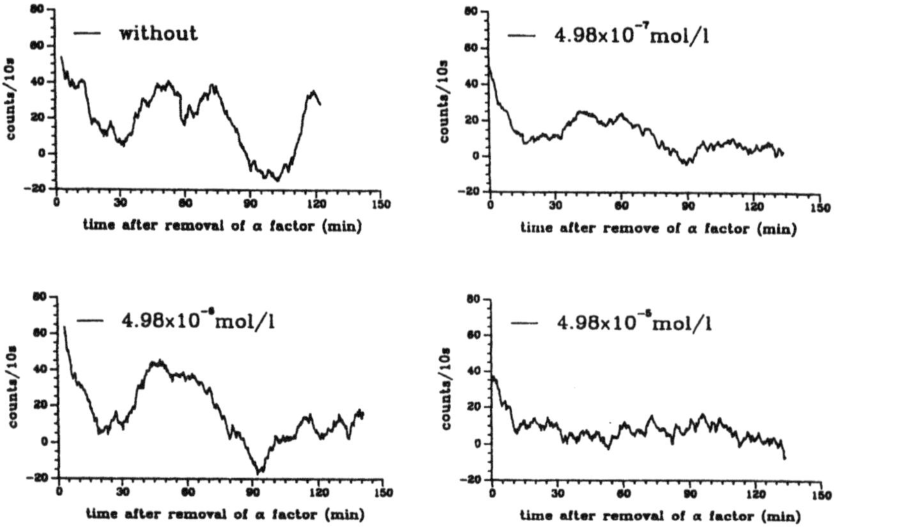

Figure 9.8: *Photon emission of synchronized* Saccharomyces cerevisiae *after removal of the α factor and the addition of oligomycin(B) in different concentrations as a function of cell division cycle.*

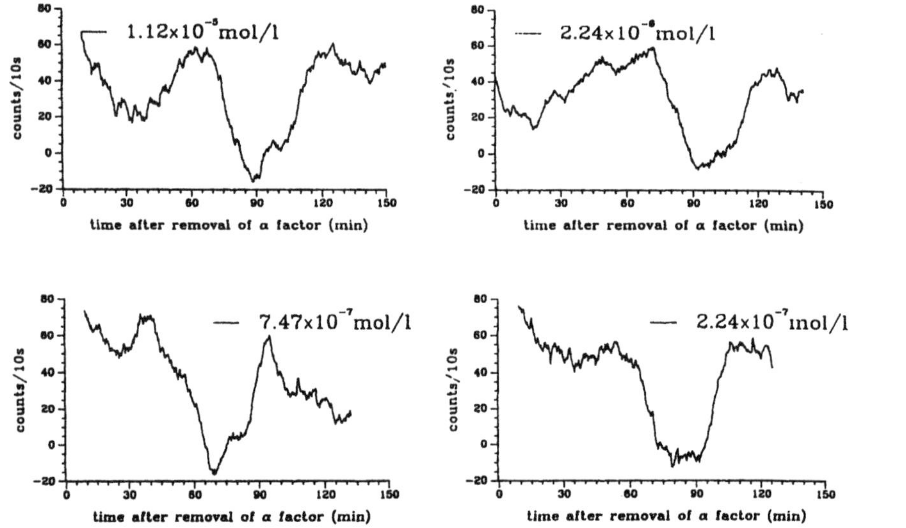

Figure 9.9: *Photon emission of synchronized* Saccharomyces cerevisiae *after removal of the α factor and the addition of cycloheximide in different concentrations as a function of the cell division cycle.*

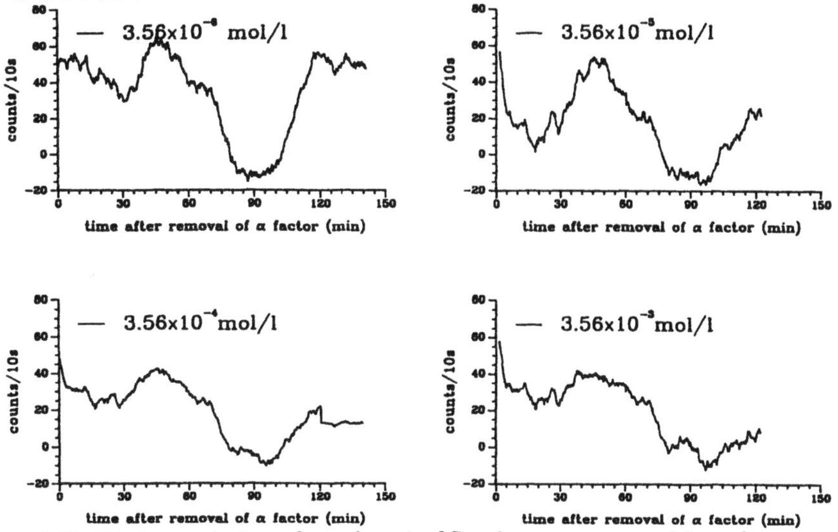

Figure 9.10: *Photon emission of synchronized* Saccharomyces cerevisiae *after removal of the α factor and the addition of ethidium bromide in different concentrations as a function of the cell division cycle.*

9.3.6 Photon emission after the addition of ethidium bromide

It has been shown that the ethidium bromide can inhibit the mitochondrial DNA synthesis but not the synthesis of nuclear DNA[24,25]. Our results show that the pattern of the photon emission is still unchanged, although very high concentrations of ethidium bromide are used (Fig. 9.10).

9.4 Discussion

The goal of this work is to confirm the correlation between the ultraweak photon emission and the cell division cycle. The results show that the photon emission is actually dependent on the cell cycle[6]. The peak in the late S phase to G2 phase indicates that photon emission reaches its highest level at the maximum of DNA synthesis. The late S phase is the time at which the mitotic spindle enlongates and one SPB (spindlebody) goes from the mother cell into the bud[26]. That is where the chromosome is beginning to separate. It is on account of this opening (or conformation change) of the DNA that the high intensity of the ultraweak photon emission occurs[27,28].

The experiment with uncouplers of DNP and FCCP[29] show that the uncouplers, i.e. the agents at appropriate concentrations inhibiting ATP formation, always lead to changes in the H^+ permeability[30]. This results show that the peaks in the late S phase to the G2 phase disappear when the uncouplers exceed the concentrations of 10^{-5}M. This means that the cells are unable to continue DNA sythesis and finish the SPB separations. By counting the cells we found that the *Saccharomyces cerevisiae* cannot double its number when the energy source ATP cannot be synthesized in the presence of the uncouplers.

On the contrary, the results with the inhibitors of mitochondrial oxidative phosphorylation show that there are no changes in the pattern of the photon emission from synchronized yeast cells. This means that there is no influence of mitochondrial oxidative phosphorylation on the photon emission. It suggests that in the nucleus there is some mechanism coupled to the ultraweak photon emission and supplying the cell with sufficient energy[31,32]. Ethidium bromide, which inhibits mitochondrial DNA synthesis[24,25], and the α factor, which inhibits only the nuclear DNA synthesis but not the mitochondrial DNA synthesis, RNA synthesis and protein synthesis[2,3], all these give no change in the pattern of the photon emission. That leads us to conclude that the mitochondrial DNA and the mitochondria do not significantly contribute to the photon emission[33].

It is clear that the ultraweak photon emission is connected with the activity of the cell nucleus[32] and that DNA plays a very important role[28,34–37,40], but further research is needed to investigate this problem even more.

9.5 Acknowledgments

I would like to thank Xing Xu, J. Eikemeyer and P. Pries for excellent technical assistance. I am grateful to Prof. F.A. Popp, Prof. J. Schole, Prof. E.-G. Niemann, Prof. W. Nagl and Dr. M.W. Ho for the discussion and critical reading. This investigation was partly supported by the Konrad-Adenauer-Foundation F.R. Germany.

9.6 References

1. S.G. Elliott and C.S. Mclaughlin, The yeast cell cycle: Coordination of growth and division rates. *Prog. Nucleic. Acid. Res. Mol. Bio.* **28** (1983) p. 143-176.

2. L.H. Hartwell, *Saccharomyces cerevisiae* cell cycle. *Bacteriological Rev.* **38** (1974) p. 164-198.

3. B.L.A. Carter, The control of cell division in *Saccharomyces cerevisiae*, in *The Cell Cycle*, ed. P.C. John (Cambridge University Press, 1981).

4. J. Hayles and P. Nurse, Cell cyle regulation in yeast. *J. Cell Sci. Suppl.* **4** (1986) p. 155-170.

5. J.R. Pringle and L.H. Hartwell, The *Saccharomyces cerevisiae* cell cycle, in *The molecular biology of the yeast* Saccharomyces cerevisiae, *Life cycle and inheritance*, eds. J.N. Strathern, E.W. Jones and J.R. Broach (Cold Spring Harbor Laboratory, 1981).

6. S.V. Konev, T.I. Lyskova and G.D. Niesenbaum, Very weak bioluminescence of cells in the ultraviolet region of the spectrum and its biological role. *Biophysics* **11** (1966) p. 410-413.

7. J. Stauff and R. Reske, *Naturwissenschaften* **51** (1964) p. 39.

8. A. Esser and J. Stauff, Lumineszenz von Hefe. *Z. Naturforsch.* **23b** (1968) p. 1554-1555.

9. T.I. Quickenden and S.S. Que Hee, Weak luminenscence from the yeast *Saccharomyces cerevisiae* and the existence of mitogenetic radiation. *Biochem. biophys. Res. Commun.* **60** (1974) p. 764-770.

10. T.I. Quickenden and S.S. Que Hee, The spectral distribution of the luminescence emited during growth of the yeast *Saccharomyces cerevisiae* and its relationship to mitogenetic rediation. *Photochem. Photobiol.* **23** (1976) p. 201-204.

11. T.I. Quickenden and R.N. Tilbury, Growth dependent luminescence from cultures of normal and respiratory deficient *Saccharomyces cerevisiae. Photochem. Photobiol.* **37** (1983) p. 344-337.

12. T. Watanabe and T. Nakamura, Bioluminescence and cell growth of *Photobacterium phosphoreum. J. Biochem.* **88** (1980) p. 815-817.

13. E. Bücking-Throm, W. Duntze, L.H. Hartwell and T.R. Manney, Reversible arrest of haploid cells at the initiation of DNA synthesis by a diffusible sex factor. *Exp. Cell Res.* **76** (1973) p. 99-110.

14. W. Duntze, D. Stötzler, E. Büching-Throm and S. Kalbitzer, Purification and paratial characterization of α-factor, a mating-type specific inhibitor of cell reproduction from *Saccharomyces cerevisiae*. *Eur. J. Biochem.* **35** (1973) p. 357-365.

15. J. Thorner, Pheromonal regulation of development in *Saccharomyces cerevisiae*, in *The molecular biology of the yeast* Saccharomyces cerevisiae, *Life cycle and inheritance*, eds. J.N. Strathern, E.W. Jones and J.R. Broach (Cold Spring Harbor Laboratory, 1981).

16. H. Inaba, Super-high sensitivity systems for detection and spectral analysis of ultraweak photon emission from biological cells and tissues. *Experientia* **44** (1988) p. 550-558.

17. M.E. Murphy and H. Sies, Visible-rang low-level chemiluminescence in biological systems. *Methods in Enzymology* **186** (1990) p. 1021-1038.

18. A.K. Campbell, *Chemiluminescence: Principles and applications in biology and medicine* (Ellis Horwood Ltd., Chichester(England), 1989).

19. P.D. Boyer, L. Gross and W. Momsen, A new concept for energy coupling in oxidative phosphorylation based on a molecular explanation of the oxygen exchange reactions. *Proc. Natl. Acad. Sci. USA* **70** (1973) p. 2837-2839.

20. A.W. Linnane, Aspects of the biosynthesis of the mitochondria of *Saccharomyces cerevisiae*, in *Oxidases and related redox systems* **Vol. 2**, eds. T.E. King, H.S. Mason and M. Moorison (John Willey & Sons. Inc. New York-London-Sydney, 1965).

21. D. Shiuan and S.I. Tu, Fluorescent labeling of mitoplast membrane: Effect of oxidative phosphorylation uncouplers. *Biochemistry* **17** (1978) p. 2249-2252.

22. L. Popolo, M. Vanoni and L. Albergghina, Control of the yeast cell cycle by protein synthesis. *Exp. Cell Res.* **142** (1982) p. 69-78.

23. S.A. Moore, Kinetic evidence for a critical rate of protein synthesis in the *Saccharomyces cerevisiae* yeast cell cycle. *J. Biol. Chem.* **20** (1988) p. 9674-9681.

24. M.T. Künzi, M.A. Tingle and H. O. Halvorson, Sporulation of *Saccharomyces cerevisiae* in the abesence of a functional mitochondrial genome. *J. Bact.* **117** (1974) p. 80-88.

25. S.F. Cottrell, A correlation between ethidium bromide uptake and petite mutagenesis during the yeast cell cycle. *Mutation Res.* **105** (1982) p. 59-63.

26. B. Byers, Cytology of the yeast life cycle, in *The molecular biology of the yeast* Saccharomyces cerevisiae, *Life cycle and inheritance*, eds. J.N. Strathern, E.W. Jones and J.R. Broach (Cold Spring Harbor Laboratory, 1981).

27. M. Rattemeyer, F.A. Popp and W. Nagl, Evidence of photon emission from DNA in living systems. *Naturwissenschaften* **68** (1981) p. 572-573.

28. W.B. Chwirot, Ultraweak photon emission and anther meiotic cycle in Larix europaea (Experimental investigation of Nagl and Popp's electromagnetic model of differentiation). *Experientia* **44** (1988) p. 594-598.

29. W.G. Hanstein, Uncoupling of oxidative phosphorylation. *TIBS* **1** (1976) p. 65-67.

30. P.P. Puglisi and A.A. Algeri, Interaction of biological mitochondrial protein synthesis on the regulation of gene activity in *Saccharomyces cerevisiae*, in *The Biological Gnesis of Mitochondria*, eds. A.M. Kroon and C. Saccone (Academic Press. Inc., 1974).

31. J. Schole, Theory of metabolic regulation including hormonal effects on the molecular level. *J. theor. Biol.* **96** (1982) p. 579-615.

32. J. Schole and W. Lutz, *Regulationskrankheit: Versuch einer fachübergreifenden Analyse* (Ferdinand Enke Verlag Stuttgart, 1988).

33. J. Stauff and J. Ostrowski, Chemilumineszenz von Mitochondrien. *Z. Naturforschg.* **22b** (1967) p. 734-740.

34. F.A. Popp, K.H. Li, W.P. Mei, M. Galle and R. Neurohr, Physical aspects of biophotons. *Experientia* **44** (1988) p. 576-585.

35. R. Van Wijk and D.H. Schamhart, Regulatory aspects of low intensity photon emission. *Experientia* **44** (1988) p. 586-593.

36. W.P. Mei, Ultraschwache Photonenemission bei synchronisierten Hefezellen (*Saccharomyces cerevisiae*) in Abhängigkeit vom Zellteiliungszyklus. *Dissertation* (Universität Hannover, 1991).

37. W. Nagl and F.A. Popp, A physical (electromagnetic) model of differentiation, 1 Basic considerations. *Cytobios* **37** (1983) p. 45-62.

38. F.A. Popp and W. Nagl, A physical (electromagnetic) model of differentiation, **2** Applications and examples. *Cytobios* **37** (1983) p. 71-83.

39. H.J. Niggli and F.A. Popp, Ultraweak-photon re-emission in cloudman S91 mouse melanoma cells: Evidence for Chromatin as Source of Radiation (to be published).

Chapter 10

Ultraweak Luminescence Studies of Microsporogenesis in Larch

Barbara W. Chwirot

Abstract

This paper presents a summary of the first complete investigation of the ultraweak luminescence of living systems carried out for the cells undergoing a cell cycle. The material under study were fractions of generative cells of larch being in the same, well determined stages of a male meiosis and pollen grain formation as in anthers of *Larix europaea Mill.*

The separation procedure employed in this work enables one to obtain highly homogenous cell fractions at a minimum perturbation of the physiological state of the cells. This feature is especially important because at a well known, high sensitivity of the ultraweak luminescence it allows for data to be obtained about every factor capable of influencing homeostasis of a living system. This data can be considered as real characteristics of the system of interest.

10.1 Introduction

This paper presents the results of the research initiated in our group in 1983 through which we have tried to find and, if possible, to characterise a relationship between properties of the so-called ultraweak luminescence (ultraweak photon emission) of living systems and processes of cell division and to some extent also of cell differentiation. The two main goals of our work have been firstly, to investigate possibilities for using the measurements of the characteristics of the light emitted by cells as a new tool for studies in the field of cell division and differentiation and, secondly, to attempt an experimental verification of some of the predictions of the model of differentiation published by Nagl and Popp in 1983[1,2]. The authors of the model assign a crucial role in the processes of differentiation to the mechanisms of ultraweak luminescence and to photons.

As the general problems related to the ultraweak luminescence of living systems and both to Nagl and Popp's model and its theoretical background are dealt with extensively in other chapters of this book we shall not discuss them here. It should be noted, however, that neither the problem of the possible sources of the observed emissions nor the question of a possible biological function of the processes leading to a generation of excited molecular states and of photons arising at their decay have yet been solved and both are still subject to intensive research and discussion.

There is, however, a large body of experimental results collected up to now confirming the hypothesis that the ultraweak luminescence can be an important source of information on the physiological state of organisms and on some features of biological processes taking place in living systems. It has been shown, for instance, that in yeast both the intensity and the spectral composition of the light emitted by the cells depend significantly on their division activity[3–7].

Similar results were also obtained for bacteria[8] and plant cells[9,10]. It was also shown for nerve as well as for muscle cells that their activation is accompanied by an increase in the ultraweak emission intensity[11,12] and that the correlation between the light emission and phagocytosis is so profound that it could be used as a direct measure of phagocytosis itself and of the influence of different chemical compounds on this process[13].

There are groups investigating the problem of possible optical interactions in biological systems and reports on the existence or absence of the so-called mitogenic[14,15] and cytopathic effects[16,17] or communications about optical interactions between separated populations of mitochondria[18,19] can be found. Related to that field are also experiments on an optical stimulation of cell divisions[20] and on a regulative action of light on selected metabolical processes such as photoactivation[21] or photoinactivation[22] of enzymes.

Many data point to the existence of significant correlations between characteristics of the ultraweak photon emission and fundamental biological processes as proliferation and death of cells, changes of metabolical processes (especially oxidative ones) and to a very high sensitivity of responses, often nonlinear, of those characteristics to the action on biological systems of external stimuli such as temperature, oxygen and other chemical and physical factors[24–32]. In some cases it was found that the ultraweak luminescence was so clearly related to the state of organisms – tissues – cells that attempts have been made to apply its measurements to medical diagnostics, food quality studies and the monitoring of environmental pollution (see for instance[31–38] and also the other chapters of this book and the references therein).

A significant advantage of the ultraweak luminescence studies is that, contrary to most of the currently used techniques, they do not require application of perturbing procedures (fixatives, dyes etc.) to the biological system under study and allow gaining of information about the object just by extensive studies of its natural activity, i.e. its ultraweak luminescence.

The extremely low intensities and broad spectral range of the observed emissions have been considered by many authors as evidence supporting the opinion that the ultraweak luminescence is a phenomenon of an accidental origin and of no importance to organisms. A representative example of such an approach is the so-called "theory of biological imperfections" sometimes referred to as "leakage theory"[27,28].

Such an approach to the question of ultraweak luminescence is, however, far from being obvious and well justified. Is it really so that the so-called "imperfections" occuring often in the same periods of life of a given organism are indeed only an expression of its imperfection? A significant support to an altenative view has been provided by a relatively new and dynamically developing field of the "photochemistry without light"[39,40].

It has been shown that the excited states of many biomolecules can not only be generated in living organisms via efficient and controlled enzymatic reactions but there is also evidence that the enzyme catalysing the reaction can also protect the excited molecular state against its noncontrolled quenching in accidental collisions with, for instance, cellular oxygen[40,41]. It was also demonstrated that such systems are capable of triggering and controlling many processes such as activation in darkness of enzymes which are normally subject to photoactivation[42], modifications of DNA of a potentially mutagenic character[43,44], transformation of phytochrome[46], conversion of provitamine D[39] or destruction of tryptophane components of proteins[41].

Also current estimates of threshold values of the light intensity or of the irradiation dose required for a response to a light stimulus to appear are much lower than those reported not long ago. A most spectacular result was the demonstration that a movement response could be triggered in *Chlamydomonas reinhardtii* by absorption of only one photon[45]. The irradiation dose necessary for inducing the so-called very low fluence (VLF) responses in plants is estimated to be of the order of 10^{12} - 10^{15} photons/m^2 [46,47] and, for instance, in *Phycomyces blakesleanus* a response can be observed even at a level of 10^{10} photons/$(m^2 \cdot s)$[48] (a photon flux corresponding to conditions of a sunny day is on the order of ca. 10^{17} photons/$(m^2 \cdot s)$. The values given above, if normalized to an area of $1\mu m^2$ which can be considered typical for a cellular scale, correspond to doses from 0.01 to 1000 photons.

Typically observed levels of the ultraweak luminescence are between 10^5 and 10^7 photons/$(m^2 \cdot s)$. One should, however, bear in mind that those values are characteristic of light escaping from the biological system. It is a well known fact that light is strongly scattered and absorbed in cellular systems and there are estimates that for a given measured intensity of the ultraweak luminescence the corresponding intensity of the light field within the organism may even be two orders of magnitude higher[29,49]. Thus, it seems justified to assume that, at least in some cases, the intensity of the ultraweak luminescence may be sufficient to influence some of the cellular processes.

Additionally, studies of the optical properties of biological systems indicate that light of appropriate wavelengths can be subject to active filtration by cellular structures and be carried over significant distances inside organisms. It has also been suggested that besides the usual scattering in living organisms, processes like directional reflection, focusing of light on some organella and even light piping both at the level of tissues and cellular structures may also be taking place[46,50,51].

It is also well established that the influence of light stimuli can be observed at all levels of organisation of organisms. At the level of the genetic apparatus (photodamage and photorepair of DNA, regulation of gene expression at levels of transcription and translation), at the level of changes of enzymatic activities, of the permeability of cell membranes, the activities of cells and tissues and also at the level of whole organisms.

Up to now a majority of the ultraweak luminescence studies have been carried out using either the whole biological objects, for example seeds or seedlings, or poorly defined systems such as roots, leaves or nonsychronised cell cultures as the material[25–32]. Only in a few cases attempts have been made to study sychronized cultures of yeast cells or bacteria[4–6,8], but it is well known that synchronization degrees which can be achieved for those cells are not very high and the synchronization itself is rather a short term event. On the other hand, both to investigate the possible relation of the properties of the ultraweak luminescence to the processes of cell division and differentiation as well as to try an experimental verification of the predictions of Nagl and Popp's theory it is of primary importance that the results are reproducible and obtained for a relatively simple, well determined and characterised biological system.

Such a natural biological model system is undoubtedly the cell cycle, especially meiosis. The meiotic cycle differs from mitosis in that it always represents a cycle of differentiating divisions resulting in qualitatively new haploid generative cells, very different from their maternal cell. The research we shall discuss in this paper was aimed just at an investigation of several features of the phenomenon of the ultraweak photon emission studied for the first time on cells being in successive, precisely determined stages of the cell cycle. More specifically, of the cycle of meiotic divisions and pollen grain formation in the anthers of *Larix europaea Mill.*

The material for the studies was obtained using the earlier elaborated method of acquirement of pure and homogenous fractions of cells of interest[52]. It is noteworthy that the microsporogenesis in larch lasts a few months and is one of the longest in plants. This unique longevity of the process makes it possible to study the cells in many successive stages of this cycle despite the fact that two to three days are required to obtain one sample. At the same time, however, it sets very strict requirements on the long-term (in the order of months) stability of the measuring apparatus.

The same experiments repeated in successive years showed a good reproducibility of the results. This fact not only demonstrates that adequate precautions were undertaken to ensure the long term stability of the system but, what is more important, it also indicates that the observed characteristics of the ultraweak luminescence are indeed related to the meiosis process and are not a result of, for example, using material collected from different trees or of differences in climatic conditions at the time of collecting cells in different developmental stages.

10.2 Materials and Methods

The materials for study were the intact male cones of *Larix europea Mill.* and fractions of microsporocytes in successive stages of microsporogenesis. The cones were collected directly from the trees growing under natural conditions, usually several times over a period of a few months, from November to April.

The isolation of cones and microsporocytes has been performed in a very weak green light. The method of acquirement of fractions of microsporocytes in successive stages of meiosis[52] consists of disrupting the structure of anther walls and squeezing the material through two nylon sieves (300-400 μm and 60 μm mesh) followed by a series of steps of differential centrifugation with non-linear gradients of sucrose density. Several schemes of centrifugation involving application of up to four different gradient combinations were elaborated and allowed for a separation of the generative cells at various, sometimes very close, stages of their development (Fig. 10.1). It was possible, for instance, to separate microsporocytes in four phases of the diplotene stage.

The separation procedure resulted in fractions of cells suspended in a 0.37 M solution of mannitol in Tris H_2SO_4 buffer at pH 7.2. This solution showed a very weak luminescence comparable to that of the ultraweak photon emission of the cells of interest. Since such an emission was not observed for water solutions of mannitol[53,54] the separation of cells was always followed by washing the microsporocytes three times with such a solution. The cells were then condensed by a centrifugation (5 min. at 1,500g) and placed in quartz cuvettes. Before each measurement the cuvettes were stored for 2 h in a dark chamber at +4°C.

The concentrations of cells of the microsporocyte fractions were measured in Thoma's chamber. The morphology of the microsporocytes was unchanged after the measurements and they were still capable of plasmolysis.

The measurements of the luminescence of the cones were done using samples consisting of about 80 cones devoid of the covering scales and cone basement. Before measurement the cuvettes containing the cones were stored for 1 h in darkness at +4°C.

For the details of the enzymatic assays the reader is referred to the original publications cited in the results and discussion section.

The apparatus used in the experiments for measurements of the intensity of the ultraweak luminescence has a configuration typical for a single photon counting technique and since there is a chapter in this book devoted purely to the methodology of such measurements the technical details will not be discussed here. It should be noted, however, that the apparatus has been significantly modified since we embarked on the ultraweak luminescence studies. In its final version it allows for computer controlled measurements with the use of two photomultipliers detecting light propagating in mutually perpendicular directions, and for a constant monitoring of dark count rate, temperature of the measuring chamber, positions of shutters etc.

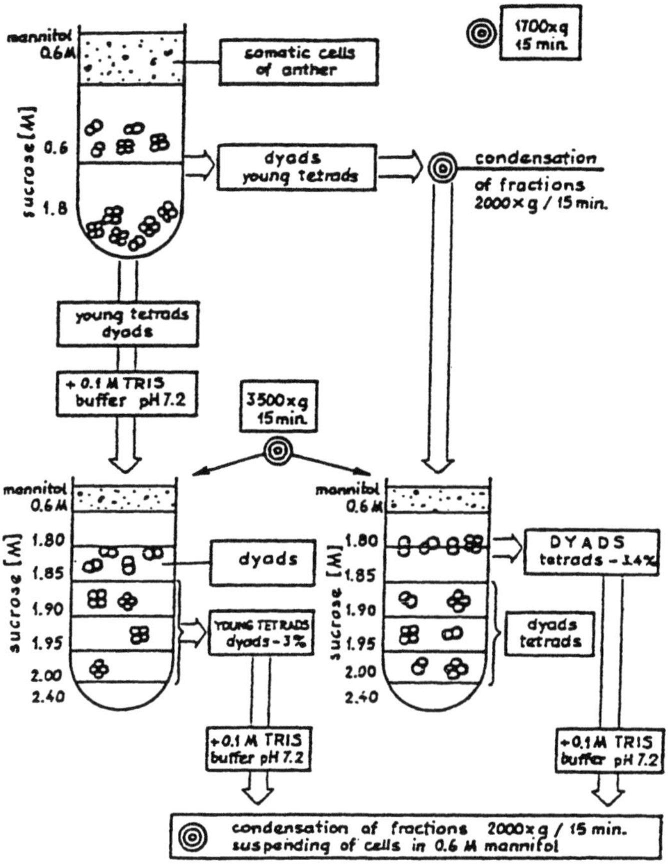

Figure 10.1: *Schematic representation of a procedure for separating the fractions of the generative cells in stages of a dyad and of a young tetrad.*

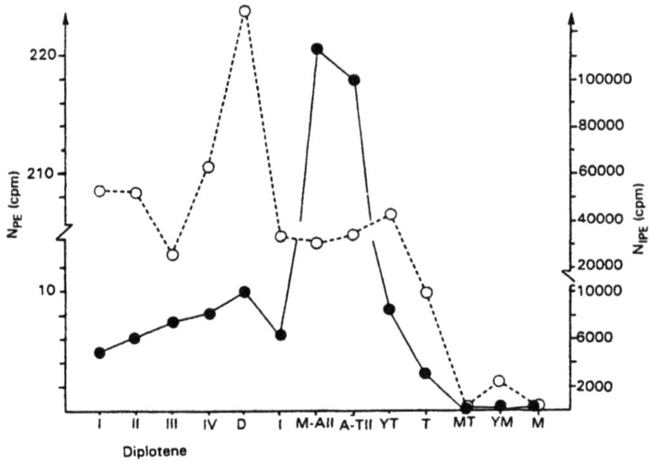

Figure 10.2: *Ultraweak (N_{PE}, •) and induced (N_{IPE}, ○) photon emission from microsporocytes during microsporogenesis in larch. D, diakinesis; I, interkinesis; M-AII, metaphase-anaphase of the second meiotic division; A-TII, anaphase-telophase of the secound meiotic disision; YT, young tetrad; T, tetrad; MT, mature tetrad; YM, young microspore; M, microspore. The measured intensity of the luminescence was normalized to the constant concentration of cells equal to 2.5 $\times 10^6$ cells/ml.*

The photomultiplier tubes currently used are EMI 9558QB and all the optics are made of quartz to allow for measurements in the UV region.

10.3 Results and Discussion

Our first experiments[53] were devoted to measurements of the intensity of the ultraweak and white-light-induced luminescence of the microsporocytes of larch during their development. They demonstrated that these cells do indeed emit light. The emission is located somewhere between 200 and 700 nm (determined by the spectral sensitivity of the photomultiplier) and changes in the intensity of the ultraweak luminescence of the microsporocytes are of a clearly different character than those observed for the whole male inflorescences (Fig. 10.2) (the latter composed of many different cells and tissues, were considered a control sample material).

The same lack of similarity was also found for the white-light-induced luminescence (Fig. 10.3) both in the case of the ultraweak and of the induced luminescence. The

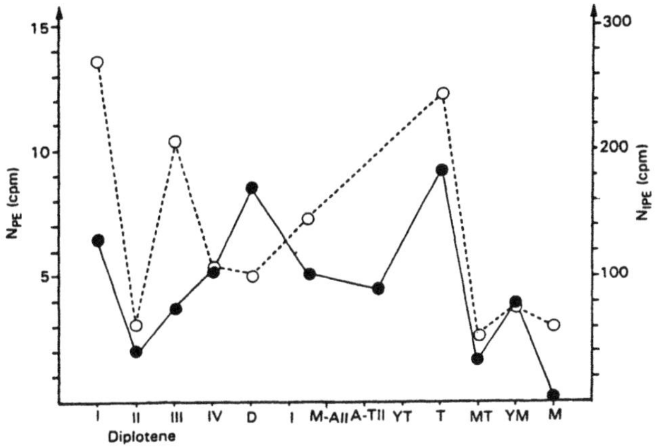

Figure 10.3: *Ultraweak (N_{PE}, •) and induced (N_{IPE}, ○) photon emission from the cones during microsporogenesis in larch. The measured intensities of the luminescence were normalized to one cone.*

range of the observed variations in the intensity was much larger than the experimental errors which for the applied experimental procedures were always smaller than 5%. Those results as well as the other data obtained during the first experiments encouraged us to continue the study.

The high levels of the white-light-induced luminescence, two or three orders of magnitude higher than those of the natural ultraweak luminescence, and its long decay times have been postulated by Nagl and Popp's model of differentiation as a consequence of the so-called "active photon storage" in biological systems.

However, crucial experimental evidence supporting this concept would be, as the authors also suggest, a demonstration of the existence of a coherent component in the light emitted by the organisms. On the other hand, according to the model, the light emitted by a biological system represents, to a large extent, chaotic losses of the biophoton field and thus a contribution of the coherent component to the ultraweak luminescence must necessarily be very small.

Additionally, the very low intensity of this luminescence and its multimode make direct measurements of its degree of coherence very difficult, if not impossible. The authors suggested, however, that the existence of mechanisms leading to a coherence of the ultraweak luminescence should be reflected in a specific decay behaviour of the induced luminescence, namely one described by behaviour hyperbolic rather than a multiexponential function. It also follows from the theoretical considerations of the authors that in some cases such decays may display oscillations of the intensity overlapping with the decay.

Thus, in the next phase of our studies we made an attempt to determine the character of the decays of the white-light-induced luminescence of the microsporocytes. The whole male cones of larch were again considered a control material (see Fig. 10.4)[56]. A very surprising result of this analysis was that for eight of the thirteen stages of development investigated, the decays could be classified with a high accuracy as hyperbolic, but in the case of the whole cones the decays could not be classified either as hyperbolic or as two-exponential (Tab. 10.1).

According to the theoretical analysis of Popp and Li[55], the occurrence of the hyperbolic decays of the white-light-induced luminescence can be considered a consequence of processes of a coherent scattering of light taking place in the cells under study and thus the fact that such decays could be observed only for developmentally homogenous fractions of the microsporocytes agrees well with the predictions of Nagl and Popp's model. It should be noted, however, that the decays classified as hyperbolic could also result from a superposition of many different exponential decays. Since such a possibility cannot be thoroughly tested at the present stage of experiments. The fact of the observation of the hyperbolic decays can only be interpreted as a first very clear indication, but only an indication, that a coherent light emission may be taking place in biological systems.

The experiments devoted to the characterisation of the decays of the induced luminescence were continued in the next season (larch trees providing the material only grow in natural conditions). To narrow down the field for the possible interpretation of the data it was decided in that phase to perform measurements of the luminescence induced by the irradiation with a white light devoid of its UV component and with a monochromatic light of two different wavelengths (525 and 660 nm)[57].

The developmental changes of the luminescence intensity observed in the case of the white-light (without UV) excitation were similar to those obtained when white light of a complete spectrum was used but both were considerably different than those of the monochromatic light excitation. The most significant result of those experiments was, however, the first observation of clear oscillations of the intensity during the decay of the monochromatic-light-induced luminescence (Fig. 10.5).

Local maxima of the induced luminescence occurring immediately after the end of the irradiation were previously observed after a monochromatic light excitation of the etiolated seedlings of cucumber and *Bryophyllum*[26]. Such local maxima could

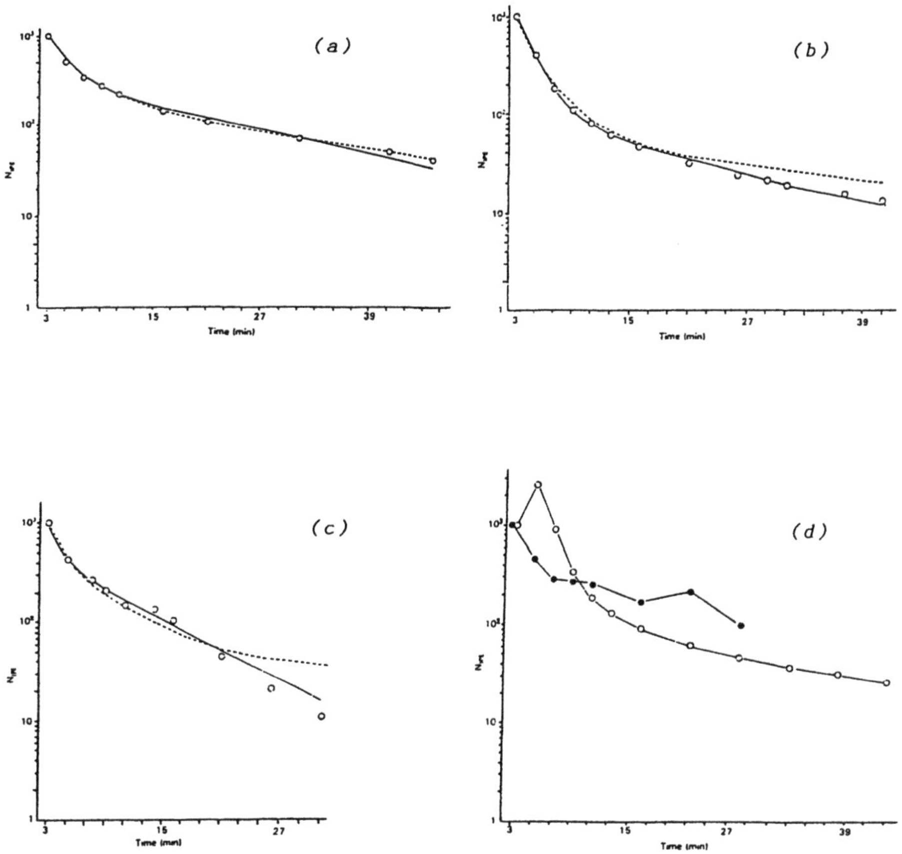

Figure 10.4: *Decay of IPE from the microsporocytes in: (a) diploten III stage (broken line – hyperbolic, solid line – two-exponential function;* ○ *– measured values normalized to 1000 at the beginning of a measurement); (b) diakinesis; (c) stage of mature tetrad; (d) Other examples of the decays which could not be described by either the hyperbolic or two-exponential functions (*○*, microsporocytes in the stage of metaphase-anaphase of the second meiotic division;* ●*, whole cones at the time of the diplotene stage).*

Table 10.1: Types of decays of the white-light-induced luminescence identified for the larch microsporocytes and whole cones in successive stages of microsporogenesis. The functions used for the least squares fitting of the observed decays were the two-exponential: $I(t) = c_1 e^{-c_2 t} + c_3 e^{-c_4 t}$ and the hyperbolic: $I(t) = A(t + B)^{-\beta}$.The estimated errors of the time constants determined were less than or equal to 4% for the decays classified as hyperbolic and 5% for the two-exponential.

Stage	Microsporocytes	Whole cones
Diplotene I	hyperbolic	?
Diplotene II	hyperbolic	?
Diplotene III	hyperbolic	?
Diplotene IV	hyperbolic	?
Diakinesis	two-exponential	?
Interkinesis	hyperbolic	?
Metaphase-anaphase of the second meiotic division	?	?
Anaphase-telophase of the second meiotic division	hyperbolic	?
Young tetrad	hyperbolic	?
Tetrad	?	?
Mature tetrad	?	?
Young microspore	?	?
Microspore	hyperbolic	?

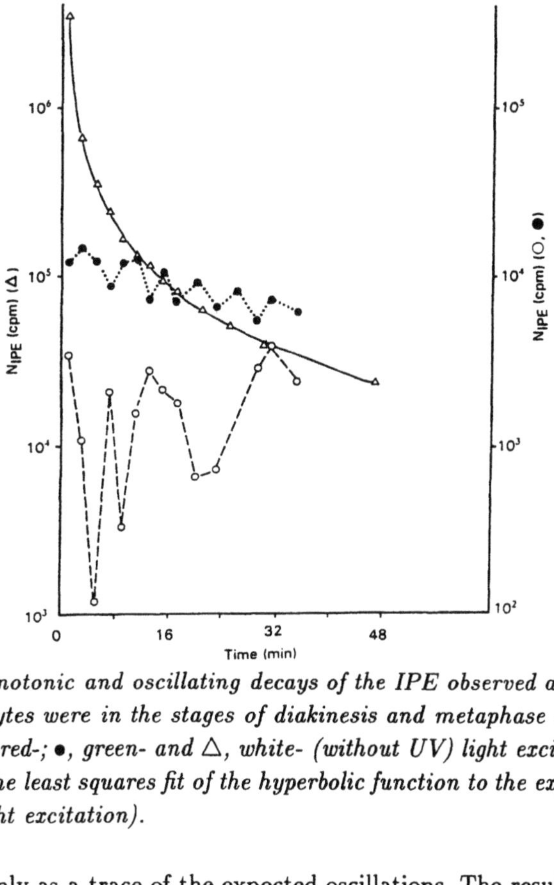

Figure 10.5: *Monotonic and oscillating decays of the IPE observed at the time when the microsporocytes were in the stages of diakinesis and metaphase of the first meiotic division (○, red-; ●, green- and △, white- (without UV) light excitation; the solid line represents the least squares fit of the hyperbolic function to the experimental data for the white-light excitation).*

be interpreted only as a trace of the expected oscillations. The results obtained for the microsporocytes of larch give a clear picture of long term oscillations and seem to demonstrate the existence of a dependence of both the amplitude and the period of the modulation of the decay on the developmental stage of the cells and on the spectral compositon of the exciting light.

Such a dependence might account for the fact that similar oscillations had not been observed earlier either in the case of irradiating the cells with a white light or in experiments with monochromatic light irradiation applied to whole organisms composed of cells of different tissues and being at different stages of differentiation or to nonsynchronous cell cultures.

The developmental variations in the intensity of both the ultraweak and induced luminescence of the microsporocytes are, as shown in[53], significantly different. Additionally, there is a strong positive correlation between the changes of the ultraweak luminescence of the microsporocytes and of the induced luminescence of the whole buds. An interesting possibility is that at least at some stages the anther wall cells could be a primary or a secondary (transmitting transformed light from the environment) source of photons regulating the development of the microsporocytes.

Studies of the problem of light conditions inside plants have been carried out relatively recently and the literature on the subject is very sparse. In such a situation, to find out what are the light conditions the microsporocytes are subject to during their development, it was necessary to perform measurements of the external transmittance of scales covering the male inflorescences and of the leaf buds of larch[58]. Those measurements were also stimulated by the experiments of Giertych et al.[59] who had found evidence that light can directly influence the development of flower buds in pine and spruce.

An interesting result of our measurements for the larch buds was that they demonstrated very clearly that the external transmittance of the covering scales undergoes significant developmental changes. Moreover, those variations are different in the two kinds of buds investigated (Fig. 10.6, 10.7). The character of the observed differences shows that the light conditions inside a developing flower bud may indeed be important for the course of the microsporogenesis.

Thus, the scales covering the flower buds not only provide a mechanical protection for the anthers developing at a time of very difficult climatic conditions (winter) but also significantly and in a dynamic manner modify the intensity and spectral composition of the light they transmit. At this stage it is impossible to say whether the light reaching the inside of the flower bud interacts primarily with the anther wall cells or influences the developing microsporocytes directly. The problem could be solved only by direct characterisation of the light measured inside the inflorescences. Such measurements, however, are not yet feasible and therefore both this problem and the questions regarding a receptor for the regulating light stimuli are still open to further investigation.

As already mentioned, the possible sources of light emitted by cells in processes of the ultraweak luminescence still have to be identified. Usually, the emission spectra are investigated to gain such information but due to the extremely low intensities involved such an approach can rarely be adopted in studies of ultraweak luminescence. Also in the case of the microsporocytes it was not possible to collect such data.

The essential feature of our studies is that the measurements are made with the cells

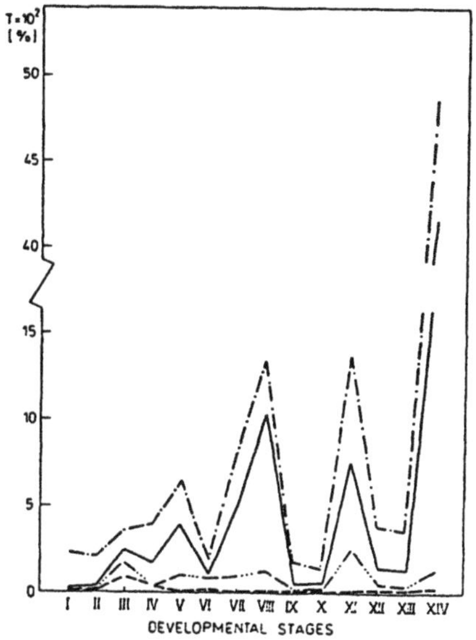

Figure 10.6: *Variations of the external transmittance of male inflorescence scales of larch for - - - - - 420 to 490 nm, — · · · — 500 to 590 nm, — — — — — 600 to 690 nm and — · — — · — 700 to 780 nm during the development of the buds; diplotene stages: I-13.XII., II-19.XII, III-3.I., IV-10.I., V-17.I., VI-23.I., VII-31.I., VIII-7.II (second contraction of chromosomes as defined by [67]), IX-14.II., X-20.II., XI-25.II. (diplotene - diakinesis), diakinesis: XII-4.III., metaphase I - anaphase I - telophase I - interkinesis:XIII-15.III., tetrad: XIV-19.III., tetrad-microspore: XV - 25.III.*

being in exactly the same well determined stages of development. To obtain a fraction of such cells of a volume of ca. 1 ml about two days are needed. A determination of the intensity of their luminescence with a reasonable accuracy (usually with a relative error of 5% at a 95% confidence level) requires from one to several hours of a continuous measurement.

The measurements of the emission in selected, even wide, spectral bands would have to be much longer and would of course also require a proportionally increased amount of the material. Additionally, it would be impossible to perform the measurements in a sufficiently short time interval to ensure the full developmental comparability of the cells investigated. It was, however, possible to determine for the same fractions of the microsporocytes the intensities of their total ultraweak luminescence and of the luminescence in the visible part of the spectrum[60]. Those experiments demonstrated

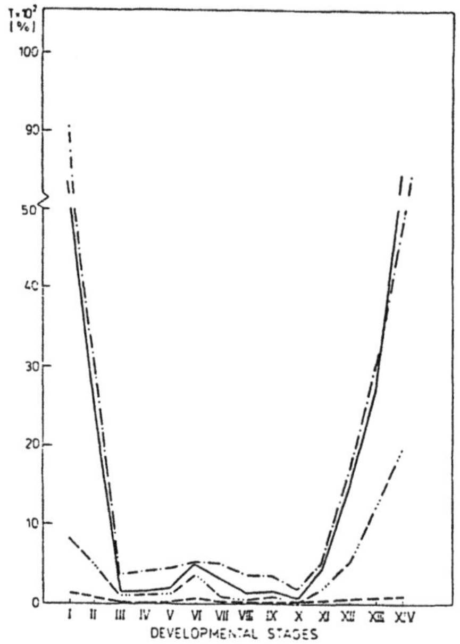

Figure 10.7: *Developmental variations of external transmittance of the leaf bud scales (all the designations as in Fig. 10.6).*

for the first time, using a physical method of detection, that the plant cells do indeed emit UV light below 310 nm.

It was also found that both regarding absolute intensity and relative participation in the total ultraweak luminescence of the microsporocytes, the UV component exhibits a clear dependence on the development stage of the cells (Fig. 10.8). The demonstration of the UV emission from the microsporocytes together with the earlier reports of such an emission from lower organisms like bacteria and fungi points to a possible general occurrence of this phenomenon in living systems. It also shows the need for further studies aimed at identification of the emitters of the UV photons which may eventually bring about an explanation of a possible biological role of this radiation.

The goal of our next experiments was to try to learn more about a possible source of the observed ultraweak luminescence and at the same time to verify a hypothesis of Popp and Nagl suggesting that DNA might play an important role in processes of light absorption and emission by living systems. The properties of the chromatin, optical ones included, are very different *in vivo* and *in vitro* and depend on many

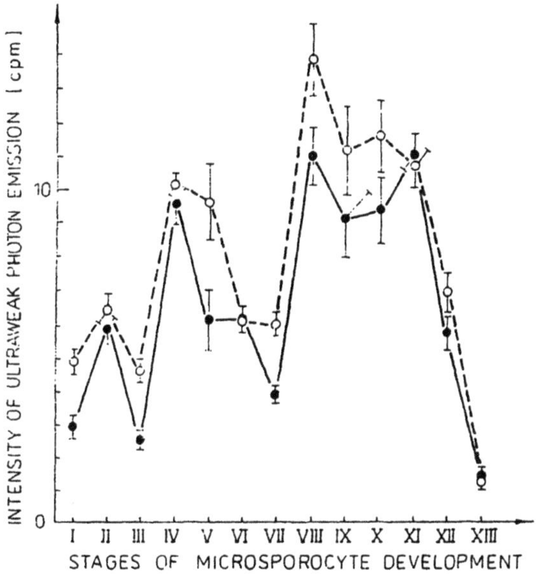

Figure 10.8: *Intensity of the photon emission from microsporocytes of larch during microsporogenesis detected in the whole (○) and in the visible (●) part of the spectrum. The broken line gives a measure of the emission in the UV region ($\lambda < 310$ nm). The vertical bars are standard deviations of the measured signals. Stages 1 to 3, successive phases of diplotene; stage 4, diakinesis, stage 5, metaphase-anaphase of the first meiotic division; stage 6, anaphase-telophase of the first meiotic division; stage 7, interkinesis; stage 8, metaphase-anaphase of the second meiotic division; stage 9, anaphase - telophase of the second meiotic division; stage 10, young tetrad; stage 11, tetrad; stage 12, mature tetrad and stage 13, microspore.*

factors which have not yet been fully understood.

For that reason, we cannot investigate a possible involvement of the DNA of the cell in the processes of the ultraweak photon emission using a well determined material, such as isolated nuclei or chromatin, not to mention pure DNA. However, the microsporocytes we used in our studies are very well suited for this kind of study. Their development is accompanied by very deep or even dramatic transformations of their genetic material typical for the meiotic divisions. The employed method of separation allows for obtaining fractions of cells of very different conformations of the chromatin.

The first assay of the developmental changes of the activity of DNase and RNase during a complete course of meiosis in plants was performed using the microsporocytes and the anther wall cells of larch (Fig. 10.9). The results, when discussed with regard to ultraweak luminescence, showed that perhaps the changes of its intensity might be related to transformations of the cellular DNA [61]. This conclusion arises because one of the results of that work was the demonstration of a positive correlation between the variations in the DNase activity in the microsporocytes and those in the intensity of their induced luminescence (Fig. 10.10).

Following the indication of a possible involvement of DNA in the process of ultraweak photon emission, it has been decided to investigate the changes of the emission intensity induced by treating the microsporocytes with ethidium bromide[62]. This compound is well known as an agent capable of modifying the conformation of DNA. Moreover, the changes of the DNA structure caused by ethidium bromide depend on its concentration.

Increasing concentrations of ethidium bromide have been shown to induce first the unwinding and then the rewinding with opposite helicity of the DNA superhelical structures. If the condensation degree of the chromatin is indeed, as suggested by Nagl and Popp, somehow dependent on the existence of long living excited states in this macromolecule, then the ethidium bromide-induced changes of its condensation should be accompanied by correlated variations in the intensity of the ultraweak photon emission of the cells under investigation.

Such a correlation has previously been observed in experiments with whole cucumber seedlings and suspension cultures of soy bean cells[63,64]. In the work we discuss here the measurements were performed on cells exhibiting very different conformations of their chromatin. The cells in the stages of anaphase I - telophase I, metaphase II - anaphase II and anaphase II - telophase II are characterised by an extremely condensed chromatin and a very low activity of metabolic processes while the chromatin of the cells in the stages of interkinesis and microspore is decondensed and a considerable increase in activity of many metabolic processes was observed in these stages[65].

A surprising result of the measurements was that a general pattern of changes of the ultraweak luminescence intensity after the ethidium bromide treatment was very similar for the microsporocytes in all investigated stages and thus was independent of the initial condensation of the chromatin (Fig. 10.11). Moreover, the main maximum of the emission was observed at the same concentrations of ethidium bromide as in the case of cucumber seedlings and soy cells.

The occurrence – at the same concentrations of ethidium bromide – of the maxima of the intensity of the ultraweak emission from cells so much different with respect to

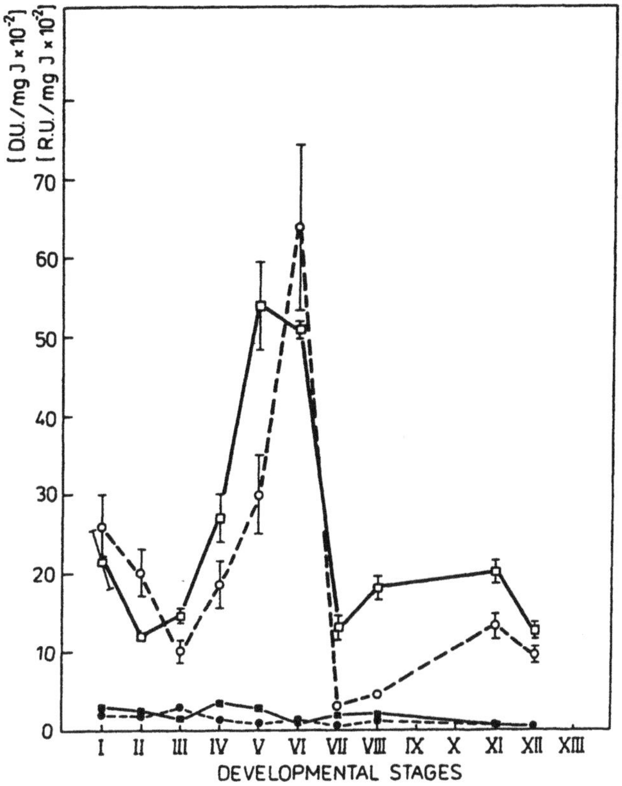

Figure 10.9: *DNase (—–) and RNase (- - -) activities in microsporocytes (•, □) and anther wall cells (○, □) during meiosis in larch. Enzyme activities are expressed as units per mg protein*[61]. *Development stages: I,II,III,IV - diplotene, V - diakinesis, VI - dyad, VII - metaphase-anaphase of the second meiotic division, VIII - anaphase-telophase of the secound meiotic division, IX - young tetrad, X - tetrad, XI - splitting tetrad, XII - microspore.*

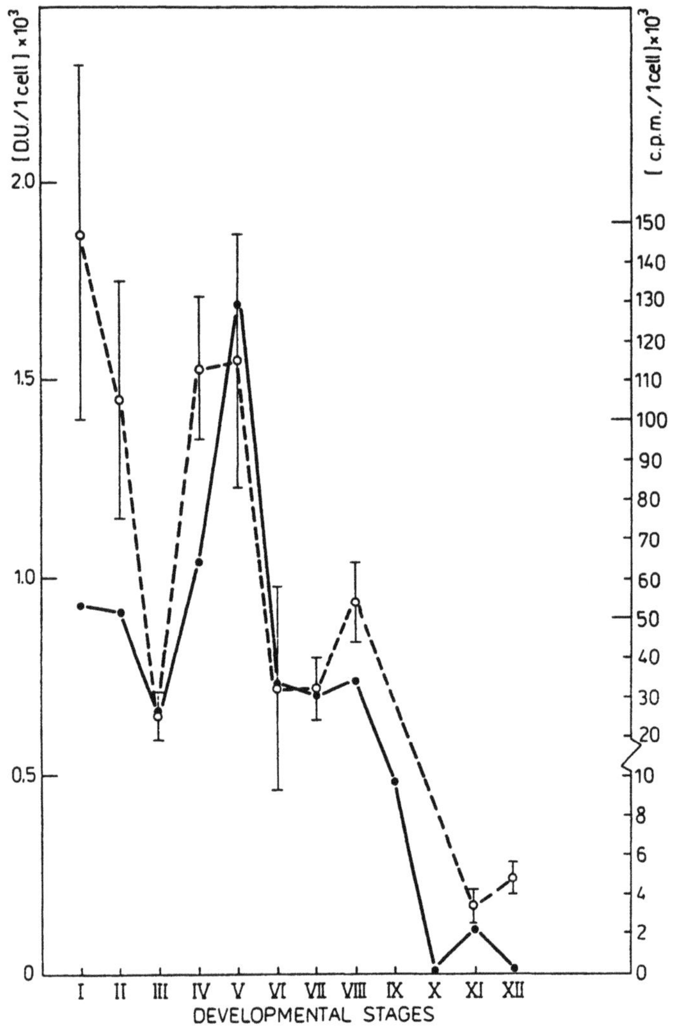

Figure 10.10: *Developmental variations of DNase activity (○) and of intensity of white light induced photon emission (•) of microsporocytes in successive stages of meiosis in larch. Both enzyme activities and photon count rates are normalized to the concentration of the cells. For a description of the developmental stages see the legend for Fig. 10.9.*

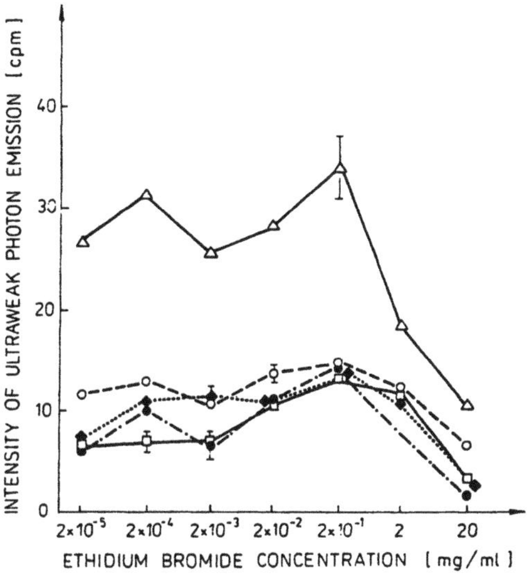

Figure 10.11: *Ethidium bromide induced changes of the ultraweak luminescence of microsporocytes of larch in different stages of their development (○ - anaphase I – telophase I, □ - interkinesis, △ - metaphase II-anaphase II, ● - anaphase II – telophase II, ◇ - microspore). Luminescence intensity was normalized to 2.5 $\times 10^5$ cells/ml.*

their origin, function and stage of the cell cycle (in this way also with respect to the initial conformation of the genetic apparatus) can hardly be related to disturbances of metabolic processes such as syntheses of the nucleic acids and/or proteins. Thus, the process common to all the systems investigated so far is the interaction between ethidium bromide and DNA. This points to the resulting changes in the conformation of the chromatin as a possible cause of the observed variations in the intensity of the ultraweak luminescence of the systems studied.

10.4 Conclusions

The experiments presented here have been the first attempt to perform a systematic research on the possible relations between the changes of the properties of the ultraweak luminescence of living systems and the occurrence of the successive stages of the cell cycle, in our case of a meiotic cycle. Such experiments could be carried out only because the method of obtaining fractions of cells being in the same well determined stages of the meiotic divisions had been elaborated earlier.

The important feature of the employed method of the cell separation is that it enables one to acquire highly homogenous fractions of intact cells with a minimum perturbation of their physiological state. The importance of this fact is clear if one takes into account the extreme sensitivity of the ultraweak photon emission to every factor influencing the homeostasis of biological systems.

Both the results discussed here, as well as the other presented in the original publications allow us to say that there is no doubt that the properties of the ultraweak as well as of the induced luminescence do change in a way clearly correlated with the occurrence of the successive stages of the meiotic divisions and of the microsporocyte development. It should be noted that there is a qualitative difference between this conclusion and the earlier reports by other authors of the existence of a correlation between the properties of the utraweak luminescence and the occurrence of periods of increased division activity of the cells of the systems they had investigated.

The importance of the material chosen for the study is also underlined by other first results such as the observation of oscillations of the monochromatic-light-induced luminescence, a clear correlation between the developmental changes in the level of the induced luminescence and the activity of the DNase or the demonstration that the influence of ethidium bromide on the intensity of the ultraweak luminescence does not depend on the initial degree of a condensation of the chromatin.

It is clear in the light of the above discussion that the measurements of the ultraweak and induced luminescence of biological systems can become a very effective tool of research in a field of cell biology. A great progress has been made in recent years in the field of the regulatory action of light on living organisms and of excited states of biomolecules, and new techniques have been successfully applied in studies of ultraweak luminescence.

There is also a renewed interest in studies of the processes of ultraweak photon emission and thus, there is a chance that it will not take long before we achieve a better understanding of a biological role of these phenomena. There is, however, no doubt that, even without a complete understanding of ultraweak luminescence as a whole, there are already areas where its measurements can be successfully applied as a noninvasive and nondestructive method of diagnostics of biological systems and it is in this direction that we are going to extend our future work in the field.

As already mentioned in the introduction to our study, we have also aimed at the experimental verification of some of the predictions of the new model of differentiation originated by Nagl and Popp. Those predictions are, however, at present of a rather general character and therefore none of the experiments performed until now does not provide a base for a clear acceptance or rejection of this interesting hypothesis.

In conclusion it can be said that many of the results support the predictions of the concept of a biological role of the excited states and of the biophotons resulting from their decay suggested by the authors. The meiotic divisions and the development of the microsporocytes are indeed accompanied by a considerable changes of the properties of both the ultraweak and induced luminescence which, according to the model, should occur in the case of the dividing and differentiating cells.

The correlation between the changes in the level of the induced luminescence and variations in the activity of DNase and a similar character of the ethidium bromide induced changes of the ultraweak luminescence of cells of a very different origin can be interpreted as an indication of an involvement of the DNA macromolecule in the processes of the emission of light by the living systems.

Also the existence of the hyperbolic and oscillating decays of the luminescence induced by an illumination are predicted by the model as a consequence of an active storage of photons in DNA dependent on the stage of the development. At the same time it should be noted, however, that all those conclusions are conditional since at the extreme complexity of the living systems each of the results can be separately interpreted on the basis of more conventional theories on different aspects of the structure and function of a living cell.

The phenomenon of the developmental variations of the properties of the luminescences of interest could be also related for example, to changes in metabolic processes and/or to transformations of cellular structures. The effects of the ethidium bromide might be discussed in terms of its action on the metabolic processes of the cells (not fully understood yet), especially on the metabolism of the mitochondria. Oscillations of the luminescence induced by the illumination are, for instance, observed in photosynthetizing cells and interpreted as resulting from changes of the thylacoidal pH[66] and one could try to look for a similar interpretation in the case of the microsporocytes.

It must be noted here, however,that even at a level electron microscopy no chloroplasts could be found in the cells. One can also argue that the observed hyperbolic decays might be fitted by a combination of more than two exponential ones.

Final conclusions regarding all the above discussed problems can be most probably reached only when new experiments with the use of different but at the same time well characterised cells and tissues will yield enough new data to allow for a rejection of those of the observed correlations which are incidental and for a selection of those reflecting general relationships. In this way they could orient the future research towards an investigation of the mechanism underlying the observed true correlations.

10.5 References

1. W. Nagl and F.A. Popp, *Cytobios* **37** (1983) p. 45.
2. F.A. Popp and W. Nagl, *Cytobios* **37** (1983) p. 71.
3. T.I. Quickenden and S.S. Que Hee, *Biochem. Biophys. Res. Comm.* **60** (1974) p. 764.
4. S.V. Konev, T.I. Lyskova and G.D. Niesenbaum, *Biophysics* **11** (1966) p. 410.
5. T.I. Quickenden and S.S. Que Hee, *Photochem. Photobiol.* **23** (1976) p. 201.
6. T.I. Quickenden and R.N. Tilbury, *Photochem. Photobiol.* **37** (1983) p. 337.
7. T.I. Quickenden and R.N. Tilbury, *J. Photochem. Photobiol.* **8** (1991) p. 169.
8. R.N. Tilbury and T.I. Quickenden, *Photochem. Photobiol.* **47** (1988) p. 145.
9. T. Ichimura, M. Hiramatsu, N. Hirai and T. Hayakawa, *Photochem. Photobiol.* **50** (1989) p. 283.
10. R.Q. Scott, M. Usa and H. Inaba, *Appl. Phys. B* **48** (1989) p. 183.
11. V. V. Artemey, A.S. Goldobin and L.N. Guskov, *Biophysics* **12** (1967) p. 1278.
12. I.H. Shtrankfeld, L.L. Klimenko and N.N. Komarov, *Biophysics* **13** (1968) p. 1082.
13. M.A. Trush, M.E. Wilson and K. van Dyke, *Meth. Enzymol.* **57** (1978) p. 462.
14. T.I. Quickenden and R.N. Tilbury, *Radiat. Res.* **102** (1985) p. 254.
15. A.A. Gurwitsch, *Usp.Biol.* **101** (1986) p.390.
16. V. P. Kaznacheev and L.P. Mikhailova, Ultraweak luminescence in intercellular interactions (in Russian), (Nauka, Nowosibirsk, 1981).

17. E. Celan, D. Gradinaru and B. Celan, in *Photon Emission from Biological Systems*, eds. B. Jezowska Trzebiatowska, B. Kochel, J. Slawinski and W. Strek (World Scientific, Singapore, 1987).

18. A.P. Batyanov, *Bull. Eksp. Biol. Med.* **XCVII** (1984) p. 675.

19. A.P. Batyanov, *Biophysics* **33** (1988).

20. O. Tiphlova and T. Karu, *Photochem. Photobiol.* **48** (1988) p. 467.

21. N. Volpi, M. Cecchi , E. Davolio and L. Bolognani, *Rep.* **5** (1987) p. 3.

22. J. Feierabend and S. Engal, *Arch. Biochem. Biophys.* **251** (1986) p. 567.

23. M. Usa, M. Kobayashi, R.Q. Scott, T. Maeda, R. Hiratsuka and H. Inaba, *Protoplasma* **149** (1989) p. 64.

24. S. Suzuki, M. Usa, T. Nagoshi, M. Kobayashi, N. Watanabe, H. Watanabe and H. Inaba, *J. Photochem. Photobiol. B* **9** (1991) p. 211.

25. F.B. Abeles, *Ann. Rev. Plant Physiol.* **37** (1986) p. 49.

26. F.A. Popp, B. Ruth, W. Bahr, J. Böhm, P. Grass, G. Grolig, M. Rattemeyer, H.G. Schmidt and P. Wulle, *Collect. Phenomena* **3** (1981) p. 187.

27. D. Slawinska and J. Slawinski, *Photochem. Photobiol.* **37** (1983) p. 709.

28. D. Slawinska and J. Slawinski, in *Chemi- and Bioluminescence*, ed. J.G. Burr, in a series Clinical and Biochemical Analysis, **16**, (Marcel Dekker Inc., New York-Basel, 1985).

29. J. Slawinski, *Experientia* **44** (1988) p. 559.

30. F. A. Popp, K.H. Li, W.P. Mei, M. Galle and R. Neurohr, *Experientia* **44** (1988) p. 576.

31. R. Van Wijk and D.H.J. Schamhart, *Experientia* **44** (1988) 586.

32. T. G. Mamedov, Biochemiluminescence of cells and tissues (in Russian), (Elm, Baku, 1982).

33. H. Inaba, *Experientia* **44** (1988) p. 550.

34. T.V. Veselova, V.A. Vaselovsky , A.B. Rubin and P.Z. Bochvarow, *Physiol. Plant.* **65** (1985) p. 493.

35. J.W. Dobrowolski, A. Ezzahir and M. Knapik, in *Photon Emission from Biological Yystems*, eds. B. Jezowska Trzebiatowska, B. Kochel, J. Slawinski and W. Strek (World Scientific, Singapore, 1987).

36. R. Zheng, *J. Luminescence* **40-41** (1988) p. 825.

37. F.A. Popp (private communication).

38. Y.Y. Wang, Y.Q. Wang, D. Li, S.X. Li, Y. Zhang and X.L. Song, *Bull. Environm. Cont. Toxicol.* **46** (1991) p. 456.

39. G. Cilento, *Experientia* **44** (1988) p. 572.

40. G. Cilento and W. Adam, *Photochem. Photobiol.* **48** (1988) p. 361.

41. G. Cilento, *Pure Appl. Chem.* **56** (1984) p. 1179.

42. R.C. Venema and D.H. Hung, *J. Biol. Chem.* **260** (1985) p. 1290.

43. K.C. Smith and N.J. Sargentini, *Photochem. Photobiol.* **42** (1985) p. 801.

44. L. A. Guillo, A. Faljoni-Alario and G. Cilento, *Biochem. Biophys. Acta* **884** (1986) p. 39.

45. P. Hegemann and W. Marwen, *Photochem. Photobiol.* **48** (1988) p. 99.

46. D.F. Mandoli and W.R. Briggs, *Cell Environm.* **5** (1982) p. 137.

47. E. de Petter, L. van Wiemeersch, R. Rethy, A. Dedonder, H. Fredericq, J. de Graaf, H. Steyaert and H. Stevens, *Photochem. Photobiol.* **42** (1985) p. 697.

48. P. Galland, M. Orejas and E.D. Lipson, *Photochem. Photobiol.* **49** (1989) 493.

49. F.A. Popp, in *Electromagnetic Bio-Information*, eds. F.A. Popp, G. Becker, H.L. König and W. Peschka (Urban & Schwarzenberg, München - Wien - Baltimore, 1979) p. 107.

50. D.F. Mandoli and W.R. Briggs, *Proc. Natl. Acad. Sci. USA* **79** (1982) p. 2902.

51. T.C. Vogelmann, *Photochem. Photobiol.* **50** (1989) p. 895.

52. W.B. Chwirot, *Acta Physiol. Plant.* **2** (1980) p. 103.

53. W.B. Chwirot, R.S. Dygdala and S. Chwirot, *Cytobios* **44** (1985) p. 95.

54. M.L. Salin and S.M. Bridges, *Plant Physiol.* **67** (1981) p. 43.

55. K.H. Li and F.A. Popp, *Phys. Lett. A* **93** (1983) p. 262.

56. W.B. Chwirot, R.S. Dygdala and S. Chwirot, *Cytobios* **44** (1985) p. 239.

57. W.B. Chwirot, R.S. Dygdala and S. Chwirot, *Cytobios* **47** (1986) p. 137.

58. W.B. Chwirot and R.S. Dygdala, *J. Plant Physiol.* **125** (1986) p. 79.

59. G. Kosinski and M. Giertych, *Planta* **155** (1982) p. 93.

60. W.B. Chwirot and R.S. Dygdala, *Cytobios* **65** (1991) p. 25.

61. W.B. Chwirot, *J. Plant. Physiol.* **122** (1986) p. 81.

62. W.B. Chwirot and R.S. Dygdala, *J. Plant Physiol.* **134** (1989) p. 762.

63. M. Rattemayer, F.A. Popp and W. Nagl, *Naturwissenschaften* **68** (1981) p. 572.

64. F.A. Popp, W. Nagl, K.H. Li, W. Scholz, O. Weingartner and R. Wolf, *Cell Biophys.* **6** (1984) p. 33.

65. W.B. Chwirot, *Postepy Biologii Komorki* **11** (1984) p. 581.

66. L.G. Sundblad, K. Palmqvist and G. Samuelsson, *FEBS Lett.* **209** (1986) p. 28.

67. I. Klasterska and C. Ramel, *Genetica* **51** (1979) p. 15.

Chapter 11

Light Emission and Rescattering in Synchronously Developing Populations of Early *Drosophila* Embryos:

Evidence for Coherence of the Embryonic Field and Long Range Cooperativity

M.W. Ho, X. Xu, S. Ross and P.T. Saunders

Abstract

Populations of synchronously developing *Drosophila* embryos emit light at a high level shortly after oviposition (when development begins), diminishing linearly in the first two hours to a steady rate that is maintained for at least the next 16 hours. On being stimulated briefly with white light, they re-emit strongly, the rescattered light decaying hyperbolically as typical of all living systems previously examined. The rate of light rescattering increases sharply between 15 to 40 min. from the start of development, then follows a more gentle rise to a plateau at around three to three and a half hours. Self-emission and light rescattering are most intense and dynamic during early development when pattern determination processes are taking place.

When the embryos are briefly exposed to light within 40 min. of development, an entirely new phenomenon is observed in the form of intense luminescent flashes which can appear any time from one to 20 min., and up to 8 hr. after light stimulation. The total number of flashes differs between experiments, and the duration of individual flashes also varies from less than 1 s up to 30 minutes. We discuss the possibility that these superdelayed luminescent flashes may result from cooperative interactions among the embryos in the entire population, which may have biological significance in the synchronization of development to external light as **Zeitgeber**.

Our results suggest that biophoton emission can be used as a noninvasive probe of the physical state of the embryos which is inaccessible by conventional chemical or biochemical means.

11.1 Introduction: the Embryonic Field

How does a relatively featureless fertilized egg turn into a shapely, highly differentiated organism? In attempting to answer this question, a disproportionately large amount of scientific effort to-day is devoted to isolating morphogenetic mutants, cloning genes and identifying the molecular species involved in "controlling" pattern formation. No other organism has been the subject of more intense investigation in this regard than *Drosophila*[1]. The summary of findings indicate that the products of maternal genes, laid down in spatial gradients, determine the spatial (and temporal) expression of primary controlling genes in the egg, which in turn determine the expression of other genes, and so on in hierarchical manner[2]. By dint of special interactions between adjacent gene products, the body is divided and subdivided finally into segments.

Segmentation, or metamerism, is nearly universal among living organisms; it is common to arthropods, vertebrates, and annelids, and in a somewhat different form, to molluscs, coelenterates, echinoderms and most other phyla. This suggests fundamental, generic processes rather than "instructions" encoded in specific genes. Genes can have a stabilizing or modifying influence on these processes[3]; so it is not surprising that mutations in certain genes can result in segmentation defects just as physical environmental perturbations can "phenocopy" not only the segmentation defects, but practically all abnormal phenotypes due to gene mutations[4–7].

Further circumstantial evidence in favour of generic physical processes in pattern formation is that a wide variety of physicochemical systems are known to produce forms similar to those in biology[8-12]. Bateson (1913), who "rediscovered" Mendelian genetics, was among the first to suggest that rhythmic mechanisms of a general sort – analogous to the breaking of waves on the seashore may be involved in segmentation.

The first generic mechanism to account for metamerism was put forward by Turing (1952)[13] who showed how chemical reaction and physical diffusion through an initially homogeneous domain could generate repeated parts with definite "chemical" wavelengths. Turing's idea was taken up and elaborated by the present generation of theorists[14,15] in accounting for many biological forms including segmentation and animal coat colour patterns.

One difficulty in applying Turing's idea is that chemical wavelengths are dependent on the precise diffusion coefficients of the chemical species involved, and hence embryos of different sizes would be expected to have different numbers of segments; which is not the case. The number of metameric parts is characteristic of the species and quite independent of the size of individual eggs. Another difficulty for those who attempt to identify diffusing "morphogens" as gene products is that the latter are either RNAs or proteins which do not, as a rule, diffuse passively through cells[16].

Certainly, the spatial pattern of the typical "morphogenetic" gene expression – as revealed by *in situ* hybridization techniques – changes quite dynamically and abruptly during early embryogenesis, placing severe constraints on the diffusion rates of putative morphogens[17]. They tend to favour inorganic ions and small chemical molecules[18] ubiquitous to living systems; pointing once again to universal, or generic properties of development, as opposed to those associated with specific genes[19].

One of the earliest generalizations to emerge from the study of biological development is the notion of the embryonic field[20] which gives a dynamic global order to the embryo at the very outset. It is a field of graded potential for developing the future organism which has a strong tendency to form organized wholes despite gross disturbances. Thus, when the early embryo is separated into two halves, each part reconstitutes itself and develops into an adult half the size but normal in every other respect. Similarly, when two early embryos are combined together, a single adult is formed which is twice the normal size.

As development proceeds, the global embryonic field becomes differentiated into sub-fields, each of which possesses a graded potential for the formation of single organs, such as the eye, the limb and so on, in a given orientation and polarity. In many organisms, this field is still evident in the adult, and is responsible for the correct regeneration of organs and limbs after amputation[21]. There has been a recent revival of interest in embryonic fields among geneticists and developmental biologists, as the expression of various morphogenetic genes and gene products seem to follow closely the domains of classically defined gradient-fields[22]. However, as the reviewers themselves emphasize, there is no direct evidence linking the gradients of gene expression and the behaviour of embryonic fields in a causal way. And the nature of the embryonic field remains something of a mystery.

Driesch's (1891) concept[20] of the embryonic field was originally inspired by analogy to a magnetic field. Years later, Burr and Northrup (1935)[23] detected electric fields around both developing and adult organisms, from which they postulate the electrodynamical field as the basis of biological organization. Becker (1962)[24] confirmed their findings, demonstrating the existence of a DC field on the surface of intact organisms which changes dramatically as the result of both anaesthesia and injury. He also identified DC currents that are specifically associated with limb regeneration in salamander[25]. Using the more sophisticated vibrating probe technique, Jaffe (1982)[26] was able to detect transembryonic ionic currents in all developing systems examined.

These currents typically appear before any overt sign of development, suggesting that they may play a role in setting up major body axes and patterns in the embryo, which somehow mirror the electrical field gradients and polarities. However, no direct causal connection has yet been established[27]. The possibility remains that the currents are themselves manifestations of a perhaps more fundamental process.

A further clue to the nature of biological organization comes from a series of reports suggesting that organisms including humans are sensitive to very weak electromagnetic fields – both natural and artificial in origin – within the entire frequency spectrum from the extra low to the microwave range[25,28,29]. As the energies involved are often below thermal background, the reports are regarded with a great deal of skepticism by the orthodox scientific community. They conclude that no known mechanism could account for the effects, even though there is now indisputable evidence that chronic exposure to weak electromagnetic fields is associated with increased incidence of cancer and other illnesses[30].

As regards a possible mechanism for the biological effects of weak electromagnetic fields, Fröhlich[29,31] had earlier suggested that, given the dielectric nature of all biological molecules – especially those constituting biological membranes – metabolic energy could well be stored in the form of electromechanical vibrations. Under certain conditions, those vibrations will condense into a coherent or collective mode – which may be a stable or metastable strongly polarized state (perhaps not unlike the DC fields observed by Becker[24]) or an oscillating one – when organisms will exhibit extreme sensitivity to external electromagnetic fields. If these "coherent excitations" are involved in biological functioning, that would explain the biological effects of weak electromagnetic fields.

The idea of coherence appears to be the key to explaining many of the most characteristic properties of the living state. These include extreme sensitivity to external cues, amplification of incoming signals, rapid and efficient transfer and transformation of energy, and long-range dynamic order[12]. Independent evidence for the existence of coherence comes from biophoton research carried out by Popp and his coworkers (see Chapters 1-5, 15), who conclude that the living system may indeed be coherent in a broad range of frequencies which are coupled together effectively to give a single degree of freedom.

This coupling of a band of modes was also predicted by Fröhlich[29] from theoretical considerations, and in his model, this is accomplished via a "heat bath" at a specified temperature which enables instantaneous transfer of energy between modes. This has prompted us to consider the possibility that global dynamic coherence may be particularly instrumental in generating the first patterning during development, and that the organizing properties of the embryonic field may be due to an underlying *coherent* electrodynamical field[12].

A number of observations offer circumstantial evidence for the involvement of global coherence in pattern formation.

1. First, all embryos begin development with a series of cell or nuclear divisions that occur simultaneously and synchronously over the entire embryo.

2. Second, the size of the embryo at which pattern is determined is of the order of 1mm in all organisms, regardless of the size of the adult or its rate of development.

This suggests a limit imposed, perhaps, by the requirement for global coherence. Significantly, growth occurs only *after* body pattern is determined, presumably registered by the expression of specific genes, best documented in the case of *Drosophila*. This also coincides with the decay of global synchrony in the embryo. As growth and differentiation proceed, organized structures are laid down sequentially which facilitate both the maintenance of coherence on a larger and larger scale, and the evolution of domains of local dynamic autonomy, which could either be effectively decoupled from the whole or coherently vibrating with it, albeit with different local frequencies[29].

The living system may owe both its dynamic flexibility and stability to an alteration of its effective degree of freedom according to functional state. Finally, Ho et al.[32] have recently found that brief exposure of early *Drosophila* embryos to weak static magnetic fields results in a high proportion of body pattern abnormalities in the larvae which hatch twentyfour hours later. The observations indicate that

1. the embryonic field involved in pattern determination does include electrodynamical components and
2. as the energies involved are below thermal threshold, there can be no effect unless a high degree of cooperativity or coherence exists during pattern determination.

In this paper, we report observations on biophoton emission characteristics of synchronously developing populations of *Drosophila* embryos, which give further evidence of coherence and cooperativity that may extend over the entire population.

11.2 Methods

Light emission and rescattering measurements were made as previously described in this book.

The Oregon strain of *Drosophila melanogaster* employed in this study was maintained as previously described[6]. *Drosophila* eggs are fertilized at oviposition, when development begins. It is therefore possible to collect large numbers of embryos that are synchronous to within one minute of development. Three to seven-day old flies, cleared of retained eggs, were anaesthesized with carbon dioxide and transferred directly into a quartz cuvette ($2\times2\times4cm^3$) where they were confined and encouraged to lay eggs over all the available surface for 3 to 5 min. (depending on how long they take to recover from anaesthesia), after which the flies were removed. The cuvette containing the freshly laid eggs – which remained *in situ* where they were deposited by the females – was transferred to the photon-detecing apparatus and maintained in a humid atmosphere for the duration of the measurements. All measurements were performed at 25°C.

Figure 11.1 is a diagrammatic representation of the developmental events and approximate time-table in *Drosophila* embryogenesis, based on Foe and Alberts, Foe, and our own observations[33,34]. The nucleus of the fertilized egg undergoes a series of synchronous divisions without cell divisions. After six to eight cycles, most of the nuclei migrate to the layer of cytoplasm just beneath the plasma membrane (the periplasm) to form a syncytial blastoderm where two more nuclear division cycles take place. At the posterior pole, a number of the nuclei become pinched off to form pole cells – the future germ cells. The rest of the nuclei undergo 4 more divisions before cellularization takes place over the entire blastoderm. The resultant cellular blastoderm gastrulates almost immediately. All the major pattern determining events have occurred by cellularization[6].

11.3 Results

11.3.1 Self-emission as a function of developmental time

Self-emission starts at a relatively high level, diminishing linearly to a steady rate by 120 min. of development, then remains unchanged for the next 16 hours (see Fig. 11.5, p. 297). This pattern of self-emission is completely reproducible, regardless of the time at which the measurement begins. The level of self-emission depends on the developmental time as well as linearly on the total number of embryos present, ranging from 60 to 289; the correlation coefficient being 0.97 (see Fig. 11.2). The mean and standard deviation of the starting rate (approx. 5-15 min. a.o.), averaged over 14 experiments is 0.24±0.03 counts per second per embryo. It can be seen from Fig. 1 that within two hours of development at 25°C, the first 12 synchronous nuclear division cycles have taken place, and the embryo is a syncytial blastoderm with a layer of about 2^{11} nuclei in the clear cytoplasm (periplasm) just beneath the plasma membrane.

11.3.2 Short term light rescattering after stimulation as a function of developmental time

The characteristics of light rescattering in *Drosophila* embryos shortly (0.2 to 10s) after stimulation for 1 min. with white light resemble those of all other systems examined so far in its typically hyperbolic decay kinetics (Fig. 11.3). It is unusual in two respects. First, the intensity relative to biomass is possibly the highest ever recorded. Second, it changes markedly with developmental time.

The rescattering rate increases steeply between 15 to 40 min., then follows a more gentle rise to a plateau between 3h to 3.5h. The rescattering rate as a function of developmental time is plotted for a number of typical experiments starting and

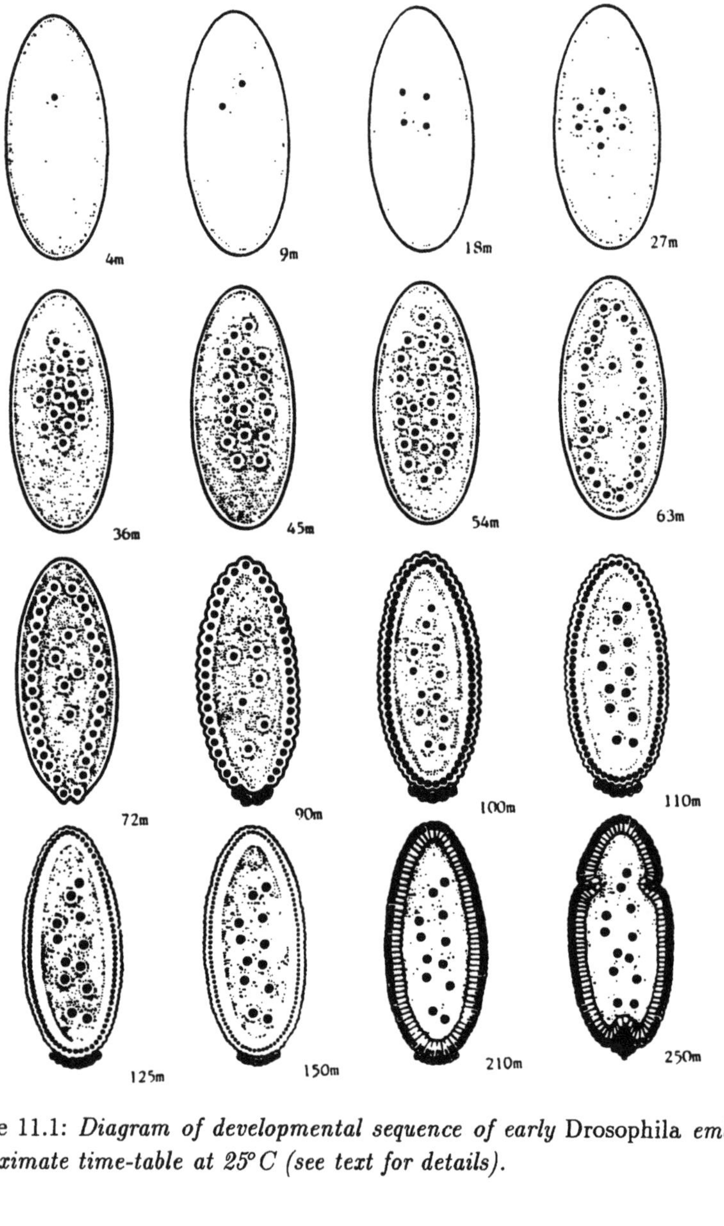

Figure 11.1: *Diagram of developmental sequence of early* Drosophila *embryos with approximate time-table at 25° C (see text for details).*

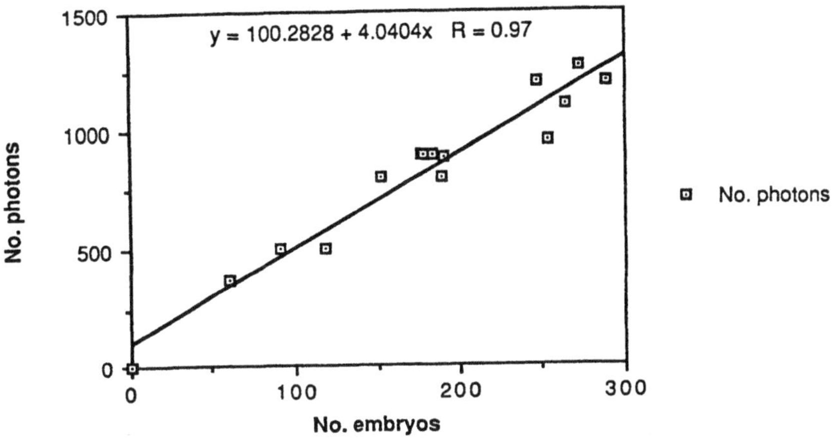

Figure 11.2: *Self-emission as a function of the number of embryos present. Self-emission is the total starting photon count in the first 20s.*

ending at different times, in order to ascertain that the increase and levelling out is not due to the previous history of light stimulation, nor dependent on the absolute rate of rescattering (Fig. 11.4). It can be seen that the period of rapid rise coincides with nuclear division cycles 2 to 6, the transition between the slopes corresponds to the mass nuclear migrations to the periplasm, and the plateau at the end corresponds to cellularization.

Although different populations of embryos show the same changes in light rescattering in the course of development, the absolute rate of rescattering bears no simple relationship to the numbers of embryos present (see Fig. 11.5). This is reminiscent of the response in the self-emission of *Daphnia* (see Chapter 14), suggesting that there may be long range interactions between embryos within the population, although much more data is required here, both in terms of numbers and the average spacing between embryos, which can vary appreciably between experiments involving similar embryo numbers.

11.3.3 Superdelayed luminiscent flashes

By far the most surprising response is elicited when the embryos are stimulated by exposure to white light (4 W/m^2) for 1 min. within the first 40 min. of development. A new kind of luminescence makes its appearance in the form of superdelayed,

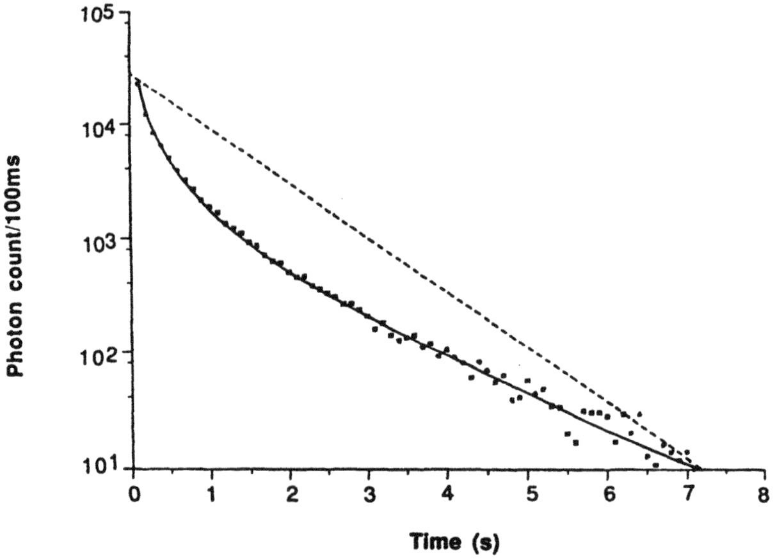

Figure 11.3: *Typical hyperbolic decay kinetics of short-term rescattered light. Light re-emission was recorded between 0.2 to 10s after 1 min. white-light stimulation; asterisks, experimental data; solid line, curve fitted on hyperbolic decay model; dotted line, exponential decay.*

intense flashes. The delay time for the first flash varies from 1 min. for single sharp bursts (duration < 1s), to about 20 min. for sustained bursts (> 20s). Multiple flashes typically appear and either follow in quick succession or after further delays, minutes or even hours apart. Each flash may be sustained for minutes to tens of minutes, and the longest delay time to the last flash is 8 hours (see Fig. 11.6). These prolonged bursts are most dramatic; and they are apparently unaffected by periodic 5s pulses of red light stimulation (see Fig. 11.7); indicating a remarkable degree of stability once the process has been initiated.

Certain conditions have to be satisfied for the observation of superdelayed luminescence:

1. The embryos must be synchronous to within 1 min. at the start of development. Populations with asynchronous development, such as those containing more than 20% retained eggs or unfertilized eggs never exhibit any superdelayed luminescence.

2. The time at which white-light stimulation is applied must fall within 40 min.

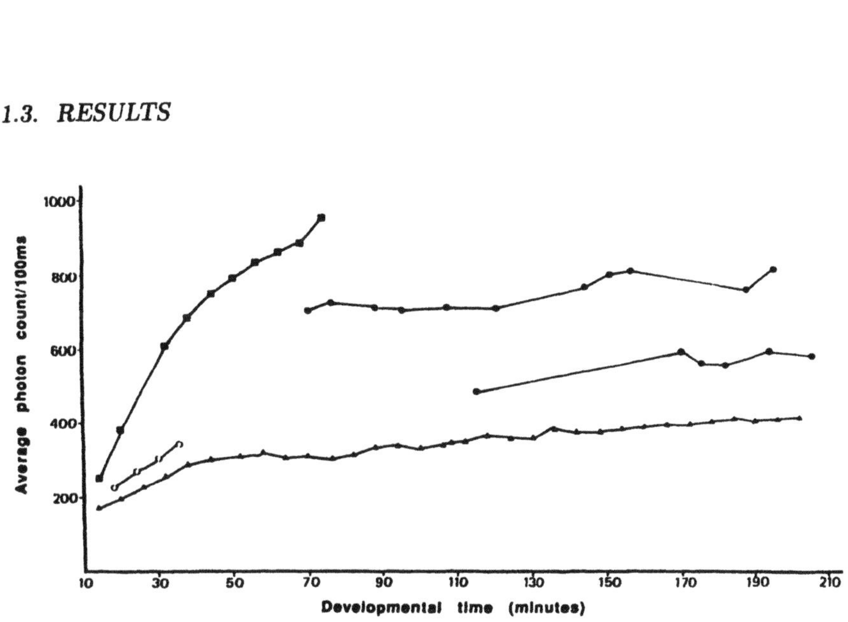

Figure 11.4: *Rate of short-term light rescattering as a function of developmental time. Different lines represent different batches of embryos; rescattering rate is expressed as photon count per 100ms averaged over the first 10s.*

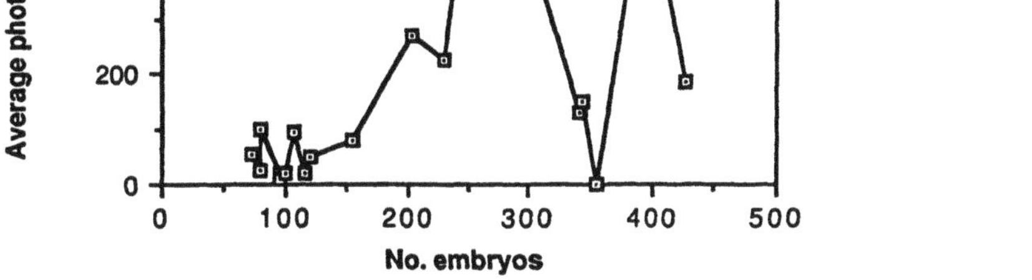

Figure 11.5: *Short-term rescattering rate as a function of the number of embryos present. The rescattering rates for all experiments are those at 50-60 min. from the start of development, expressed as photon count per 100ms averaged over the first 10s.*

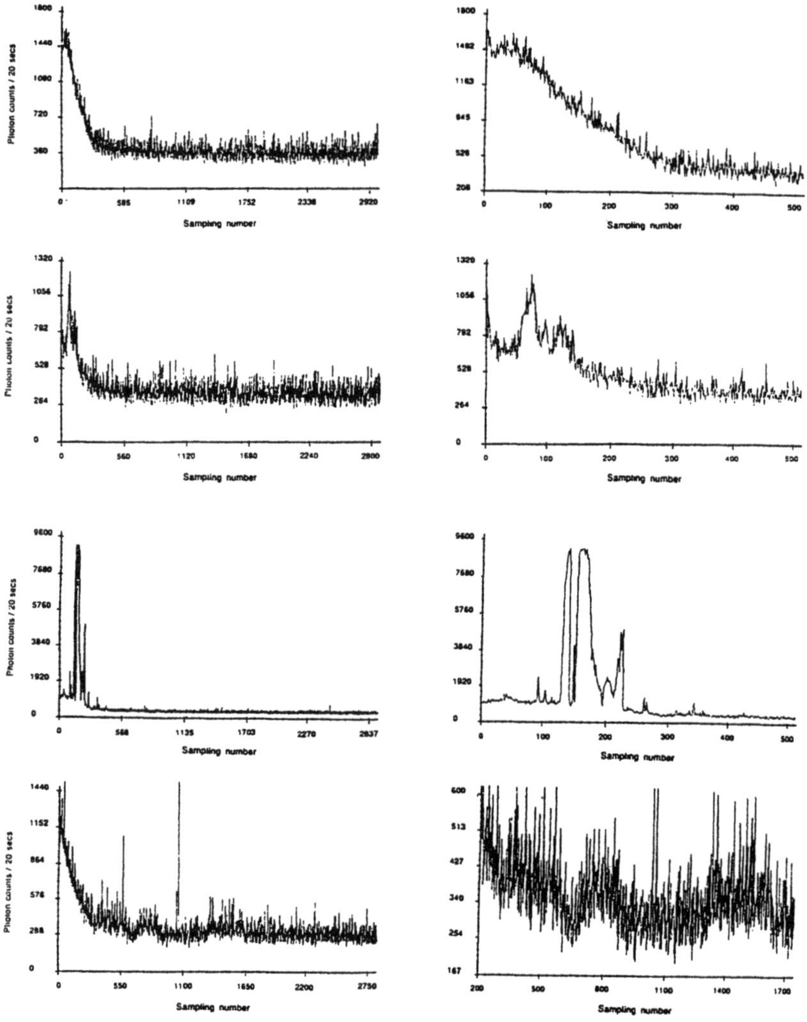

Figure 11.6: *Superdelayed luminescence in synchronously developing populations of* Drosophila *embryos. Top to bottom: control batch of 289 embryos not exposed to white light stimulation, exhibiting typical pattern of self-emission; batch of 107 embryos exposed to white-light for one minute at 10 min. of development; batch of 232 embryos exposed to white light at 5 min. development; batch of 172 embroys exposed to white light at 5 min. of development. The traces on the right are expanded versions of those on the left.*

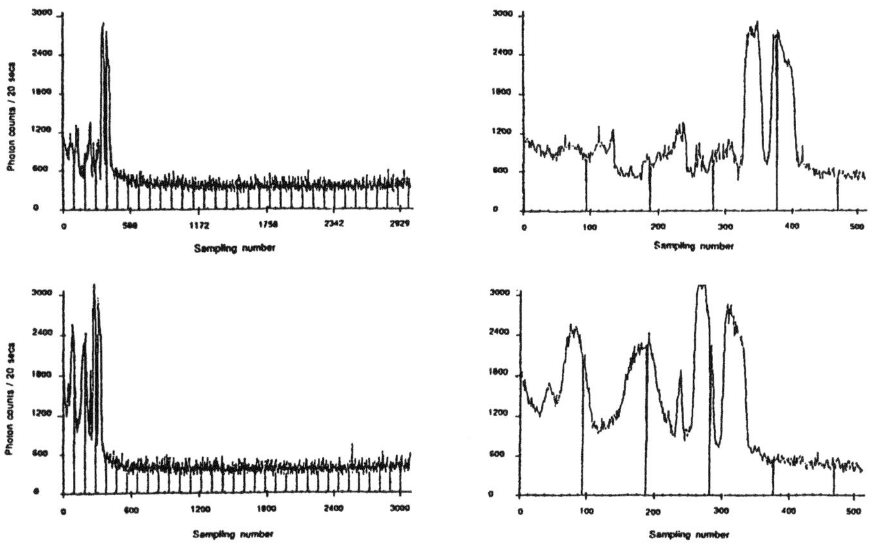

Figure 11.7: *Superdelayed luminescence unperturbed by periodic exposure to red light. Top, batch of 274 embryos exposed at 20 min. of development; bottom, batch of 245 embryos exposed at 5 min. of deveopment. Five second pluses of red light were delivered 30 min. apart during the entire period of measurement.*

from the start of development.

3. The intensity and duration of light stimulation must not be less than about 4W/m^2 for 60s. Light of limited spectral composition (red, green, or blue) was ineffective, or much less effective. But this may be because the intensity of light is lowered substantially by the filters used.

4. The embryos must be fairly evenly distributed over the inside of the quartz cuvette.

Even given the above conditions, however, the multiple prolonged flashes do not always occur. Instead, either single flashes, or multiple sharp bursts may appear; or there may be no delayed flashes at all. In a total of 29 experiments where all the above conditions have been satisfied, 13 gave no flashes at all, 9 gave single flashes or multiple sharp bursts, and only 7 gave multiple prolonged flashes (see Table 11.1). At the moment, we do not know whether uncontrolled geometrical factors such as

Table 11.1: Parameters of superdelayed luminescence with multiple prolonged flashes

N	N^2	t_0	t_D	$t_{D(max)}$	P_e	P_{max}
172	29584	7	20	186	47504	750
107	11449	10	20	20	19362	582
232	53824	5	31	45	356000	8160
245	60025	5	20	92	259168	2000
274	75076	20	57	132	205479	2100
159	25281	15	34	128	87556	2800
100	10000	10	10	35	5743	1000

N, number of embryos; t_o, developmental time at light exposure; t_D, delay time to first flash; $t_{D(max)}$, delay time to last flash: P_e, total superdelayed luminescence (photon count) in excess of baseline; P_{max}, maximum photon count achieved in the superdelayed luminescence.

the distribution of eggs in the cuvette and/or other uncontrollable stochastic effects may be responsible for the variability.

A suggestive relationship exists between the total superdelayed re-emission, P_e, (in excess of the self-emission background), and the number of embryos, N, in all cases involving multiple prolonged flashes. There is a significant correlation between P_e and N or N^2, the correlation for N^2 being somewhat better than for N (Fig. 11.8); the correlation coefficient R, for N^2 is 0,85 vs 0,81 for N (significant at 1% and 5% level respectively); and the F value for linear regression, 15.54 vs 11.51 (significant at 1% and 2.5% level respectively).

Other possible correlations are:

1. between the delay time to the first flash, t_D, and the time a.o. at which the light stimulation is applied, t_0. ($R = 0.82$, $F = 12.19$).

2. between t_D and N ($R = 0.8$, $F = 11.04$).

3. between P_{max} , the maximum photon count achieved in the superdelayed luminescence, and P_e ($R = 0.85$, $F = 13.30$).

All of the above correlations are significant at the level of 5%.

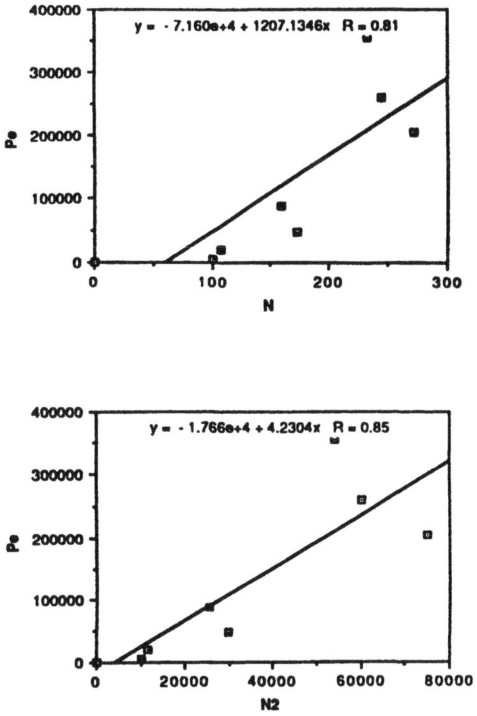

Figure 11.8: *Regression of P_e on N and N^2.*

There does not seem to be any consistent relationship between the timing of individual flashes and the precise embryonic events occurring in the developing embryos, which clearly indicates that the flashes are not correlated to specific developmental or metabolic chemical events occurring at specific times.

The question arises as to whether single embryos may exhibit similar flashes, which only build up to the intense bursts recorded if many embryos flash in synchrony. In preliminary experiments carried out in Astromed (with the help of Andrew Wilkinson and Rob Haggart, Cambridge Science Park, Cambridge, U.K.) we have succeeded in obtaining an image of the single embryo 15 to 20 minutes after it has been stimulated with the microscope light (Fig. 11.9). This opens a way to imaging single embryos that has many important applications.

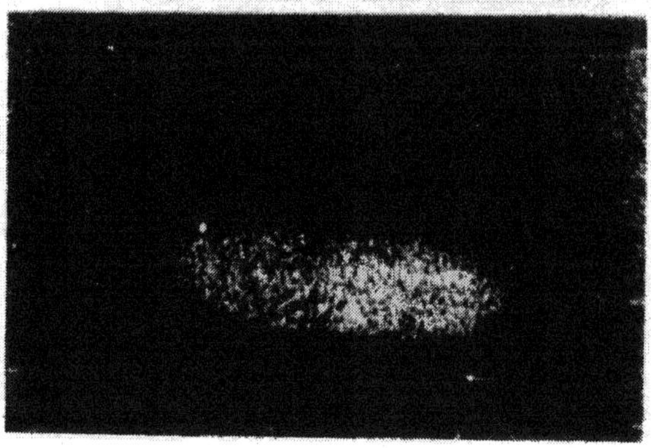

Figure 11.9; *Single* Drosophila *embryo imaged by its own superdelayed luminescence. The embryo was stimulated for several minutes with light from the microscope at 10 min. of development, and the image was made up of light re-emitted between 15-20 min. afterwards.*

11.4 Discussion

Our results show that light emission and rescattering change most markedly in the course of early development when major pattern determination processes are taking place. Short-term rescattering after light exposure follows the hyperbolic decay kinetics which is universal to all living systems and is a sufficient condition for coherence (see Chapter 2). The intensity of rescattering achieved is higher than all systems previously examined, and furthermore, it changes in parallel with specific embryonic events.

We have also demonstrated, for the first time, a superdelayed luminescence that is dependent, neither on detailed metabolic events nor on particular embryonic stages. Taken all together, the characteristics of superdelayed luminescence suggest that it is a cooperative phenomenon dependent on both synchrony within the population and some general physical state of the embryos within the first forty minutes of development. This is when nuclear divisions are occurring relatively rapidly and synchronously over the entire embryo, such that a high degree of global coherence may be inferred.

The dependency of the total superdelayed luminescence on N^2 (see Chapter 3), the variable delay time, and the forms of the flashes (particularly the multiple peaks or ringing effects), the dependence on geometry and the stochasticity and variability of the effect, are all reminiscent of the phenomenon of superradiance in a population of emitters which can interact nonlinearly to build up a coherent emission involving the whole population[35].

Drosophila females typically lay eggs just before sunrise, so the external light source may be used as an initial synchronizing signal (**Zeitgeber**) which maintains the circadian and other biological rhythms. The superdelayed re-emission could then be a means of maintaining communication and synchrony among individuals in the population. On the other hand, since the females do not deposit their eggs in nature in quite the same configuration as they are made to in our experiments, the effects we observe may simply be a consequence of our measurement conditions, and serve to inform us of the highly coherent state of the embryos at the time when light stimulation is applied. They indicate that synchronously developing embryos can interact nonlinearly to generate light emission which is coherent over the entire population, and often orders of magnitude higher than the self-emission rate.

These experiments demonstrate that photon-emission can be used to probe the physical state of the embryo during pattern determination which cannot be accessed by conventional biochemical or molecular biological means. We have barely scratched the surface of this fascinating area, and a lot more research needs to be done. Nevertheless, our observations already strongly suggest that the embryonic field is a physically coherent domain which not only gives global order to the embryos but enables them to interact cooperatively with other embryos, giving us new insights into the sociality that is at the basis of all life[36].

11.5 Acknowledgements

We would like to thank Phil Callahan for suggesting the best method for measuring photon-emission from *Drosophila* embryos, based on his sympathetic resonance with the insect world, for which he is justly renowned. We are very grateful to F.A. Popp for his inspiration, encouragement and support for the project. Thanks are also due to all in the International Institute of Biophysics for their many and varied contributions to this work, in particular, to Beate Hemmer and Corina Benkel, without whose excellent technical skills and goodwill, this work would have been impossible. A small part of this work was performed while M.W. Ho was supported by a Royal Society study visit grant to F.A. Popp's laboratory in Kaiserslautern.

11.6 References

1. V. French, P. Ingham, J. Cooke and J. Smith, eds. Mechanisms of Segmentation. *Development* **104** (Suppl.) (1988) p. 1-254.

2. H. Meinhardt, Models for maternally supplied positional information and the activation of segmentation genes in *Drosophila* emboyogenesis. *Development* **104** (Suppl.) (1988) p. 95-111.

3. M.W. Ho, Environment and heredity in development and evolution, in *Beyond Neo-Darwinism: Introduction to the New Evolutionary Paradigm*, eds. M.W. Ho and P.T. Saunders (Academic Press, London, 1984) p. 267-289,.

4. M.W. Ho and P.T. Saunders, Beyond Neo-Darwinisman epigenetic approach to evolution. *J. Theor. Biol.* **78** (1979) p. 573-591.

5. M.W. Ho and P.T. Saunders, Adaptation and natural selection: mechanism and teleology, in *Towards a Liberatory Biology*, ed. S. Rose (Allison and Busby, London, 1982) p. 85-102.

6. M.W. Ho, A. Matheson, P.T. Saunders, B.C. Goodwin and A. Smallcombe, Ether-induced disturbances in segmentation in *Drosophila melanogaster. Roux Arch. Devl. Biol.* **196** (1987) p. 511-521.

7. M.W. Ho, An exercise in rational taxonomy. *J. Theor. Biol.* **147** (1990) p. 43-57.

8. D'A. W. Thompson, *On Growth and Form* (Cambridge University Press, Cambridge, 1914).

9. P.T. Saunders, Development and evolution, in *Beyond Neo-Darwinism: Introduction to the New Evolutionary Paradigm*, eds. M.W. Ho and P.T. Saunders (Academic Press, London, 1984) p. 243-265.

10. M.W. Ho, Where does biological form come from? *Rivista di Biologia* **77** (1984) p. 147-179.

11. M.W. Ho, How rational can rational morphology be? A post-Darwinian rational taxonomy based on a structuralism of process. *Rivista di Biologia* **81** (1988) p. 11-55.

12. M.W. Ho, Coherent excitations and the physical foundations of life, in: *Theoretical Biology; Epigenetic and Evolutionary Order from Complex Systems*, eds. B.C. Goodwin and P.T. Saunders (Edinburgh University Press, Edinburgh, 1989) p. 162-76.

13. A. Turing, The chemical basis of morphogenesis. *Phil. Trans. Roy. Soc. B* **237** (1952) p. 37-72.

14. A. Gierer and H. Meinhardt, A theory of biological pattern formation. *Kybernetik* **12** (1972) p. 30-39.

15. J. D. Murray, *Mathematical Biology* (Springer-Verlag, Berlin, 1980).

16. D.N. Wheatley, On the possible importance of an intracellular circulation. *Life Sciences* **36** (1985) p. 99-307.

17. T.C. Lacalli, Modelling the *Drosophila* pair-rule pattern by reaction-diffusion: gap input and pattern control in a 4-morphogen system. *J. Theor. Biol.* **144** (1990) p. 171-194.

18. B.C. Goodwin and L.E.H. Trainor, Tip and whorl morphogenesis in Acetabularia by calcium-regulated strain fields. *J. Theor. Biol.* **117** (1985) p. 79-106.

19. M.W. Ho and P.T. Saunders, How to segment generically and reliably (submitted) (1991).

20. H. Driesch, Entwicklungsmechanische Studien I-II. *Z. Wiss. Zool.* **53** (1891) p. 160-182.

21. P. Weiss, *Principles of Development* (Holt and Co. New York, 1939).

22. E.M. De Robertis, E.A. Morita and K.W.Y. Cho, Gradient fields and homeobox genes. *Development* **112** (1991) p. 669-678.

23. H.S. Burr and F.S.C. Northrup, The electro-dynamic theory of life. *Quart. Rev. Biol.* **10** (1935) p. 322-333.

24. R.O. Becker, The direct current control system: A link between environment and organism. *New York State Journal of Medicine* **62** (1962) p. 1169-76.

25. R.O. Becker, Cross Currents: The Promise of Electromedicine, The Perils of Electropollution. (Jeremy P. Tarcher, Inc., Los Angeles, 1990).

26. L.F. Jaffe, Developmental current, voltages and gradients, in *Developmental Order: Its Origin and Regulation* (Alan R. Liss, Inc., New York, 1982) p. 183-215.

27. R. Nucitelli, Ionic currents in morphogenesis. *Experientia* **44** (1988) p. 657-666.

28. A.S. Presman, *Electromagnetic fields and Life* (Plenum Press, New York, 1970).

29. H. Fröhlich, The biological effects of microwaves and related questions. *Adv. Electronics and Electron Phys.* **53** (1980) p. 85-152.

30. S. Shulman, Cancer rises seen in electromagnetic fields. *Nature* **345** (1990) p. 463.

31. H. Fröhlich, Long range coherence and energy storage in biological systems. *Int. J. Quantum Chem.* **2** (1968) 641-9.

32. M.W. Ho, T.A. Stone, I. Jerman, J. Bolton, H. Bolton, B.C. Goodwin, P.T. Saunders and F. Robertson, Brief exposure to weak static magnetic fields during early embryogenesis causes cuticular pattern abnormalities in *Drosophila* larvae. *Phys. Med. Biol.* (1991) (in press).

33. V.E. Foe and B.M. Alberts, Studies of nuclear and cytoplasmic behaviour during the five mitotic cycles that precede gastrulation in *Drosophila* embryogenesis. *J. Cell Sci.* **61** (1983) 31-70.

34. V.E. Foe, Mitotic domains reveal early commitment of cells in *Drosophila* emboyos. *Development* **107** (1989) p. 1-22.

35. M. Gross and S. Haroche, Superradiance: An essay on the theory of collective spontaneous emission. *Phys. Rep.* **93** (1982) p. 301-396.

36. M.W. Ho, Natural being and coherent society, in *Evolutionary Theory and Social Values*, ed. R. Wesson (Oxford University Press, New York, 1991).

Chapter 12

Experimental Evidence on Ultraweak Photon Emission from Normal and Tumour Human Tissues

F. Musumeci, A. Triglia and F. Grasso

Abstract

Among the many subjects of the research on low level luminescence from biological matter, one of the most important applicative features, the possibility of using the parameters connected to photon emission from human tissues to detect cancer pathologies, is discussed. A survey of the experimental work of the last years in this field is presented together with the experimental data collected at the University of Catania. The problems of the conditions under which the measurements should be carried out, those of the normalization of the emission rate and of the choice of the measurement technique, and then of the characteristic parameters of the involved processes, are discussed. On the basis of the results shown it should be possible to discriminate between normal and tumour tissues.

12.1 Introduction

The importance of the role of electromagnetic waves within biological systems has been more and more acknowledged in the last decades. In some papers[4,7,11,13,16,19,21] the strong connection between a wide range of electromagnetic waves and the state of living matter has been demonstrated, together with the double role that such fields may play, namely as:

1. indexes of the functional status and
2. promoters or inhibitors of specific processes.

Significant interactions of electromagnetic waves with biological systems seem to be present in a wide range of frequencies, although there are several problems connected with their reproducibility and the lack of refined and suitable experimental techniques. These interactions are ranging from the radio waves[13], known in orthopedic therapy and connected to phenomena suffered by hyperallergic people[20], to the microwaves[7], effective at an intensity so low as to exclude thermal modification in the biological samples, to the visible waves[4,11,16,19,21], where the therapeutic application of He-Ne lasers in dermatology is well known, up to near the ultraviolet waves, the range in which the existence of the mitogenetic radiation[8,9] has been postulated.

The range between near infrared and near ultraviolet has been the object of interesting research, giving rise to several experimental results and suggesting stimulating theoretical implications[5,12,14,15,22]. The first investigations in this field were performed at the beginning of this century by A.G. Gurvich[9], to explain some experimental observations on duplicating cells. The presence of this radiation was revealed in Italy forty years ago by Facchini et al.[1,2], who discovered that living systems emit light at a very low level.

Nowadays the detection and the analysis of the interaction between biological systems and electromagnetic waves of this range has become more feasible due to the reliability of the experimental apparatus devoted to single photon counting detection and to the production of electromagnetic fields with stabilized frequency and amplitude.

The original Gurvich hypothesis suggested that living systems transmit and receive information by means of electromagnetic waves. This statement is at present controversial and neither the experiments nor their interpretation are conclusive[5]. It is widely accepted, however, that the intensity of the emitted radiation and its statistical properties is closely connected to the biochemical processes occurring in a system, as well as its physiological state. This suggests the possibility of using the emitted radiation as a non-destructive tool to investigate the "status" of a living system.

Attempts have already been made to use the measurement of ultraweak radiation in agriculture[21], biology and medicine[10,18,23,24] as a qualitative diagnostic method for different pathologies including cancer.

Experiments on tumour cell cultures have been made in Popps group (University Marburg) by Bernhard Ruth[16]. The radiation emitted from Yoshida and Ehrlich tumour cells, grown respectively in the abdomen of rats and mice, was measured. The measurements were performed after the cells were taken from the animals. The spontaneous emission was not sufficient to be detected; but a slight increase occurred with the addition of a cytostatic substance (e.g. Cyclo-Phosphamide). However, the addition of 4-hydroperoxicyclophosphamide produced a remarkable increase in emission with a maximum in about an hour.

The integrated rate of radiation was calculated and connected to the quantity of chemical substance added to the cells; the quantum yield thus obtained was not proportional to the quantity of added substance, but varied by two orders of magnitude. This property could lead us to postulate that the emitted radiation was not chemiluminescence, but a biological phenomenon in which the produced radiation itself induces radiation of the same type. In this case a physical mechanism should be considered in order to explain the experimental data.

In Cracow, Jan W. Dobrowolski, Ali Ezzahir and Michal Knapik performed research into the chemiluminescence from cancer and normal cells, particularly from leukemic and normal blood cells[3]. Two different types of measurements were carried out: the spectral distribution and time counting series. In the wavelength range between 340 and 600 nm, the photon emission from cells of breast cancer exposed to monochromatic light of a wavelength of 295 nm, and from leukocytes of patients suffering from different leukemic pathologies was measured; the spectral distributions obtained were different from the control ones, either in shape or intensity, mainly for the control leukocytes that have shown a very low intensity, in comparison with the ones from leukemic people.

Time counting series were carried out on blood serum and on lymphocytes of patients suffering from chronic lymphatic leukemia; the spontaneous emission from controls was lower than from the leukemic samples and, after irradiation with UV rays, the emission increased in both cases, but control blood serum, in contrast to the lymphocyte one, emitted less than leukemic samples. Due to the fact that the emitted radiation should be connected to the recombination of free radicals, the yield could give a sensitive indicator of the amount of anti-oxidants in the biological system that depends on the stage of the development of the cancer.

In Kaiserslautern and Heidelberg, W. Scholz, U. Staszkiewicz, F. A. Popp and W. Nagl studied the ultraweak photon emission from normal human amnion cells and malignant transformed Wish cells[18]. No spontaneous emission from the samples was measured, at least in the limits of the dark count rate of the photomultiplier; whereas after irradiation with visible light, they exhibited a great intensity followed by a time decay. The decays were better fitted by an hyperbolic approximation than by an exponential one and the hyperbolic decay constant was used as a characteristic parameter.

Different measurements of cell density were performed and the trends obtained for the two different types of cells were easily discriminated: the decay constant of the Wish cells showed slightly variable values, whereas the one of the amnion cells showed a strong dependence on the cell density; this last result was interpreted as a prolonged capacity for storage of the excitation energy with increasing amnion cell density. This technique, if the results are confirmed in other cell systems, could be used to discriminate between normal and malignant cells.

Denis H. J. Schamhart and R. van Wijk studied the photon emission from different tumour cells characterized by a different degree of differentiation[17]; the samples were normal hepatocytes and in vitro cultured hepatoma cells. The integral of emitted photons, after the illumination with visible light, was measured for different cell densities: the emission of poorly differentiated tumour cells showed a non-linear increase with increasing cell density, while the emission of well differentiated tumours showed no dependence on cell density; the normal hepatocytes gave rise to a moderate decreasing trend. So, at high cell density, there was a great difference in the emission between normal and tumour cells that could be discriminated; in addition, it should also be possible to correlate the emission to the degree of differentiation.

Despite this research on cancer, there are not many experimental data on the photon emission from tumours in vivo[23] or from surgically removed tumour tissues[6], so the correlation with the pathological characterization of the examined samples was quite incomplete. Moreover some of the available data[23] have been obtained by means of biological detectors whose use and application limits are not yet well established.

Nevertheless, it is evident that the possibility of performing analysis directly on surgically removed tissues or on the patients should have a great applicative interest, because this type of analysis is relatively easy, quick and non-destructive.

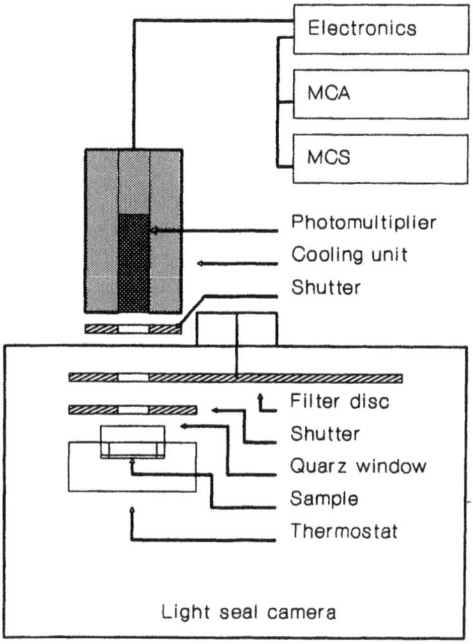

Figure 12.1: *Schematic block diagram of the experimental setup.*

In the following, the results of measurements of low luminescence from tumour and normal tissues will be reported. The measurements have been made in the Institute of Physics in collaboration with the Institute of the Ist General Surgery Clinic of the University of Catania; the aim of this research was to establish a measurement and analysis technique suitable to discriminate the tumour pathologies on the basis of their ultraweak emission in the visible range.

12.2 Materials and Methods

To measure the photon emission from biological systems a special experimental setup has been designed and assembled, as illustrated in Fig. 12.1. It consists of a dark steel chamber, where the sample to be analyzed can be maintained at a constant temperature by means of a thermostatic sample holder, the temperature of which can be set between 5°C and 80°C and maintained within ± 0.1°C.

The sample is put into a pyrex or plastic cup, which is covered by a Suprasil quartz window, transparent to UV light up to 190 nm; the window prevents the evaporation of liquids contained in the sample. To avoid condensation on the window, it is heated to a temperature 1 – 2°C higher than that of the sample.

A light shutter is placed just above the window. The shutter can be closed by an electromagnetic actuator to measure the chamber emission background; in the absence of sample, the background has been shown to be the same whether the shutter is open or closed. Just over the shutter the light source is mounted, which has a circular shape in order that the photons emitted from the sample can reach the photomultiplier; the light source is made of interchangeable LED groups, emitting different wavelength radiation centered on 565, 583 and 657 nm.

Above the light ring lies an aluminum disk which acts as a filter-holder for 15 different filters. The disk can be rotated electrically so that different filters are interposed between the sample and the photomultiplier; this allows the spectral distribution of the emission from the sample to be measured in the range between 200 nm and 700 nm. The dark chamber is closed by a second light shutter above which the photomultiplier is fastened.

The photomultiplier (Thorn EMI type 9635 QA or 9558 QA) is outside the main chamber, and is cooled to −30°C by means of a Peltier cooling system. This temperature-reduction makes it possible to decrease the dark current by three orders of magnitude.

Pulses coming from the photomultiplier are amplified, shaped, discriminated by an appropriate electronic chain, and sent to a multiscaler which counts the number of pulses arriving during each preset time interval; up to 4096 equal time intervals can be fixed, in the range from 2 microseconds to 30 minutes each. Under operating conditions, the background count rate is only about 8 pulses/ sec as is necessary if the low rate emission from samples is to be detected.

The intensity of the emitted light depends on several factors. These include the physiological condition of the sample itself and its preparation and handling, and also the physiological and pathological conditions of the living system from which the sample has been taken, i.e. of the patient in the present case.

To reduce masking effects due to sample handling, a standard procedure for sample preparation and treatment was followed. Samples were surgically removed human tissues. As soon as they had been taken from the living body, the samples were put into a dark container which was thermostatically maintained at 37°C, and immersed in Ringer solution to allow the metabolic processes to be maintained. Samples were put into the measuring chamber 30 – 100 minutes after removal. After the photon emission measurements the size was measured and it was of the order of one or a few cm^3.

In order to study the influence of the sample preparation on the photon emission, some tests were performed under different conditions; some samples were not immersed in the Ringer solution. The measurements were always carried out on different freshly cut pieces belonging to the same tissue.

12.3 Results

It has been found, in the preliminary observations, that the spontaneous emission was characterized by very low intensity (from ten to hundred counts per second per cm^2 of sample surface) and that the intensity decreased with the increasing time from the surgical removal; the decreasing rate depended on the conditions under which the samples were maintained.

To determine the best conditions in which the samples should be maintained, some measurements have been performed on tissues taken from the same patient and stored as they were or immersed in Ringer solution. The effect of the Ringer solution was to increase the emission by about 30% and to keep the signal decay time constant up to about 1 hour[6], whereas without Ringer the decay time was about half an hour. Moreover some samples were cut and the photon emission was quickly measured; the results obtained showed enhanced intensity and became almost independent on the time elapsed since the surgical removal.

The problem with the normalization of the signals was due to the fact that it was impossible to obtain samples of the same mass and volume, so it was necessary to find some rule for normalization. Making the hypothesis that the measured quantity was an extensive one, the normalization should be relative to the area or to the mass of the samples. In order to discriminate between these two possibilities, measurements were performed on four samples, taken from two different tumour tissues and having different mass and area facing the photomultiplier surface. The results are shown in Fig. 12.2 and suggest that the normalization to the facing surface area should be, for the spontaneous emission, the best one.

The reported measurements belong to two series of samples which have been analyzed in different periods of time and with different purposes.

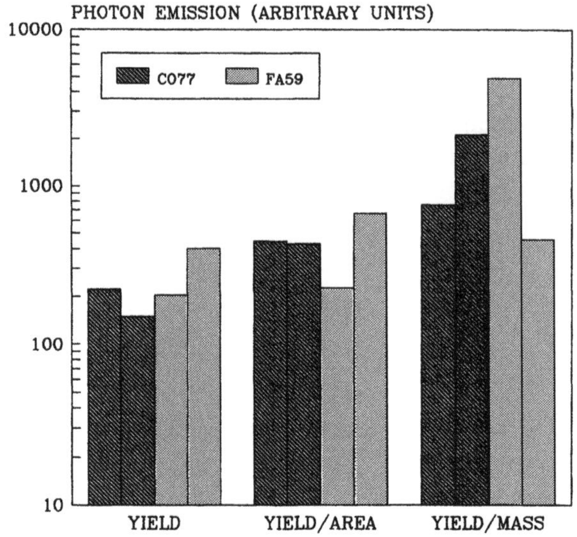

Figure 12.2: *Different normalization methods of photocounts from two tumour samples. Each sample has been cut in two pieces with different mass and area.*

12.3.1 Spontaneous emission

The first group concerns the spontaneous photon emission from human tissue samples; the aim of the measurements was to establish the existence of the ultraweak radiation, to measure it and to exploit the possibility of discriminating tumour from normal tissues.

Among the 25 samples examined in the first series of experiments, 16 were from tumours and 9 were normal (non-tumour) tissues. The characteristics of the samples, including their size and weight, and the pathological description, are shown in Table 12.1, 12.2. The column 3 "sample size" reports the average dimension of the sample in mm; the first and second figures are the dimensions of the side facing the photomultiplier, while the third is the sample thickness.

The Table 12.1 also shows the intensity of the light emitted (number of photons/($cm^2\cdot$ min) after subtraction of background), the estimated error in the counting rate, and the time that elapsed between the surgical removal of the sample and the emission measurement. From Table 12.1 it is evident that all the normal samples emitted light with negligible intensity, whereas the tumour samples emitted light with an intensity which could be as high as 1400 photons/($cm^2\cdot$ min). All of

Table 12.1: Measured photon emission from tissues

Sample #	Patient code	Sample size mm	Sample weight g	Emission photons $/cm^2 \cdot$ min	Elapsed time min
1	sa46	30×15×5	1.64	266±3	30
2a	fm56	5×3×2	.02	66±50	100
2b	"	4×2×2	.02	0*±12	110
3	xx	-	-	348±12	45
4	pg60	5×3×2	.03	1440±80	120
5	1r69	-	-	228±12	70
6	va58	9×6×2	.18	510±24	90
7	cl74	20×16×8	2.92	819±15	55
8	ma57	40×30×10	9.75	24±1	30
9	g	25×20×3	1.29	34*±2	30
10	fr	35×25×4	2.68	13*±1	45
11	na	10×20×5	1.23	84±6	50
12	mr70	12×15×3	.30	84±6	350
13	rs67	30×20×8	3.66	24±3	35
14	dc	10×15×5	1.14	138±8	40
15	s	45×12×4	2.92	34*±2	30
16	fr	30×15×6	2.67	108±12	30
17a	cr68	10×10×1	.24	24±24	95
17b	"	20×10×4	.95	12*±12	85
18	at	10×6×4	.24	320±24	20
19a	tn61	13×8×4	.35	306±36	55
19b	"	30×20×7	4.31	40*±6	85
20a	vm39	17×8×4	.41	40*±9	90
20b	"	14×10×4	.52	0*±9	100
21	m66	20×20×6	2.15	26*±3	130

The symbol * indicates non-tumour tissues. The pathological description of the tissues is displayed in the next table.

Table 12.2: Pathological description of the tissues

Sample	Description of pathologies
1	metastasis from infiltrating ductal carcinoma
2a	thymoma with preminent lymphocytic component
2b	perothymic normal lymph node
3	gastric adenocarcinoma
4	mild differentiated adenocarcinoma
5	light cell kidney carcinoma
6	lymphnodal metastasis from infiltrating ductal carcinoma
7	omentum metastasis from liver carcinoma
8	infiltrating ductal carcinoma
9	chronic appendicitis
10	lithiasic chronic cholecystitis
11	poorly differentiated adenocarcinoma
12	diffuse malignant lymphoma w/ small & great cells
13	epidermoid well differentiated carcinoma
14	pleomorphic sarcoma
15	chronic appendicitis
16	mild differentiated adenocarcinoma
17a	ductal carcinoma with peritumoral lymphocytic infiltrate
17b	normal breast tissue
18	melanoma
19a	lobular infiltrating carcinoma
19b	normal breast tissue
20a	chronic appendicitis
20b	lithiastic chronic cholecystitis
21	normal thyriod tissue

Table 12.3: The ratio of intensities of photon emission from tumour sample to emission from normal tissue.

Patient code	Emission ratio tumor/normal	Statistical error
MA57	1.5	0.3
RG66	10.0	0.4
CO77	1.9	0.3
FA59	8.6	0.8

the 9 samples belonging to the sub-set of non-tumour tissues have an emission rate lower than 40 photons/($cm^2\cdot$ min), with an average value of 22 $\pm$ 6 photons/($cm^2\cdot$ min). On the other hand, the 16 samples belonging to the sub-set of tumour tissues showed a much wider distribution of emission rates. The average value was 300 $\pm$ 90 photons/($cm^2\cdot$ min).

The results indicate that the two sub-sets have different populations: if they actually belong to the same distribution, the measured intensities of emission by the tumour samples should be distributed with equal probability below and above the average value as those of the non-tumour sub-set (22 photons/($cm^2\cdot$ min)). However, the rates for only 3 samples out of 16 fell below the average, whereas the rates for 13 samples were above. This difference remained even when the most unfavorable condition was considered, i.e. when the statistical error for the photon counting is subtracted from the measured value.

By using standard statistical methods it is possible to show that this conclusion has a reliability of 99% .

In the second series of measurements the spontaneous photon emission was compared between the tumour tissue sample and a contiguous normal tissue one belonging to the same patient. The results are shown in Tab. 12.3 as the ratio of the emission from tumour sample to that from normal one: the yield relative to tumour tissues is always higher outside the experimental errors.

12.3.2 Photoinduced emission

The second series of measurements was performed on the radiation emitted from tissues after the excition by a light source.

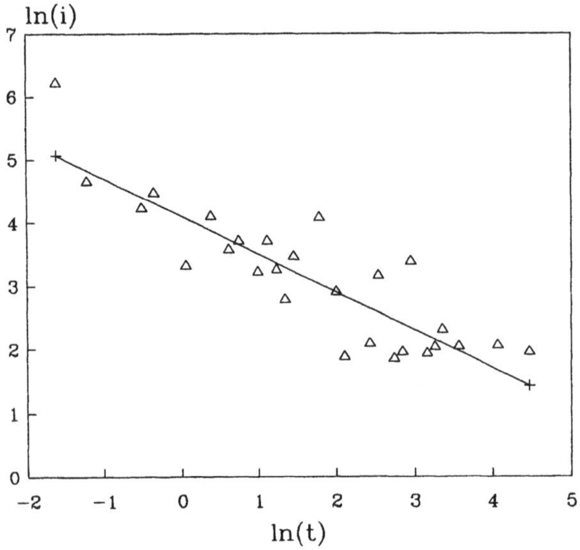

Figure 12.3: *Time decay of photoinduced emission from a tumour sample. The line is the hyperbolic fit of the experimental data.*

The samples have been illuminated for 10 sec by means of three matrices of LEDs. Each one centered on a different wavelength: 657 nm (red), 583 nm (yellow) and 565 nm (green); the light intensity on the samples was about 7.8×10^{-15} photons/$(s\cdot cm^2)$ for red light and 4.1×10^{-16} photons/$(s\cdot cm^2)$ for yellow and green light.

The analysis of the results obtained shows that the time decrease of the photoinduced emission of all the samples, calculated by subtracting the relative photoinduced emission of the chamber, is well described by a simple hyperbolic decay characterized by:

$$i(t) = \frac{e^{b_o}}{t^{b_1}}$$

This trend (see Fig. 12.3) has been shown at intervals going from 0.1 to about 100 seconds starting from the switching-off of the lamp. The lower limit is imposed by the presence of a time lag from the interruption of lighting to the complete opening of the shutter of the photomultiplier. The higher limit is determined by the presence of the background noise of the experimental setup.

Plotting this trend on a log-log diagram, it has the shape of a straight line, whose intercept with Y axis is b_0 and whose slope is b_1; so it becomes very easy to determine,

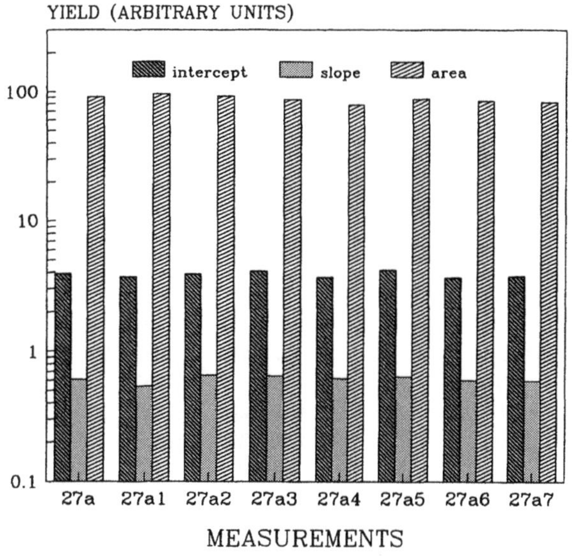

Figure 12.4: *Hyperbolic decay parameters of the photoinduced emission relative to the tumour sample MC55, measured different times in the same experimental conditions.*

by a linear regression, the above parameters.

This analysis was performed on the sample of the second series and the results are shown in Table 12.4; in this table the pathological description of the samples, the intensity of spontaneous light emitted (number of photons/($cm^2\cdot$ min) after subtraction of background) normalized to the area of sample, the estimated error, the wavelength of the stimulating light source, the intercept b_0 and the slope b_1 of the linear regression in the log-log plots and the relative integrals are reported. When the light stimulation was repeated on the same sample, only the parameters of the hyperbolic decays are reported in Table 12.4, 12.5.

The lacking figures in the b_0 and the b_1 columns are due to the impossibility to fit the experimental data with a correlation coefficient greater than 0.7.

The measurements with light stimulation were repeated several times on some samples, in order to check the reproducibility of the calculated parameters; in Fig. 12.4 one example of these measurements is reported. One can see that the ability to reproduce is very good for all the above parameters which can be so considered as the characterizing status of the sample.

This measurement reproducibility is good for the emission induced by stimulation

Table 12.4: Decay parameters (The pathological description of the tissues is displayed in the next table).

Sample #	Patient code	Photon counts/cm^2· min	λ nm	b_0	b_1	intlog
22a	MA57	192±30	657	2.3	0.319	57
22b	MA57	294±66	657	2.8	0.262	93
22b1			657	2.9	0.363	69
22c	MA57	0±42	657			57
23a	RG66	1200±180	657			60
23b	RG66	132±30	657	2.3	0.386	55
23b1			657			27
24a	CO77	420±120	657	4.1	0.546	85
24b	CO77	420±120	657			72
24c	CO77	168±24	657			43
24c1			657	2.3	0.591	38
25a	FA59	228±60	657			14
25b	FA59	252±36	657	1.5	0.194	3
25c	FA59	54±24	657	1.8	0.389	9
25c1			657	1.8	0.399	6
26a	CL71	49.8±18	583	4.4	1.081	108
26a1			583	4.3	1.072	104
26a2			583	4.3	1.027	106
26b	CL71	54±6	583	2.7	0.983	52
26b1			583	2.7	1.114	47
26b2			583	2.9	1.065	56
26c	CL71	78±30	583	5.1	1.042	126
26c1			583	5.2	1.231	126
26c2			583	5.2	1.114	129
27a	MC55	774±42	583	3.9	0.603	91
27a1			583	3.7	0.542	96
27a2			583	3.9	0.659	93
27a3			583	4.1	0.649	87
27a4			583	3.7	0.622	79
27a5			583	4.2	0.647	88
27a6			583	3.7	0.604	85
27a7			583	3.8	0.601	83
28a	FG45	414±30	583	3.4	0.449	96
29a	DG61	360±24	565	2.9	0.542	16
29a1			565	2.1	0.443	19
29a2			565	1.7	0.254	34
30a	FA63	276±30	565	2.6	0.371	54
30a1			565	3.9	0.393	100

Table 12.5: Pathological description of the tissues

Sample	Description of Pathologies
22a	normal glande tissue
22b	breast cancer
22b1	—
22c	axillary tissue
23a	rectum cancer
23b	normal mucous tissue
23b1	—
24a	gastric antrum tumor
24b	gastric antrum tumor
24c	perigastric normal tissue
24c1	—
25a	colon cancer
25b	colon cancer
25c	colon normal tissue
25c1	—
26a	colon cancer
26a1	—
26a2	—
26b	colon normal tissue
26b1	—
26b2	—
26c	colon cancer
26c1	—
26c2	—
27a	rectum cancer
27a1	—
27a2	—
27a3	—
27a4	—
27a5	—
27a6	—
27a7	—
28a	stomach cancer
29a	rectum cancer
29a1	—
29a2	—
30a	gastric sarcoma
30a1	—

with red and yellow light, whereas it is not apparent when the stimulation is performed with green light. Therefore red light stimulation was used in order to study the possibility of discriminating between tumour and normal tissues.

Fig. 12.5 shows the values of the parameters describing the hyperbolic decays of some tumour tissues and the contiguous normal one belonging to the same patient. It seems that b_0 and intlog should be suitable to discriminate tumour samples from normal ones (Fig. 12.5a and Fig. 12.5b), whereas the slope b_1 is not effective for this purpose (Fig. 12.5c). However, the relation between b_0 and b_1 (Fig. 12.5d) is strongly different for tumour and normal tissue and this could furnish a good criterion of discrimination.

12.4 Discussion

On the basis of the reported results one can affirm that it is possible to discriminate between tumour and normal tissues by measuring their spontaneous low-level luminescence.

This possibility has a great importance for practical application since it could permit a fast analysis of human tissue samples (each measurement requires about ten minutes).

Nevertheless, there are still many problems to be solved. For instance, the experimental data show that some tumour samples (e.g. #8, #13 and #17a) have a low emission rate which cannot be distinguished, within the limits of statistical error, from that of normal tissues. At present there is no evidence to show whether or not this is due to anomalous behaviour.

Moreover the present data are not sufficient to show whether there is a significant correlation between the degree of malignancy and the emission intensity. However, some tumour samples that emitted at a very low intensity (e.g. #13) are definitely characterized by a low degree of malignancy, whereas some of those that emitted at high intensity are from highly malignant tumour tissues.

With regard to the photoinduced emission, the choice of the parameters suitable for the description of the phenomenon seems to play a fundamental role. The reported data suggest that it is possible to use the intercept and the area relative to the log-log plot of the photoinduced decay as discriminating parameters, whereas the slope seems to be ineffective.

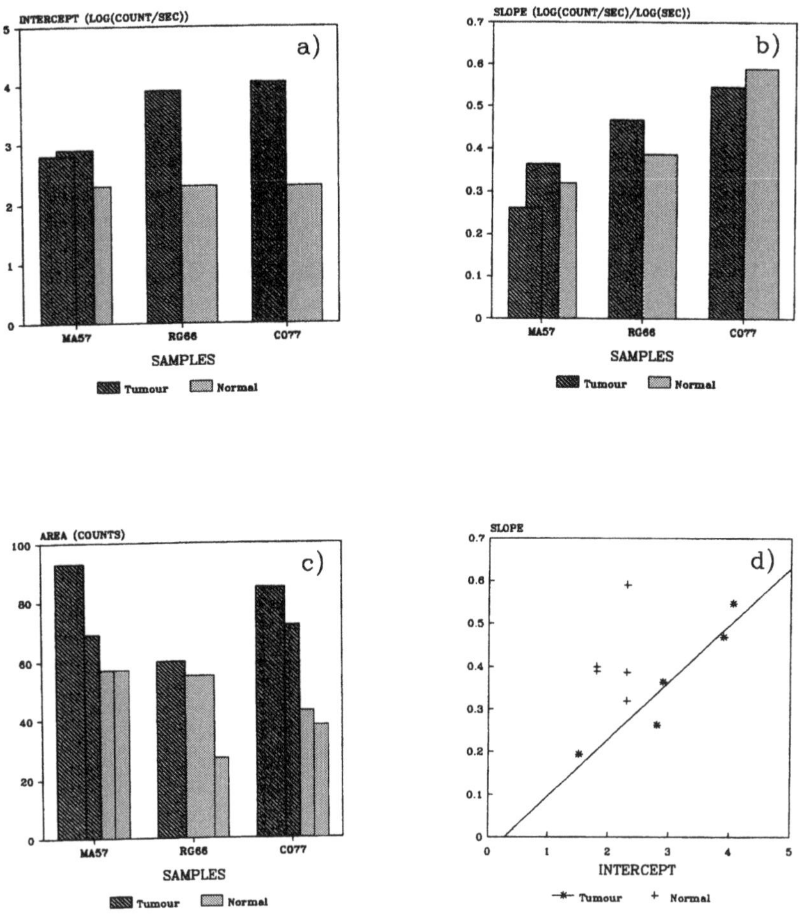

Figure 12.5: *Comparison between normal and tumour photoinduced emission as described by different parameters of the hyperbolic decay.*

These considerations seem to be in contrast with the results of other researchers[18], who considered just the slope as the significant parameter to distinguish tumour from normal cell cultures. However, their experimental conditions were quite different: the measured samples were cell cultures instead of tissues, the light source was not the same and the intensity was higher. Nevertheless, as the slope and the intercept depend on the intensity of the stimulating radiation, and the correlation between the two parameters is different for tumour and for normal samples (Fig. 12.5), the contrast could disappear by stimulating the tissue emission by a higher intensity light source.

In conclusion, the experimental data produced at the Institute of Physics of the University of Catania and those which are reported in the cited references permit to establish that the spontaneous and the photoinduced ultraweak radiation emitted from tumour tissues are different from the ones emitted from normal tissues.

However,before this technique can be used for medical application, it is necessary to improve research in this field in order to correlate the biological parameters describing the tumour pathologies with the physical parameters which characterize the emitted ultraweak radiation.

If this goal is reached, this technique could become a powerful, quick and non-destructive analysis method for the pathological status of the patients and of the effectiveness for the adopted therapies.

12.5 References

1. L. Colli and U. Facchini, Light emission by germinating plants. *Il Nuovo Cimento* **12** (1954) p. 150-153.

2. L. Colli, U. Facchini, G.Guidotti, R. Dugnani-Lonati, M. Orsenigo and O. Sommariva, Further measurements on the bioluminescence of the seedlings. *Experientia* **11** (1955) p. 479-484.

3. J.W. Dobrowolski, A. Ezzahir and M. Knapik, Possibilities of chemiluminescence application in comparative studies of normal and cancer cells with special attention to leukemic blood cells, in *Photon Emission from Biological Ystems*, eds. B. Jezowska-Trzebiatowska, B. Kochel, J. Slawinski and W. Strek (World Scientific, Singapore, 1986) p. 170-183.

4. G.E. Fedoseyeva, T.I. Karu, V.S. Letokhov, V.V. Lobko, N.A. Pomoshnikova, T.S. Lyapunova and M.N. Meissel, Effect of He-Ne Laser Radiation on the Reproduction Rate and Protein Synthesis of the Yeast. *Laser Chem.* **5** (1984) p. 27-33.

5. H. Fröhlich, Theoretical Physics and Biology, in *Biological Coherence and Response to External Stimuli*, ed. H. Fröhlich (Springer Verlag, Berlin, 1988) p. 1-24.

6. F. Grasso, C. Grillo, F. Musumeci, A. Triglia, G. Rodolico, F. Cammisuli, C. Rinzivillo, G.Fragati, A. Santuccio and M. Rodolico, Photon Emission from Normal and Tumour Human Tissues. *Experientia* **48** (1992) p.10-13

7. W. Grundler, U. Jentzsch, F. Keilmann and V. Putterlik, Resonant Cellular Effect of Low Intensity Microwaves, in: *Biological Coherence and Response to External Stimuli*, ed. H. Fröhlich (Springer-Verlag, Berlin-Heidelberg, 1988) p. 65-85.

8. A.A. Gurvich, A historical review of the problem of mitogenetic radiation. *Experientia* **44** (1988) p. 545-550.

9. A.G. Gurvich, Die Natur des spezifischen Erregers der Zellteilung. *Arch. Entw. Mech. Org.* **100** (1923) p. 11-40.

10. M.J. Harber, Application of luminescence in medical microbiology and hematology, in *Clinical and Biochemical Luminescence*, eds. L. J. Kricka and I. J. Carter (Marcel Dekker Inc., New York, 1982) p. 189-218.

11. T.I. Karu, G.S. Kalendo, V.S. Letokhov and V.V. Lobko, Biostimulation of HeLa Cells by Low-Intensity Visible Light. *Il Nuovo Cimento D* **1** (1982) p. 828-835.

12. K.H. Li, F.A. Popp, W. Nagl, and H. Klima, Indications of Optical Coherence in Biological Systems and its Possible Significance, in *Coherent Excitation in Biological Systems*, eds. H. Fröhlich and F. Kremer (Springer-Verlag, Berlin-Heidelberg, 1983) p. 117-122.

13. J.K. Pollock and D.G. Pohl, Emission of radiations by active cells, in *Biological Coherence and Response to External Stimuli* ed. H. Fröhlich (Springer-Verlag, Berlin-Heidelberg, 1988) p. 65-85.

14. F.A. Popp, Biophoton emission. *Experientia* **44** (1988) p. 543-544.

15. F.A. Popp, K.H. Li, W.P. Mei, M. Galle and R. Neurohr, Physical aspects of biophotons. *Experientia* **44** (1988) p. 576-585.

16. B. Ruth, Experimental Investigations on Ultraweak Photon Emission, in *Electromagnetic Bio-Information*, 2nd ed., eds. F.A. Popp et al. (Urban & Schwarzenberg, München, 1989) p. 128-143; 1. ed. (1979) p. 107-122.

17. D.H.J. Schamhart, R. van Wijk, Photon Emission and the Degree of Differentiation, in *Photon Emission from Biological Systems*, eds. B. Jezowska-Trzebiatowska, B. Kochel, J. Slawinski and W.Strek (World Scientific, Singapore, 1986) p. 99-210.

18. W. Scholz, U. Staszkiewicz, F.A. Popp and W. Nagl, Light-Stimulated Ultraweak Photon Reemission of Human Amuion Cells and Wish Cells. *Cell Biophysics* **13** (1988) p. 55-63.

19. J. Slawinski and D. Slawinska, Low level luminescence from biological objects, in *Chemi- and Bioluminescence*, ed. J.G. Burr (Marcel Dekker Inc., New York, 1985) p. 495-531.

20. C.W. Smith, Coherent Electromagnetic Fields and Biocommunications, in *Electromagnetic Bio-Information*, 2nd ed., eds. F.A. Popp, U. Warnke, H.L. König and W. Peschka (Urban & Schwarzenberg, München, 1989) p. 1-17.

21. S. Tryka and R. Koper, Luminescence of cereal grain subjected to the effect of mechanical loads, in *Photon Emission from Biological Systems*, eds. B. Jezowska-Trzebiatowska, B. Kochel, J. Slawinski and W.Strek (World Scientific, Singapore, 1986) p. 248-254.

22. R. Van Wijk and D.H.J. Schamhart, Regulatory aspects of low intensity photon emission. *Experientia* **44** (1988) p. 586-593.

23. M.I. Yanbastiev, On some applied-medical aspects of biophotons, in *Photon Emission from Biological Systems*, eds. B. Jezowska-Trzebiatowska, B. Kochel, J. Slawinski and W.Strek (World Scientific, Singapore, 1986) p. 184-198.

24. B. Yoda, Y. Goto, A. Saeki and H. Inaba, Chemiluminescence of smoker's blood and its possible relationship to cigarette smoke components, in *Photon Emission from Biological Systems*, eds. B. Jezowska-Trzebiatowska, B. Kochel, J. Slawinski and W.Strek (World Scientific, Singapore, 1986) p. 199-210.

Chapter 13

Time Behaviour of Delayed Luminescence in Acetabularia Acetabulum

F. Musumeci, M. Godlewski, F.A. Popp and M.W. Ho

Abstract

The decay kinetics of delayed luminescence (200ms to 100s) in *Acetabularia acetabulum* were investigated with excitation sources of different intensity, wavelength and duration. The results are as follows:

1. The decay kinetics are hyperbolic, characterized by parameters that seem to be independent of excitation wavelength.

2. Clear oscillations are present, which are near periodic in a logarithmic time scale.

3. Both decay parameters and oscillations are significantly altered on poisoning the *Acetabularia.*

The possibility that the delayed luminescence is a coherent emission from the living organism is examined.

13.1 Introduction

The afterglow of green plants exposed to light illumination was first discovered by Strehler and Arnold[1].

The phenomenon is now well known as "delayed luminescence" (DL). Lavorel[2] introduced the "τ scale" which separates phases of the decay kinetics into fast ($\tau \leq 1\mu$s), medium and slow ($\tau \geq 1$ ms) relaxation dynamics.

The fast one, also called "prompt fluorescence"[3] has been assigned to the recombination of positive charges on the donor side of photosystem II (PS II) and negative charges on the acceptor side, corresponding to a partial reversal of the primary photoact in photosynthesis[2]. The slow decay which is sometimes referred to as "long time delayed luminescence" involves both PS I and II[4].

Most conclusions were drawn from similarities between known spectra of PS-compounds and the spectral distribution of DL. However, with increasing sensitivity of photocounting techniques it turned out that DL is a common phenomenon of living tissues and even of organic[5] as well as inorganic matter[6].

DL is sensitively dependent on a variety of influences, ranging from impurities, physical and chemical agents (like heat, CO_2 and oxygen) to physical properties of the exciting light and the procedure of illumination. However, the common factor is the relaxation dynamics of long-term DL. Instead of following a simple exponential law it relaxes according to a curve which has been empirically described by either a superposition of at least two exponential functions[2] or, even better, by a hyperbolic decay law[5,7,8].

This point is crucial, since exponential decay can be interpreted in terms of chaotic interaction (like random recombination processes in the matter under investigation)[9], while a hyperbolic decay under ergodic conditions is sufficient to indicate a coherent recoupling of the decay products (e.g. photons) to its source[10,11].

The analysis of the decay behaviour, which has also been used for uncovering the coupling of PS I and II in long-term DL[5], may then become a physical tool for examining collective properties of the matter under investigation, and consequently for describing the mechanism in terms of the degree of coherence of DL. In particular, it may provide new insights into the old question of whether there are several distinct forms of luminescence arising from distinct substrata and processes or whether there is only one luminescence that may be "triggered" by various external factors[2].

In order to demonstrate this, we have examined the long-term DL of *Acetabularia* under different conditions.

13.2 Material and Methods

All measurements of the luminescence were performed using the single photon counting technique[12,13]. The measuring equipment used in the present study is nearly the same as that described in reference[7]. For the photon detector, we used a Thorn EMI photomultiplier tube type 9558QB cooled to −20°C to reduce dark counts. Pulses from the photomultiplier were amplified in an Ortec pulse preamplifier and separation between noises and signal was done by means of an Ortec amplifier-discriminator.

The amplification coefficient, discrimination level and supplying photomultiplier (PMT) high tension values were optimized to achieve the optimum signal/noise ratio. Pulses were counted and stored in 8192 multiscaler channels of Nucleus MCS II type mounted in a PC AT computer. A dwelling time of 0.1 sec was used in the majority of experiments. A single individual *Acetabularia acetabulum* was placed in sterile artificial seawater in a quartz cuvette of about 15 ml capacity and put into the light-tight chamber maintained at 20°C in front of the PMT[7,12]. Between the sample in the measuring chamber and the PMT Photocatode metal interference type Schott filters were mounted to allow for measurement of emitted photons at chosen wavelengths. A slide shutter (driven pneumatically) was used to determine the dark counts of the PMT and to protect the PMT during photoexcitation of samples. The time required to open the shutter was about 250 ms.

Photoexcitation was done by illumination with white light from tungsten (150W) lamp or with monochromatic light from a halogen lamp passing through a monochromatic grating. To obtain different relative light intensities neutral grey Schott filters were used.

13.3 Analysis of Experimental Results

The analysis of the results shows that under all excitation and emission measurement conditions, re-emitted light follows a simple hyperbolic decay in the first 100 seconds after excitation (see Fig. 13.1)

It is always possible, of course, to use alternative models to describe this behaviour. For example, it can be expressed with a series of appropriately weighted exponential functions.

Table 13.1 shows parameters describing hyperbolic and multi-exponential fits of

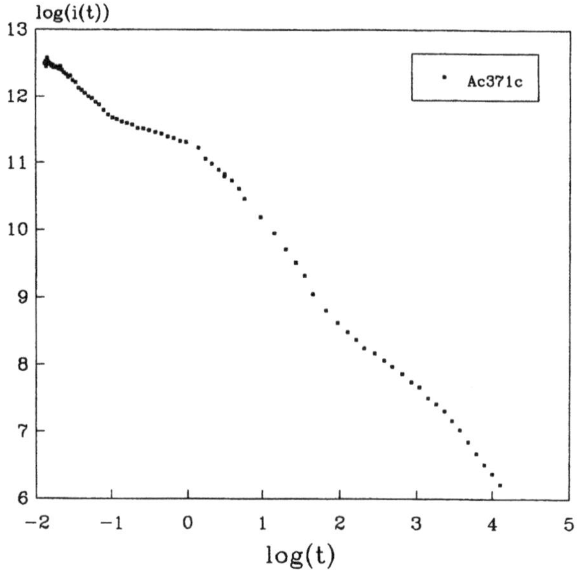

Figure 13.1: *Typical trend of delayed luminescence (DL)of* Acetabularia *reported in logarithmic coordinates.*

decay curves of the same *Acetabularia* excited with white light of two different intensities.

It is evident that totally different parameters must be used for the two multi-exponential fits, whereas in the case of the hyperbolic fits, only slight variations exist in the values of three parameters between the two cases.

This implies that the description in terms of a sum of exponentials is not realistic from a physical point of view as it requires a model in which different intensities of excitation activate mechanisms which are totally different one from another.

Furthermore, the presence of many decay mechanisms, each independent of the other and characterized by different decay times, is incompatible with the results shown in Fig. 13.2, where one can clearly see that the kinetics of decay of different spectral components are nearly identical to that of the total emission.

This result shows that specific spectral components do not have different exponential decay coefficient, as would be expected if the decay of the total emission were to be explained by a multi-exponential function.

In other words, the spectral components cannot be assigned to any specific photochemical or fluorescent reaction. On the contrary, a hyperbolic decay of a collectively

Table 13.1: Values of the parameters of hyperbolic and multiexponential function obtained from least square method fit to experimental data.

Hyperbolic function: $I(t) = \frac{I_0}{(t+t_0)^m}$			
decay curve	I_0	t_0	m
A	2.12×10^5	0.05	1.04
B	2.12×10^5	0.43	1.03

Multiexponential function: $I(t) = \sum_i I_i e^{-t/\tau_i}$								
decay curve	I_1	τ_1	I_2	τ_2	I_3	τ_3	I_4	τ_4
A	2.68×10^6	0.08	2.26×10^5	1.28	1.26×10^4	16.7	2.57×10^3	100
B	2.47×10^5	0.256	9.0×10^4	1.5	2.52×10^4	6.73	3.73×10^3	78

note – The decay curve A and B were measured for the same *Acetabularia* excited with white light. The ratio between the intensity of excitation light for curve A and B was 1000.

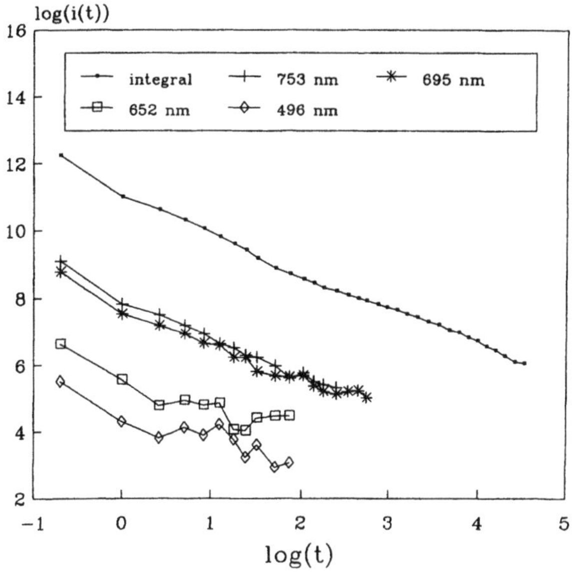

Figure 13.2: *Trend of some spectral components of DL of* Acetabularia *reported in logarithmic coordinates.*

coherent mode is, by far, the simplest and the most realistic interpretation of the data.

The counts per unit of time of the photomultiplier are proportional to the radiating flux averaged by the time. The time average of flux, in its turn, is proportional to the mean of the square of the electromagnetic field calculated at times which are of the order of the temporal resolution of the photomultiplier.

If one supposes that the electromagnetic field decays according to hyperbolic law, it is possible to write that (see Appendix):

$$I(t_n) = \frac{I_0}{(t_0 + t_n)^{2\alpha}} \cdot \frac{1}{2-\epsilon} \cdot (1 - \epsilon \sin^2(\beta \log(t_n + t_0) + \phi)), \qquad (13.1)$$

where I_0 is the intensity value at time $(t + t_0) = 0$, t_0 is a sort of "time of memory" of the system as described by Popp and Li[10], ϕ is the phase of the emitted photon, α and β, respectively, are the real and the imaginary part of the decay exponent of the electromagnetic field and ϵ is an empirical parameter that describes the damping of the variations due to partial coherence of the emitted photons.

In Fig. 13.3, the result of this analysis, performed on a certain number of samples, is shown.

The analysis has been done by calculating the best fit of the $\log(I(t))$ compared to the $log(t + t_0)$ and assuming as the value of t_0 the one which made the coefficient of determination most accurate.

Once the values of I_0, 2α and t_0 have been calculated in this way, we have determined the values of the remaining coefficients in order to reduce the sum of the square of the discards as much as possible.

In Table 13.2, the values of these parameters, for the different measurements, and the value of I_s (the I value when $t_n = Dt_s$) are shown. The value of β is not given because in all the cases that we have analyzed, $\beta \sim 2\alpha$, within the limits of experimental error. Furthermore, for many measurements it was not possible to give the values of ϵ and ϕ because in these cases the signals were too low to distinguish the variations from the background noise.

It is of interest to see whether the parameters are, in some way, correlated. At a qualitative level, this correlation can be observed between the intensity and the duration of illumination: when the intensity decreases, ϵ tends to 0 and the variations disappear.

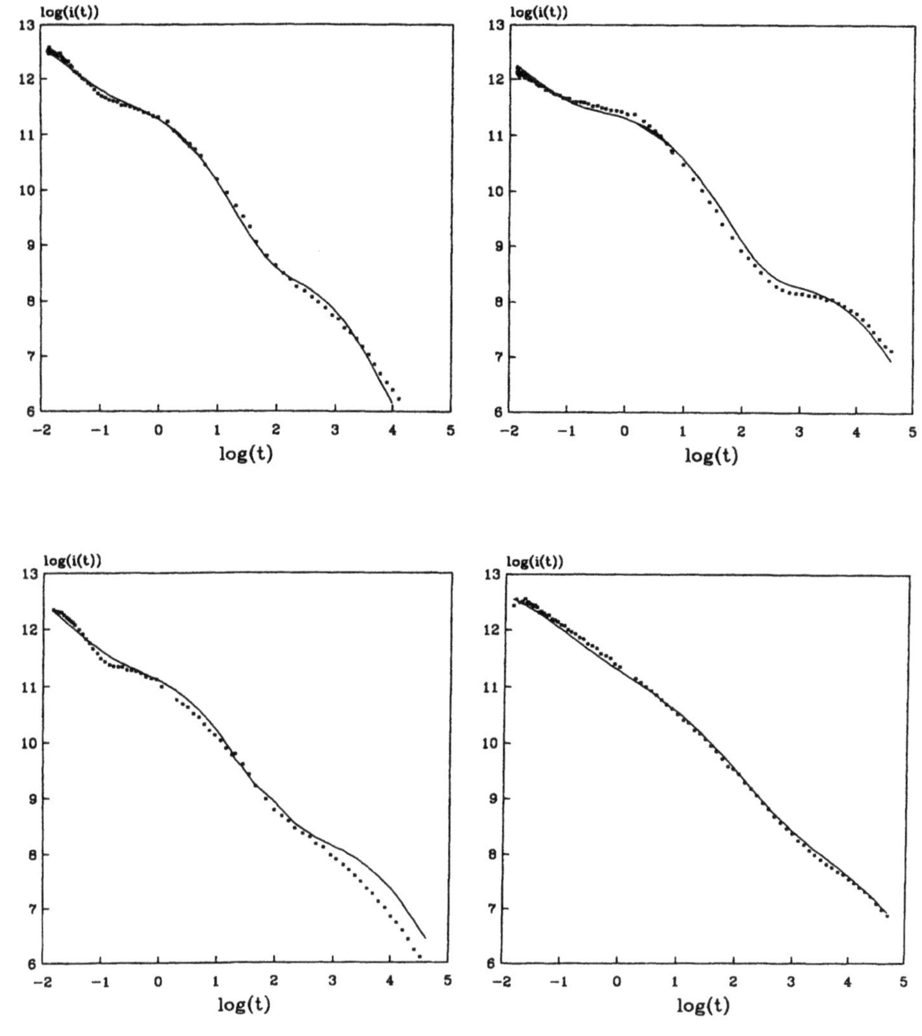

Figure 13.3: $\log i(t)$ *as functions of time parameter* $\log t$. $\cdots$ *: data of* Acetabularia, —*: theoretical curves.*

Table 13.2: Values of the parameters of Eq. (13.1) obtained from least square method fit of equation to experimental data.

#	$I(t_1)$	I_0	t_0	2α	ε	ϕ	excitation light
Ac1	4.85×10^5	7.56×10^4	0.05	1.04	0.4	0.39	white I
Ac1	1.66×10^5	9.74×10^4	0.15	1.00	0.5	0.5	white I/40
Ac1	2.36×10^5	2.37×10^5	0.25	0.97	0.4	0.75	white I/400
Ac1	9.90×10^3	4.31×10^4	1.10	0.86	0.1	2.60	white I/4000
Ac1	1.39×10^5	2.21×10^5	0.35	1.03	0.2	1.30	white I/1000
Ac2	2.22×10^5	2.24×10^5	0.38	1.17	0.4	0.79	white I/4
Ac2	1.52×10^5	1.70×10^5	0.30	0.98	0.5	0.79	white I/40
Ac2	2.18×10^5	1.80×10^5	0.37	1.27	0.4	0.79	white I
Ac3	2.29×10^5	1.42×10^5	0.30	1.10	0.4	0.79	white I
Ac3	9.14×10^3	1.94×10^4	1.10	0.83	[b]	[b]	375nm
Ac3	7.56×10^3	2.60×10^4	1.80	0.98	[b]	[b]	400nm
Ac3	2.36×10^4	3.40×10^4	0.70	0.87	[b]	[b]	422nm
Ac3	5.42×10^4	5.96×10^4	0.50	0.92	[b]	[b]	451nm
Ac3	3.32×10^4	6.80×10^4	1.00	1.00	[b]	[b]	476nm
Ac3	2.56×10^4	5.02×10^4	1.00	0.91	[b]	[b]	496nm
Ac3	7.48×10^3	1.94×10^4	1.50	0.78	[b]	[b]	526nm
Ac3	6.49×10^4	1.07×10^5	0.80	1.01	[b]	[b]	550nm
Ac3	2.74×10^4	6.80×10^4	1.30	0.90	[b]	[b]	574nm
Ac3	9.20×10^4	1.92×10^5	0.90	1.12	[b]	[b]	595nm
Ac3	6.62×10^4	8.80×10^4	0.60	0.95	[b]	[b]	626nm
Ac3	5.26×10^4	9.20×10^4	0.90	1.00	[b]	[b]	652nm
Ac3	7.24×10^4	9.20×10^4	0.60	0.99	[b]	[b]	665nm
Ac3	6.89×10^4	8.40×10^4	0.60	0.97	[b]	[b]	680nm
Ac4	2.83×10^3	9.80×10^3	2.00	0.89	[b]	[b]	451nm I/5
Ac4	1.93×10^4	2.60×10^4	0.70	0.91	0.12	0.2	451nm I
Ac4	3.59×10^5	5.13×10^4	0.40	0.96	0.08	0.6	451nm I×4
Ac5	4.26×10^4	8.00×10^3	2.00	0.82	[b]	[b]	550nm I/5
Ac5	1.58×10^5	8.02×10^4	1.00	1.01	0.1	0.6	550nm I
Ac5	2.86×10^3	1.36×10^5	0.10	1.02	0.1	0.4	550nm I×4

[a] I is the full intensity from halogen lamp impinging on the sample. [b] For this cases DL intensity was too low to determine ε and ϕ.

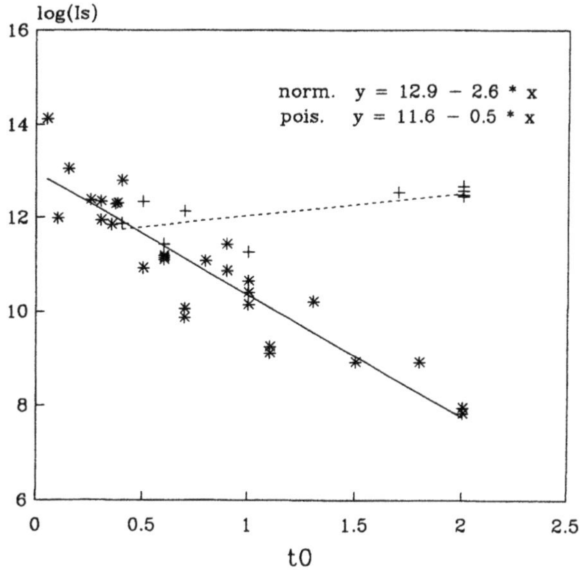

Figure 13.4: *Relation between* $\log I_0$ *and* t_0 *for normal (*) and poisoned (+)* Acetabularia. *The two equations at the high right corner represent the best fit of normal (—) and poisoned (- - -) samples.*

A more rigorous analysis can be performed for the parameters which characterize the decay. In particular, we observed (Fig. 13.4) that there is a linear relation between $\log(I_s)$ (normalized to the value of the corresponding emission of the sample considered as reference) and the value of t_0.

Another linear relation can be observed between the values of I_0, normalized in the way we described earlier, and the values of 2α but the coefficient of determination is rather low (about 0.6).

An improvement of this relation can be obtained if values of t_0 which are obtainable from the linear relation in Fig. 13.4. are chosen. In this case, we still obtain very good, best fits for the $\log(I(t_n))$ *vs. log*$(t_n + t_0)$ (in every case the coefficient of determination is > 0.95) and the coefficient of determination related to the linear relation between the $\log(I_0)$ and 2α, is about 0.64 (see Fig. 13.5).

This suggests that there is a relation between the parameters which describe the hyperbolic decay such that, once one of them is determined, it will be possible to predict the others.

Thus, the experimental curves can be described by the relations

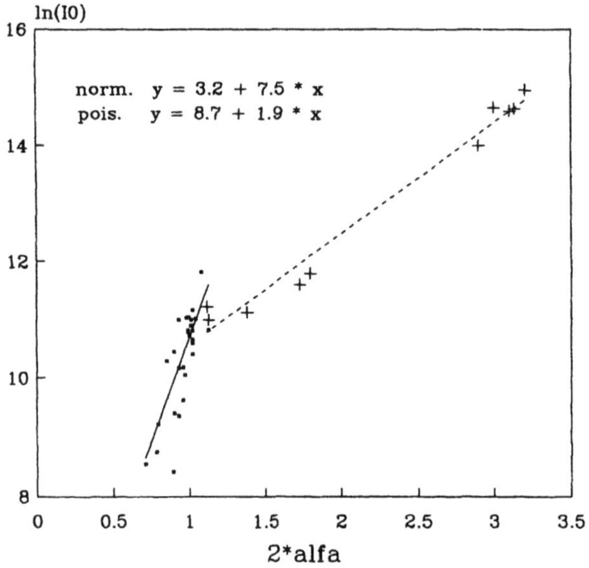

Figure 13.5: *Relation between* $\log I_0$ *and slope for normal* (□) *and poisoned* (+) Acetabularia. *The two equations at the high left corner represent the best fit of normal (—) and poisoned (- - -) samples.*

$$I_s \cong \frac{I_0}{t_0^{2\alpha}},$$

$$\log(I_s) \cong 12.95 - 2.60 \cdot t_0,$$

and

$$\log(I_0) \cong 3.17 + 7.48 \cdot 2\alpha.$$

As can be seen, the fit between predicted and experimental curves is quite good.

This is surprising if one considers that the data are obtained with different individuals, exited with light of different spectral characteristics. It constitutes a further proof that the rescattered light cannot be the sum of a large number of uncorrelated processes based on the excitation of single atomic or molecular levels. Rather,

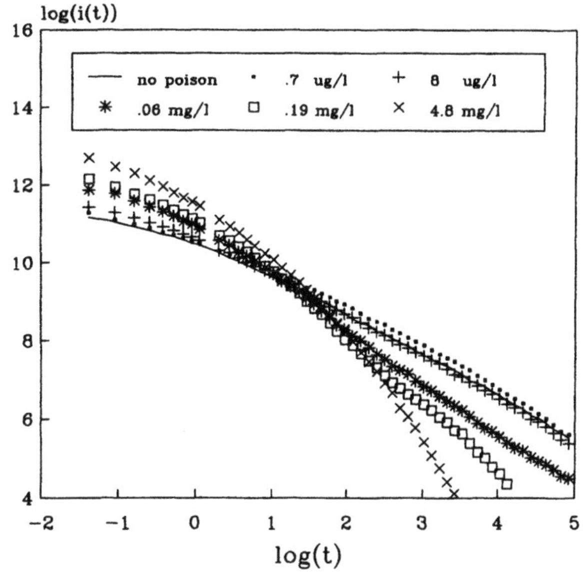

Figure 13.6: *Trend of DL of a poisoned* Acetabularia *reported in logarithmic coordinates. Different symbols are referring to different poison concentration.*

it suggests the existence of cooperative phenomena correlated with the functional state of the plant.

To test this hypothesis we measured the decay kinetics in plants poisoned with different concentrations of atrazine*.

The results shown in Fig. 13.6 refer to one plant and Table 13.3 shows the values of the parameters fiting a hyperbolic decay for two different plants.

As can be seen, the modifications of the behaviour of the two samples caused poisoning, are dramatic and extremely evident.

1. There is a marked increase in the maximum intensity of re-emission I_s which can also be measured in very low concentrations of atrazine (of the order of some μg/liter).

2. There is a marked decrease in decay time (an increase in the coefficient 2α) of the hyperbolic decay.

*2chloro-4-ethilamino-6-isopropylamino-s-triazine

Table 13.3: Values of parameters of equation obtained from least square method fit to experimental data.

#	$I(t_1)$	I_0	t_0	2α	poisoning (atrazine)
Ac6	8.16×10^4	1.48×10^5	0.9	1.10	no poison
Ac6	2.28×10^5	2.21×10^5	0.5	1.73	0.2mg/l
Ac6	2.85×10^5	2.40×10^6	1.7	2.90	1.0mg/l
Ac6	2.68×10^5	6.21×10^6	2.0	3.30	4.0mg/l
Ac7	7.03×10^4	1.11×10^5	0.7	1.08	no poison
Ac7	7.84×10^4	1.51×10^5	1.0	1.12	0.7μg/l
Ac7	9.20×10^4	1.20×10^5	0.6	1.13	8.0μg/l
Ac7	1.43×10^5	1.37×10^5	0.4	1.38	0.06mg/l
Ac7	1.87×10^5	2.72×10^5	0.7	1.80	0.2mg/l
Ac7	2.94×10^5	4.61×10^6	2.0	3.00	4.8mg/l

note – Excitation with light of 565 nm wavelength for 5 seconds.

3. The existing relation between I_s and I_0 disappears, as they seem to become independent of one another, (see Fig. 13.4, p. 335) and the relation between I_0 and 2α also changes radically (see Fig. 13.5, p. 336).

4. As the concentration of poison increases, the decay tends towards a mono-exponential mode. This is shown in Table 13.4 which contains the parameters that are estimated for the best fit of $\log(I(t_n))$ vs. t_n: As can be seen, the coefficient of determination increased to unity and the mean quadratic value of the sum of the square of the discards EMSQ tends to 0 as the concentration of poison increases.

   This could reflect the inhibition of cooperative processes produced by the poisoning.

5. The decay kinetics of selected spectral components seems to remain the same even though this analysis can be done only for a few spectral components because of the high rapidity with which the radiation decays and therefore the increasing importance of the baseline (see Fig. 13.7). These dramatic changes produced by poisoning are quite reproducible as shown in Fig. 13.8 and Table 13.3, p. 338, where the decay kinetics and parameters for similar conditions of

Table 13.4: Parameters of exponential function $I(t) = Ie^{-t/\tau}$ determined by least square method fit to experimental data.

#	I	τ	d	$emsq$	poisoning (atrazine)
Ac6	1.66×10^4	11.1	0.720	1.87	0.2mg/l
Ac6	4.39×10^4	4.1	0.868	0.971	1.0mg/l
Ac6	8.64×10^4	2.9	0.941	0.373	4.0mg/l
Ac7	2.08×10^4	8.8	0.766	1.360	0.2mg/l
Ac7	6.72×10^4	3.5	0.898	0.656	4.8mg/l

note – $d$vand $emsq$ are respectively the coefficient of determination and the sum of the squares of residuals of linear regression performed on $\ln I(t)$ vs t.

poisoning in two different plants are presented.

13.4 Conclusions

The experimental results shown in this paper give some new insights into the DL of *Acetabularia.*

In all the experiments performed on different individuals and under various conditions, DL tends to follow a hyperbolic decay with oscillations which are nearly periodic in a $\log(t)$ scale.

It is therefore no longer reasonable to interpret DL as the result of the sum of many uncorrlated decays from single atomic levels or excited metastable states, but instead, to think of the presence of more complicated mechanisms of a physical or physical-chemical nature which take the organized structure of the whole system into consideration.

Our results suggest that a knowledge of the parameters of the decay kinetics can give information about the functional state of the whole system.

The existence of mathematical relations between the different parameters of hyperbolic decay shows that, during the decay itself, the starting conditions are retained

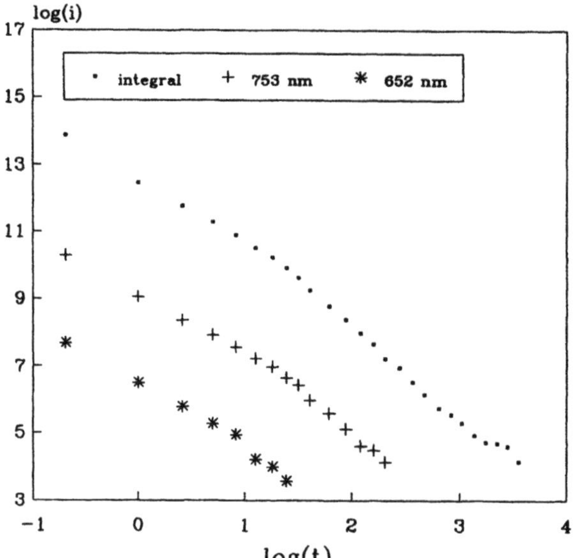

Figure 13.7: *Trend of some spectral component of DL of* Acetabularia *poisoned with 0.2mg/l of atrazine.*

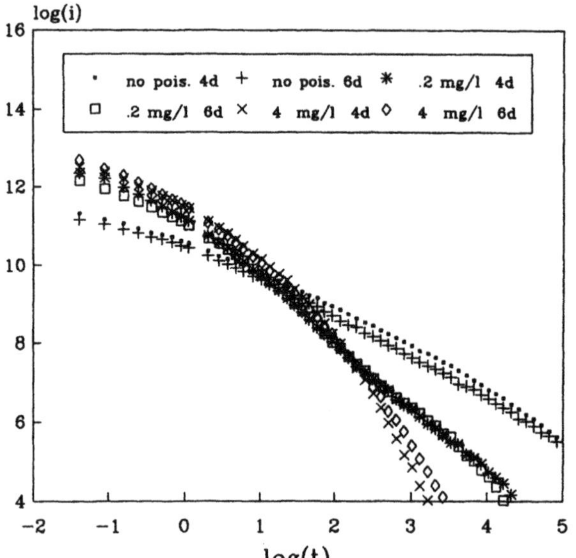

Figure 13.8: *Trend of DL of two* Acetabularia *(4d and 6d) in normal status and poisoned with atrazine at two different concentrations.*

in a memory. This is another argument against the interpretation of DL as the result of the sum of uncorrelated processes.

Furthermore, as the mathematical relations between the different parameters are either disrupted or altered and as the decay kinetics tend to a simple exponential when the organism is severely damaged, we may infer that hyperbolic decay is closely related to the normal functional state of the whole system and that it could, therefore, give us information about co-operative processes occurring under these conditions.

13.5 References

1. B.L. Strehler and W. Arnold, Light production by green plants. *J. Gen. Physiol.* **34** (1951) p. 809-820.

2. J. Lavorel, Luminescence, in *Bioenergetics of Photosynthesis*, ed. R. Govindjee (Academic Press, New York, San Francisco, London, 1975) p. 225-314.

3. S. Malkin, Delayed Luminescence, in *Primary Processes of Photosynthesis*, ed. J. Barber (Biochemical Press, Elsevier/North-Holland. ISBN 0-444-41585-8) p. 349-431.

4. W. Schmidt and H. Senger, Long-term delayed luminescence in Scenedismus obliques: Spectral and Kinetic properties. *Biochem. Biophys. Acta* **890** (1987) p. 15-22.

5. G.D. Mendenhall, Charakterisierung von Materialien durch Chemilumineszenz-Messungen. *Angew. Chem.* **102** (1990) p. 380-391.

6. S.P. McGlynn, M. Kinoshita, M. McCarville, B.N. Srinivasan and J.W. Rabalais, Delayed Luminescence of Organic Mixed Cristals-VIII, the Luminescence of Impure Crystals. *Photochem. Photobiol.* **8** (1968) p. 349-359.

7. F.A. Popp, B. Ruth, W. Bahr, J. Bõhm, P. Grass, G. Grohlig, M. Rattemeyer, H.G. Schmidt and P. Wulle, Emission of visible and ultraviolet radiation by active biological systems. *Collect. Phenomena* **3** (1981) p. 187-214.

8. W.B. Chwirot, R.S. Dygdala and S. Chwirot, Quasi monochromatic light induced photon emission from microsporocytes of larch showing oscillating decay behaviour predicted by an electromagnetic model of differentiation. *Cytobios* **47** (1986) p. 137-146.

9. L. Fonda, G.C. Ghirardi and A. Rimini, Decay of unstable quantum systems. *Rep. Progr. Phys.* **41** (1978) p. 587-631.

10. K.H. Li and F.A. Popp, Non-exponential decay law of radiation systems with coherent rescattering. *Phys. Lett.* **93A** (1983) p. 262-266, and corresponding chapters in this volume.

11. F.A. Popp, On the coherence of ultraweak photoemission from living tissues, in *Disequilibrium and Self-Organization*, ed. C.W. Kilmister (D. Reidel Publishing Company, Dordrecht, Boston, Lancaster, Tokio, 1986) p. 207-230.

12. D. Slawinska and J. Slawinski, in *Chemiluminescence and Bioluminescence*, ed. J.G. Burr (Marcel Dekker, NY, Basel, 1985) p. 494-530.

13. B. Ruth, Experimental investigation into Ultraweak Photon Emission, in *Electromagnetic Bio-Information*, 2nd ed., eds. F. A. Popp, U. Warnke, H.L. König and W. Peschka (Urban & Schwarzenberg, München, 1989).

13.6 Appendix

The counts per unit of time of the photomultiplier are proportional to the radiating flux averaged by the time. This flux, in its turn, is proportional to the mean of product $E^*(t)E(t)$, $E(t)$ being the electric field.

If the electric field is assumed to be a superposition of positive and negative frequency parts in the form: $E(t) = E(t)^+ + E(t)^-$ with an amplitude decaying according to hyperbolic law, it is possible to write for the electric:

$$E(t) = [\frac{E_0 e^{i\phi}}{(t+t_0)^{\alpha+i\beta}} + \frac{E_0 e^{-i\phi}}{(t+t_0)^{\alpha-i\beta}}] \mid e^{i\omega t} + e^{-i\omega t} \mid, \tag{13.2}$$

where E_0 is field value at the time $(t+t_0) = 0$, t_0 is a sort of "time of memory of the system" as described by Popp and Li[10], ω is the frequency of the emitted photon, ϕ its phase and i the imaginary unit.

The expression (15.2) can be written as:

$$E(t) = \frac{E_0}{(t+t_0)^\alpha} \mid e^{i(\beta \log(t+t_0)-\phi)} + e^{-i(\beta \log(t+t_0)-\phi)} \mid \cdot \mid e^{i\omega t} + e^{-i\omega t} \mid . \tag{13.3}$$

What we actually measure is the mean value of the radiating flux calculated over a time equivalent to the dwelling time of the measurement system δt, i.e. we have:

$$I(t_n) = < E^*(t)E(t) >_{\delta t} .$$

This product can be written as:

$$\begin{aligned} E^*(t)E(t) &= \frac{2E_0^2}{(t+t_0)^{2\alpha}}\{2 + e^{2i(\beta\log(t+t_0)-\phi)} + e^{-2i(\beta\log(t+t_0)-\phi)} \\ &+ e^{2i(\omega t+\beta\log(t+t_0)-\phi)} + e^{-2i(\omega t+\beta\log(t+t_0)-\phi)} \\ &+ e^{2i(\omega t-\beta\log(t+t_0)-\phi)} + e^{-2i(\omega t-\beta\log(t+t_0)-\phi)} \\ &+ e^{2i\omega t} + e^{-2i\omega t}\}. \end{aligned} \tag{13.4}$$

It is easy to verify that the time averages performed over a time interval that is of the order of the temporal resolution of the photomultiplier, (i.e., large compared with $2\frac{\pi}{\omega}$, but too small to describe the decay of the DL) are equal to 0 for all terms in the square brackets, except the first three, since either the term average is zero or it is possible to divide the term into the product of two parts, in which the quantity $e^{i2\omega t}$ varies very quickly while the rest of the expression remains at an almost constant value.

Therefore, working out the remaining terms, it is also possible to write that

$$I(t_n) = \frac{4E_0^2}{(t_0+t_n)^{2\alpha}}(1 - \sin^2(\beta\log(t_n+t_0) - \phi)) \tag{13.5}$$

with t_n as a full multiple of dt.

The values of the phase ϕ are connected to the temporal coherence of the discharged photons such that if the coherence time τ is longer than the duration of the measurement, Eq. (13.5) is entirely satisfied. Alternatively, if the coherence tmie τ of the field is shorter than the dwelling time, the argument of the $\sin^2$ changes many times in the course of a single dt, the sin^2 will be substituted by its mean value $\frac{1}{2}$, and the variations present in Eq. (13.5) will disappear.

In the intermediate case, the quantity measured could be given by the sum of a certain number of contributions which are only partially coherent to one another. It is also possible to describe the behaviour of the system by using a sum of expression (13.5) with different phase angles ϕ_i.

The partially coherent contributions tend to reduce the extent of the variations and it is nearly impossible to calculate the influence in a rigorous way. For a first approximation, they can be described by adopting a parameter ϵ which describes empirically the damping of the variations. So, setting $E_0^2 = I_0$ we get Eq. (13.1):

$$I(t_n) = \frac{I_0}{(t_0 + t_n)^{2\alpha}} \frac{1}{2 - \epsilon} (1 - \epsilon \sin^2(\beta \log(t_n + t_0) + \phi)).$$

Chapter 14

Population Density-dependence of Biophoton Emission from Daphnia

Michael Galle

Abstract

At first, the model organism *Daphnia magna* and the experimental conditions are described. Later on, three typical results of the density characteristic of the photon intensity for three stages of development (juvenile, adolescent, adult) are shown. For the adolescent stage there is evidence of a relationship between the mean distance of the animals and the intensity of the photons.

The results are qualitatively discussed with regard to the mutual interactions in swarm formation on the basis of the theory of Dicke.

14.1 Introduction

The experiments were performed with parthenogenetic females of the water flea *Daphnia magna* (Entomostraca, Phyllopoda, Cladocera, Daphnidae). These free swimming planktonic organisms use the second pair of their antennae for movement. The subitanous eggs develop in the brood pouches on the back of the animals. The neonates are released as free swimming small daphniae in the water. They grow up to maximally 6 mm in size[13,15]. *Daphnia magna* forms aggregations and swarms[9,30]. There is no evidence that the formation of aggregations and swarms is caused by chemical communication[3].

For the investigations the animals were kept in the laboratory within carbon filtrated and oxygen enriched tap water. The water was changed twice a week and temperated at 20°C. The animals lived in a seasonally changing day/night rhythm. The water fleas were fed with the unicellular green algae Scenedesmus subspicatus.

For the experiments only animals in good biological conditions were used, i.e., those being of brown colour and carrying much reserve substances.

In our opinion (see also paper by Tembrock[26]) the basic nature of communication between animals includes the regulation of distance. And as some theoretical and experimental work suggests a regulatory role of biophotons in living systems[18], we investigated the relationship between light emission and population density of *Daphnia magna*. Earlier studies indicate already some interaction by photons between these animals[19].

In more detail the question of the experimental procedure was as follows: how does the photon intensity change if we increase the density of similar animals (sex, body size, developmental stage) in a group? For identical photon sources we can expect a linear relationship between intensity and density apart from an absorption effect at high densities. Every deviation from this relationship should indicate a communicative mechanism[16,19].

For biological systems especially systematic time effects must be taken into account (see Sec. 14.4).

14.2 Experimental Procedure

For the experiments with adolescent animals, juvenile females about 1.5 mm in size were cultured in 2 liter beakers of water at a density of 30 individuals per vial. The animals were fed up to the time when they developed eggs but did not yet release the neonates.

The females then were transferred to another 2 liter beaker without food at a density of 200 to 300 individuals per cup for two hours.

The subsequent measurements, which took place during day time, i.e. from 10 am to 4 pm, in September and October, were performed according to the following procedure. Increasing in number from one to about 250, females were put into the 15 ml photomultiplier cuvette of quartz glass. The daphnia population in the cuvette was increased by addition of new animals to the population which has been measured before. After each addition of new animals, the photon emission decreases, and then reaches a stable value in 5 to 7.5 min.

Measurements of the light emitted by the animals were carried out in the dark with an EMI 9558QA photomultiplier operating in the single photon counting mode. During the 6 hours of measurements the animals did not release neonates and the water was exchanged at several intervals.

Control measurements were taken from stock water without animals.

The experimental procedure for the adult females was the same as for the adolescent ones, with the exception that the adult animals released neonates before and during the whole experiment.

For the experiments with the juvenile animals there were taken females of similar size from the original cultures.

14.3 Results

Figure 14.1 illustrates the result of a typical experiment with the adolescent animals. Figure 14.2 shows the mean values of the data obtained in six experiments. It can be seen that well defined maxima and minima of photon emission occur in dependence of the population size. Table 14.1 orders for these six experiments the numbers of the adolescent animals in the cuvette according to the minima and maxima of intensity.

The following calculation gives, in addition, evidence of a clear relationship of the intensity characteristic to the body size[8]. If we take the average number (56.7) of animals at the second maximum (N_{max2}, see table 14.1), for instance, the mean distance d over the individuals in volume V (15 ml) is, assuming an regulary spaced distribution and the animals as spheres:

$$d = \frac{1}{2} d_{max2} = \frac{1}{2} \sqrt[3]{\frac{V}{N_{max2}}} = \frac{1}{2} \sqrt[3]{\frac{15cm^3}{56.7}} = 3.21 \pm 0.23mm$$

where d_{max2} is the mean distance of the centers of the spheres (animals) at the second maximum, and the error given is calculated from the volume error and the SD in table 14.1.

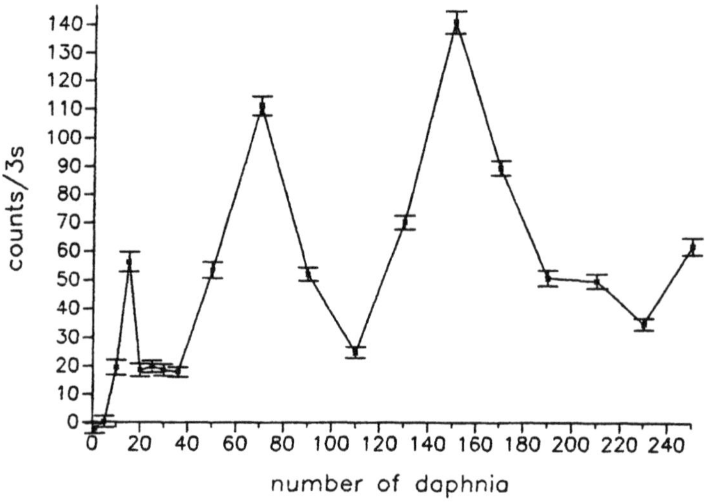

Figure 14.1: ***The intensity of photon emission as a function of the number of individuals. The number of adolescent* Daphnia magna *was raised within a 15 ml quartz cuvette between each measurement. The diagram shows the mean values and the 2-fold standard error of 150 readings 7.5 min after transfer of the animals into the equipment.***

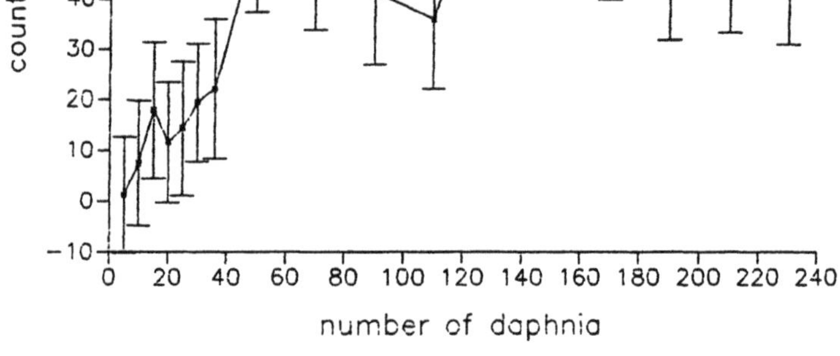

Figure 14.2: ***Mean values of the photon intensity of the six experiments with the adolescent animals in 15 ml volume with the weighted standard deviation*[23].**

Table 14.1: Relationship between the minima and maxima of intensity of photon emission (I = mean of 150 values taken 7.5 min after transfering the animals into the cuvette; counts per 3 seconds), the number (N, per cuvette and 15 ml) and the body size (B, in mm) of the daphniae as monitored in six experiments. M is mean value, SD represents standard deviation. The first maximum and minimum were not identifiable in all experimemts. The animals in these six experiments carried 1 to 4 eggs in their brood pouches.

Exp. No.	B	1st Max I(N)	1st Min I(N)	2nd Max I(N)	2nd Min I(N)	3rd Max I(N)	3rd Min I(N)
1	3.5	—	—	118.0(50)	52.7(93)	132.0(123)	72.0(153)
2	3.5	—	—	32.1(50)	10.3(90)	41.1(110)	5.6(184)
3	3.0	56.4(15)	18.0(36)	111.3(70)	24.9(110)	141.1(150)	35.1(230)
4	3.0	16.3(15)	6.0(20)	48.7(70)	38.7(90)	75.5(130)	42.6(210)
5	3.5	—	—	41.7(50)	27.4(110)	82.1(130)	47.8(150)
6	3.0	8.5(15)	4.8(20)	30.7(50)	20.9(70)	55.0(130)	30.1(170)
M	3.25	27.1(15)	9.6(25.3)	63.8(56.7)	29.2(93.8)	87.8(128.8)	38.9(182.8)
SD	0.27	25.7(0)	7.3(9.2)	40.0(10.3)	14.8(15.0)	40.6(13.0)	21.9(31.9)

The result corresponds approximately to the body size of the daphniae (3.25 ± 0.27 mm, see SD in table 14.1). The relationships between the number of animals and the resulting mean distances of maxima and minima of photon emission from the daphniae indicate integer ratios of these parameters to the body size (Table 14.2). More details and data from additional experiments are given in[7].

Figure 14.3 shows a typical experiment with adult females (body size: 3.5 ± 0.3 mm) dismissing successive neonates during the measurements. At the end of the experiment there were swimming 160 neonates with the adults in the cuvette (15 ml volume). The adults carried 3 to 6 eggs in their brood pouches.

Figure 14.4 illustrates a typical experiment with juvenile females (body size: 1.7 ± 0.2 mm).

Table 14.2: Comparison of the theoretically expected values (T) for the ratio of the mean distances of animals within the minima and the maxima of photon emission expressed in terms of integer ratios of the body size, and the corresponding ratio of the mean distances as calculated from the experimental data (E) according to the formula[a], where N_x represents the mean number of animals in the cuvette underlying the minima and maxima of intensity except the mean number underlying the peak given in the numerator (SD taken from table 1), and ρ the density (d_x=mean distances of the centers of the spheres, that means animals, at the minima and maxima of intensity). M indicates the theoretical value of the mean distance in units of body length between the animals. The minima and maxima are ordered according to their empirical appearance.

Photon Intensity	E[a]	T	M
3rd Minimum	0.667±0.058	0.625 = 5/8	1/4
3rd Maximum	0.761±0.072	0.750 = 6/8	1/2
2nd Minimum	0.845±0.095	0.875 = 7/8	3/4
2nd Maximum	Reference	1.000 = 8/8	1
1st Minimum	1.308±0.238	1.125 = 9/8	5/4
1st Maximum	1.557±0.094	1.250 = 10/8	3/2
0 Minimum	—	1.375 = 11/8	7/4
0 Maximum	—	1.500 = 12/8	2

[a] $E = \frac{d_x}{d_{max2}} = \sqrt[3]{\frac{\rho_{max2}}{\rho_x}} = \sqrt[3]{\frac{N_{max2}}{N_x}}$

14.4 Discussion

The present results indicate a correlation between biophoton emission and population of adolescent *Daphnia*.

The change of the photon emission is correlated to a density dependent formation of a group of animals. So it is part of a social process. A systematic time effect could be excluded[7].

The physical basis underlying the previous mentioned correlation between body size and photon intensity in the case of the adolescent animals can be seen in the theory of Dicke[5,6]. In brief, spontaneously emitting photon sources undergo, with increasing density and decreasing distance, couplings, and both coherence and interference phenomena arise. The resulting emission pattern (that is rather complex and only

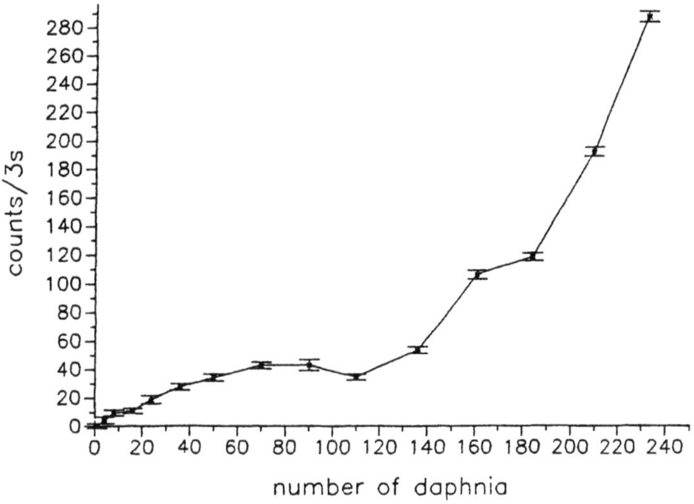

Figure 14.3: *The intensity of photon emission as a function of the number of individuals. The number of adult* Daphnia magna *was raised within a 15 ml quartz cuvette between each measurement. The diagram shows the mean values and the 2-fold standard error of 150 readings 7.5 min after transfering the animals into the equipment.*

qualitatively measurable at the moment) is very sensitive to the distance of the photon sources and may lead to respectively attractive and repulsive forces. Hence, the theory of Dicke allows to conclude the involvement of such forces in the regulation of group formation.

We can exclude self-absorption of photons as a cause for the appearance of maxima and minima, because even the highest possible value of self-absorption is much too low to account for the differences between the maxima and the minima measured[19].

The observed phenomena are also supported by observations on microaggregations of daphnidae in different laboratories, in so far as the animals involved were reported to be of the same sex, of similar size and developmental stage for about the same range of densities[1,24]. Even in nature the density of aggregations and swarms is in the same range as the densities in the cuvette. For example, Colebrook[2] observed swarms of *Daphnia hyalina* in a juvenile stage at a density of 9.1 individuals per ml (the body size of adult females in this species is about 2 mm). Ratzlaff[22] observed dense swarms of females of Moina affinis with a density of 284 individuals per ml (the body size of adult females is about 1 mm). In nature, swarm formation by

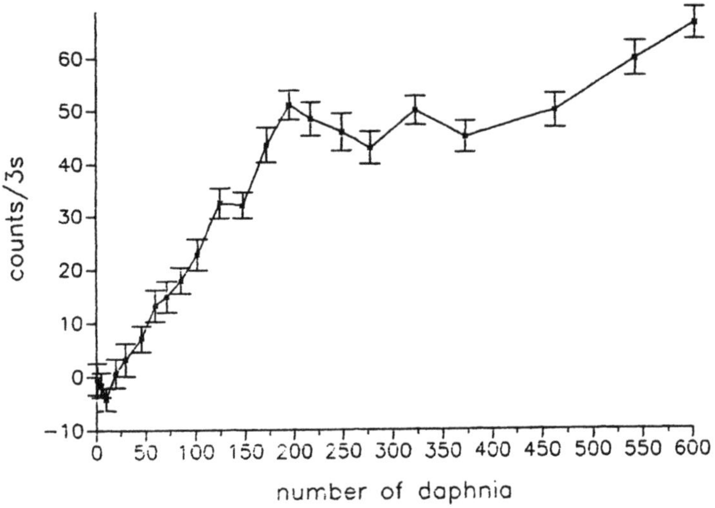

Figure 14.4: *The intensity of photon emission as a function of the number of individuals. The number of juvenile* **Daphnia magna** *was raised within a 15 ml quartz cuvette between each measurement. The diagram shows the mean values and the 2-fold standard error of 150 readings 7.5 min after transfering the animals into the equipment.*

Daphnia magna only occurs if there are high densities in the pools[3,9,30].

In addition to cooperative phenomena in groups of planktonic animals, biophotons apparently exhibit some regulatory role also in cell populations, as was recently shown for normal and malignant human amnion cells and hepatocytes of rats[25,28]. There is an increasing number of papers giving evidence that long distance cell-to-cell and organism-to-organism communications might be accomplished by transmission and reception of electromagnetic signals[20,27].

A distinct distance behaviour, such as indicated by the body size correlated photon emission characteristic of the adolescent animals, is a widespread phenomena in animal communities. In aggregations and swarms of fish and birds the units of measurements to regulate the distance between the animals are the body size or other characteristic body units[11,17,29].

The comparison among Fig. 14.1 (adolescent), Fig. 14.3 (adult with neonates) and Fig. 14.4 (juvenile) demonstrates that the density behaviour of the photon intensity of the animal group also depends on the developmental stage of the animals. Probably the distinction between Fig. 14.1 and Fig. 14.3 is caused by the mixing of adult animals with the neonates. This dependence of the photon density characteristic from the developmental stage indicates furthermore a biological meaning of the biophoton emission. For animals it is a common phenomenon that the juveniles have another distance-sensitivity-behaviour as the adult ones[26].

14.5 Final Remarks

Pattern formation in time and space represents a fundamental principle in the differentiation and evolution of living systems, and probably of all matter[10,21].

Many physical, chemical and biological factors have been studied in the broad range from the molecular to the ecological levels, and a good deal of hypotheses have been put forward.

One of the least understood phenomena, although often described, is patching and swarming of planktonic organisms[14]. No biological advantage or importance could be suggested so far for swarming behaviour[4,12], but at least there should be a common interactive force or communicative mechanism involved.

As swarms are uni-species, constant in form, dense, and behave like a "super-organism", some cooperativity must underlie swarm formation. Such a long-distance-field interaction mechanism might be the here investigated light emission.

14.6 References

1. Z. Brandl and C.H. Fernando, *Can. J. Zool.* **49** (1971) p. 775.

2. J.M. Colebrook, *J. Animal Ecol.* **29** (1960) p. 241.

3. T.J. Crease and P.D.N. Hebert, *Freshw. Biol.* **13** (1983) p. 491.

4. D.J. Crisp, *Nature* **193** (1962) p. 597.

5. R. Crubellier, S. Liberman and D. Pillet, *J. Phys. B* **18** (1985) p. 3811.

6. R.H. Dicke, *Phys. Rev.* **93** (1954) p. 99.

7. M. Galle, *Ph.D. Thesis* (University of Saarland, 1992) in preparation.

8. M. Galle, R. Neurohr, G. Altmann, F.A. Popp and W. Nagl, *Experientia* **47** (1991) p. 457.

9. J. Green, *J. Animal Ecol.* **24** (1955) p. 84.

10. H. Haken (ed.), *Dynamics of Synergetic Systems* (Springer, Berlin, 1980).

11. E. Horstmann, *Umschau* (1951) p. 494.

12. D.S. Johnson and T.E. Chua, *Crustaceana* **24** (1973) p. 332.

13. A. Kästner, *Lehrbuch der speziellen Zoologie* **Band I:** Wirbellose, 2.Teil: Crustacea (Fischer, Stuttgart, 1967).

14. B.J. Malone and D.J. McQueen, *Hydrobiologia* **99** (1983) p. 102.

15. D. Matthes, *Mikrokosmos* **77** (1988) p. 24.

16. W. Nagl and F.A. Popp, in *Energy Transfer Dynamics*, eds. T.W. Barrett and H.A. Pohl (Springer, Berlin, 1987) p. 248.

17. B.L. Partridge, *Spektr. Wiss.* **8** (1982) p. 64.

18. F.A. Popp et al. (Multi-author Review), *Experientia* **44** (1988) p. 543.

19. F.A. Popp, K.H. Li, W.P. Mei , M. Galle and R. Neurohr, *Experientia* **44** (1988) p. 576.

20. F.A. Popp, U. Warnke, H.L. König and W. Peschka, *Electromagnetic Bio-Information, 2nd Edition* (Urban & Schwarzenberg, Munich, 1989).

21. I. Prigogine, G. Nicolis, A. Babloyantz, *Physics Today* **11** (1972) p. 23.

22. W. Ratzlaff, *Limnol. Oceanogr.* **19** (1974) p. 993.

23. L. Sachs, *Statistische Auswertungsmethoden* (Springer, Berlin, 1969).

24. K.R. Santharam, P. Navaneethakrishan and S. Krishnaswamy, *Arch. Hydrobiol.* **80** (1977) p. 398.

25. W. Scholz, U. Staszkiewicz, F.A Popp and W. Nagl, *Cell Biophys.* **13** (1988) p. 55.

26. G. Tembrock, *Verhaltensbiologie* (Harri Deutsch, Frankfurt/Main, 1987).

27. T.Y Tsong, *Trends Biochem. Sci.* **14** (1989) p. 89.

28. R. van Wijk and D. Schamhart, *Experientia* **44** (1988) p. 586.

29. U. Warnke, *Umschau* (1986) p. 336.

30. J.P.W. Young, *Freshw. Biol.* **8** (1978) p. 279.

Chapter 15

Some Remarks on Biological Consequences of a Coherent Biophoton Field

F.A. Popp

Abstract

The destructive interference which, as a consequence of Dicke's theory, provides a powerful mechanism of biological regulation, is discussed in terms of the uncertainty principle. This enables to frame the fields where these tools of superradiance and subradiance can be used by living systems. Some striking examples (population formation of daphnia and tumour growth) show that this concept helps to understand yet unexplained phenomena in biology and medicine.

In order to understand essential consequences of a coherent biophoton field, let us discuss the coherence in its most simple terms.

Take, at first, the case where two radiators R_1 and R_2 each emit a quasi-monochromatic wave with about the same wavelength λ. The distance between the radiators is denoted by Δx, and the momentum uncertainty by

$$\Delta p = \frac{h}{\lambda^2}\Delta\lambda, \qquad (15.1)$$

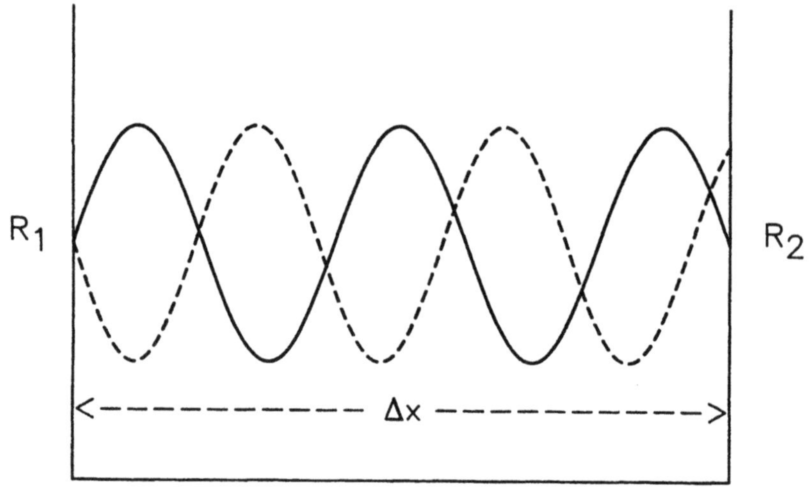

Figure 15.1: *Destructive (and constructive) interference between radiating sources of biological systems (biomolecules, organelles, cells, organs, organisms, populations) may play the most important role of biological organization. This principle combines uncertainty relation and energy conservation (see text).*

where h is the Planck constant and $\Delta\lambda$ the deviation from the exact value λ.

Fig. 15.1 shows the region between the two radiators R_1 and R_2, where the two waves interfere.

At R_1, the two waves with equal amplitudes and of wavelength λ and $\lambda + \Delta\lambda$ shall be in antiphase. We ask, over what distance they can show destructive interference such that the energy of the field over this distance is lower than the sum of the energy of the two single waves.

Obviously, destruction is obtained if, and only if, the uncertainty $\Delta\lambda$ per wavelength over a distance of n consecutive wavelengths does not add up to a value greater than $\frac{\lambda}{2}$:

$$n\Delta\lambda \leq \frac{\lambda}{2}. \qquad (15.2)$$

The value of n is determined as the ratio of the distance Δx to the wavelength λ:

$$n = \frac{\Delta x}{\lambda}. \tag{15.3}$$

Insertion of (15.3) into (15.2) results in

$$\frac{\Delta x}{\lambda}\Delta\lambda \leq \frac{\lambda}{2},$$

which, after substitution of (15.1), yields the remarkable result

$$\Delta x \Delta p \leq \frac{h}{2}. \tag{15.4}$$

From the uncertainty relation which, in contrast to (15.4), requires $\Delta x \Delta p \geq h$ we know that (15.4) can never be fulfilled. However, this simple example can be looked upon as a very basic case of understanding a manifold of important consequences, e.g.:

(1) The elementary link between energy conservation and the uncertainty principle. Actually, the violation of the uncertainty relation simultaneously violates the energy conservation, or, in turn, on a microscopic scale the energy conservation in a stationary state requires the validity of the uncertainty principle.

(2) The creation of a photochemical potential by use of a coherent field.

Of course, the two radiators are able to trap photons by the mechanism of destructive interference within the coherence volume between them. This elementary process of sucking up photons may be a basic mechanism of feeding, or, even more general, of information and communication. In turn, by constructive interference in the coherence volume between the systems, stored photons are released.

(3) The creation of long-range forces over the coherence volume between the systems, which may regulate the "social" behaviour, such as group formation, "Gestaltbildung" and similar effects.

Of course, from (15.4) it follows that by increasing the coherence length, a system is able to attract or repulse identical or similar systems over longer and longer distances. The force increases proportional to the amount of energy which can be stored by destructive interference. Like the Casimir force[1], this force does not depend on electric charges or magnetic momenta. It depends only on the coherence length, the amount of stored energy and Planck constant. (The order of magnitude of this force between cells may amount to $10^{-12}N$ or even more. Consequently, it can fairly well account for significant contributions to the adhesion between cells [2]. An example, demonstrating, how complicated the spatial structure of this force can develop, is displayed in Fig. 15.2.

(4) In view of the fairly small distances d between base-pairs the lattice system of DNA is subject to strong Casimir forces in the optical range ($d << \lambda$). Thus, a balance between exciplex-excitation interference forces and Casimir forces may be the essential cause of the stabilization of this elementary biomolecule[3], as well as its extraordinary optical properties in terms of coherence. This may provide its ability to regulate intracellular and intercellular processes by its dynamic interference pattern of biophoton emission [4].

(5) Even the doubling capacity of DNA can be understood in terms of coherence: As soon as the coherence of the biophoton field breaks down locally, the probability of absorbing a second photon once the first one has been absorbed becomes twice as high as it was before [5]. This bunching effect provides that enough energy is always available for integrating an excited base into the missing link, where the local incoherence takes place. From this point of view, repair and DNA-replication are physically interlinked phenomena, based again on the coherence of the biophoton field.

These examples may suffice to show that a coherent biophoton field invites the explanation of a variety of biological phenomena which are not understandable in purely biochemical terms. Further examples like immunological response or the trigger effect of biophoton on biological rhythms are discussed elsewhere [4,6].

A more detailed analysis of Galle's daphnia experiment [7] and of the significant differences of photon emission from normal and malignant cells shall now be given.

Fig. 15.3 (right side) displays the dependence of the average photon intensity i of daphnia on the mean distance d between two daphnia, in units of the body size l. From the measured values in Fig. 15.3 (left side, see Chapter 14) to those in Fig. 15.3 (right side) one uses the transformation

$$i\left(\frac{d}{l}\right) = i\left(n\left(\frac{d}{l}\right)\right) = i\left(\frac{1}{2l}\sqrt[3]{\frac{V}{n}}\right),$$

where $V = 15cm^3$ as the volume of the cuvette has been inserted, and n represents the number of daphnia in the cuvette.

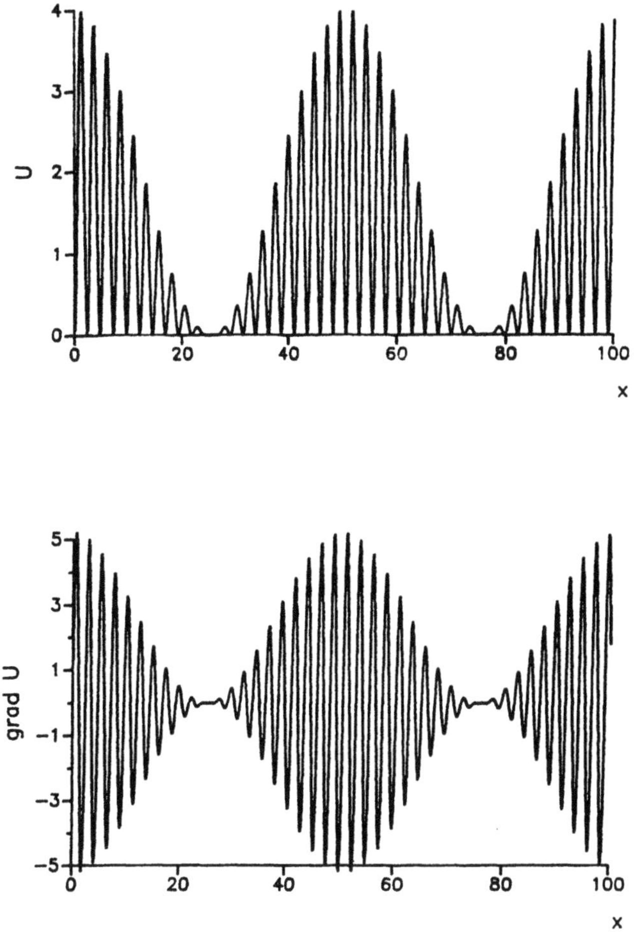

Figure 15.2: *The intensity of two interfering waves* $U = (\sin kx + \sin(k + \Delta k)x)^2$ *for* $k = 1.24$ *and* $\Delta k = 0.124$ *displays the pattern* $U(x)$ *(upper figure). The force between the radiators, or* $-grad\ U$*, determines the zones of attraction and repulsion between the radiators.*

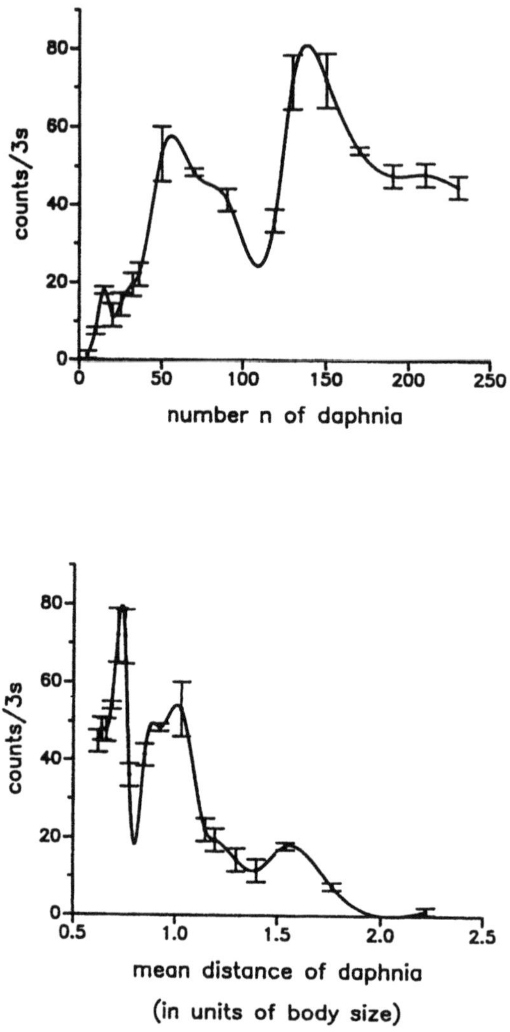

Figure 15.3: *The photon intensity of daphnia displays an interference-like dependence on the number n of daphnia (upper figure, see also Chapter 14). It turns out that this "pattern" can be described in terms of integer ratios of the distance between the animals (see text). The lower figure shows the photon intensity as it depends on the mean distance between the animals.*

In order to analyse the effect in more detail we describe the intensity i in terms of the electric field amplitudes $\vec{e}_j(\vec{r},t)$ from all the daphnia $j = 1, 2, 3, \cdots, n$, where $\vec{r}$ and t are the position and time, respectively:

$$i(\vec{r},t) \propto \left(\sum_{j=1}^{n} \vec{e}_j(\vec{r},t)\right)^2 . \tag{15.5}$$

In order to obtain the actual registered intensity i_W, this must be integrated over the volume of the cuvette and the measurement time interval.

$$i_W = \int_V \int_{\Delta t} i(\vec{r},t) dV dt \propto \int dV \int_{\Delta t} dt \left(\sum_{j=1}^{n} \vec{e}_j(\vec{r},t)\right)^2 . \tag{15.6}$$

Since

$$\left(\sum_{j=1}^{n} \vec{e}_j(\vec{r},t)\right)^2 = \sum_{j=1}^{n} (\vec{e}_j)^2(\vec{r},t) + \sum_{j=1}^{n} \sum_{k \neq j} \vec{e}_j(\vec{r},t)\vec{e}_k(\vec{r},t) , \tag{15.7}$$

by introducing

$$e_0^2 = \frac{1}{n} \int_V \int_{\Delta t} dV dt \left(\sum_{j=1}^{n} (\vec{e}_j)^2(\vec{r},t)\right) , \tag{15.8}$$

we get

$$\int_V \int_{\Delta t} dV dt \left(\sum_{j=1}^{n} \vec{e}_j(\vec{r},t)\right)^2 = n e_0^2 (1 + (n-1)\cos\psi), \tag{15.9}$$

where ψ corresponds to an average phase between always two field-amplitudes $\vec{e}_j(\vec{r},t)$ and $\vec{e}_k(\vec{r},t)$.

In the case of a chaotic field, ψ becomes exactly $\frac{\pi}{2}$, which means that the interference term $(n-1)\cos\psi$ cancels out. Since i_W does not follow a linear relation $i_W \propto n$, the interference term does actually not vanish, even for a time interval Δt as long as 3s. Fig. 15.4 (left figure) shows the dependence of this interference term $(n-1)\cos\psi$ on the number of daphnia, where the best fit of $n e_0^2$ has been used.

Its contribution changes with n, varying between about -60 to $+\ 140$ % of the non-interfering part. With increasing n, the contribution of the interference term

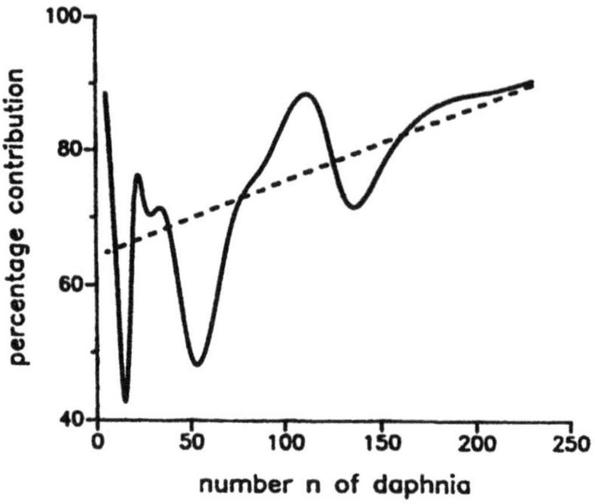

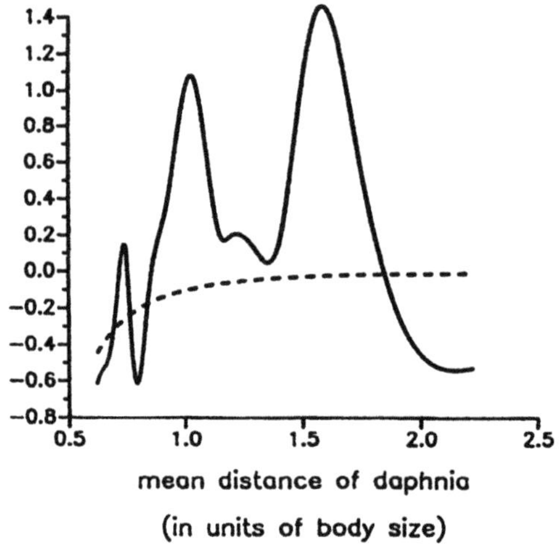

Figure 15.4: *If one describes the photon intensity of daphnia $i(n)$ in terms of a linear increase and an interference, the interference term shows systematic deviations from the linear term, varying from -60 to $+$ 140 % of the non-interfering part (see text). The upper side displays the dependence on the number n of daphnia, the lower side the dependence in units of the body size.*

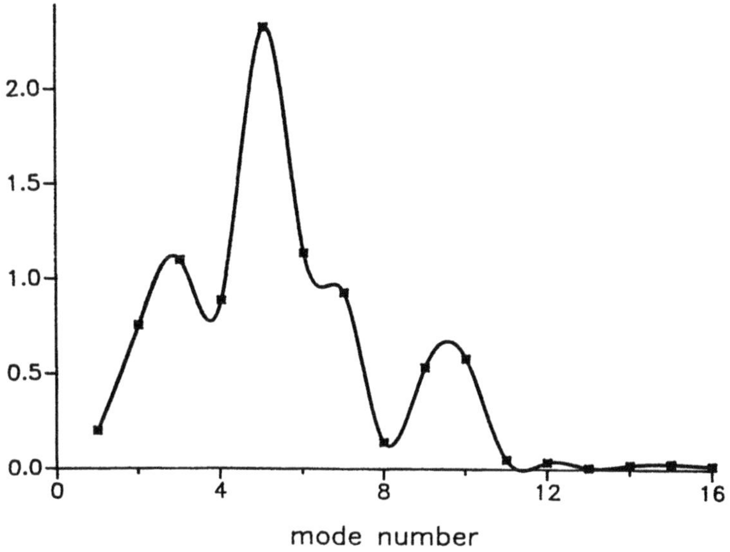

Figure 15.5: *Fourier spectrum of the distance-dependent interference term. For the mode numbers 4, 8 and about 12 there are distinct minima (see text).*

becomes flatter and less distinct. However, the dependence on the mean distance between the daphnia (Fig. 15.4) shows that with decreasing distance the interference pattern becomes sharper with a tendency to destruction. This may be due to a quicker movement of the daphnia at lower distances and at the same time sharper phase-relations between them.

The Fourier spectrum of the interference term shows a distinct pattern (Fig. 15.5), where the mode number m correspond to a k-value $k = 0.264m$ (in units of the reciprocal of the body size). The minima of the Fourier components appear at about $m = 4$, 8 and 11.5, corresponding to about 1, 1/2 and 1/3 of the body size.

This invites interpretation of the results in terms of destructive interference of standing waves $\cos(k(n)n + \varphi)$, where the Fourier analysis displays an interference structure of the following type

$$\sum_{j=1}^{n}(-1)^{(j+1)}\{\frac{1}{j}(\cos(jk_0n^{3/4}) - \sin(jk_0n^{3/4}))\}.$$

Fig. 15.6 shows the measured average intensities $i(n)$ of 6 consecutive experiments on daphnia compared to the theoretical values

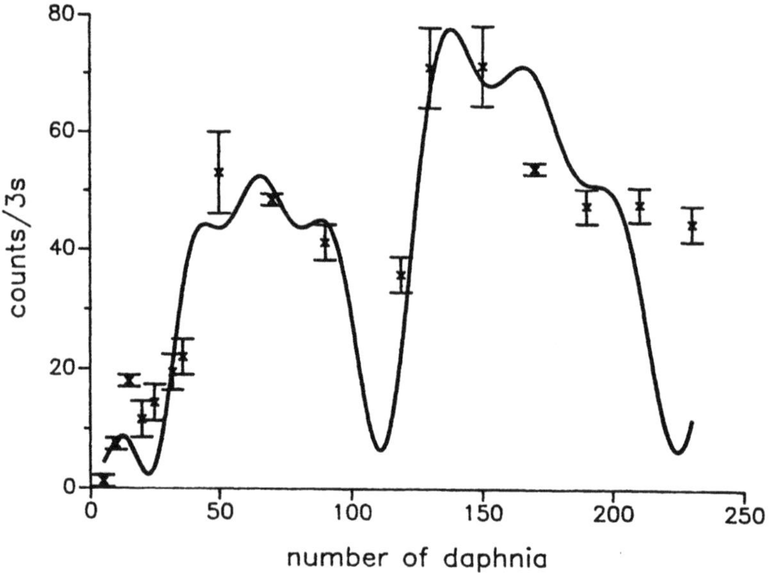

Figure 15.6: *Photon emission from daphnia as calculated from the interference model (see text) compared to the experimental values.*

$$i(n) = \alpha\, n\, exp(-\beta n)\left(1-\frac{1}{2}\left(\sum_{j=1}^{3}(-1)^{(j+1)}\frac{1}{j}\left(\cos(jk_0 n^{3/4})-\sin(jk_0 n^{3/4})\right)\right)\right), \tag{15.10}$$

where $\alpha = 0.89687$, $\beta = 0.00627$, and $k_0 = 0.264$. Qualitatively it describes the interference pattern which is registered. The deviations of the theoretical values from the experimental ones are almost within the error range.

The results confirm that destructive interference takes place. However, at present we do not know the mechanism behind. This active process of adjusting to the waves in such a way that destructive interference with a phase shift of about $\frac{\pi}{4}$ results may be traced back to Dicke's theory applied to the optical properties of the DNA including its lattice vibrations.

Of course, investigations of the internal rythms of daphnia, Thailand fireflies, Gonyaulax polyedra and even cereals indicate a close connection between extremely low frequency vibrations of the optical system and this appearance of standing waves. The figures (see Chapter 1) demonstrate some typical examples. As a consequence, these coherent long-range interactions of identical biological systems are likely to be

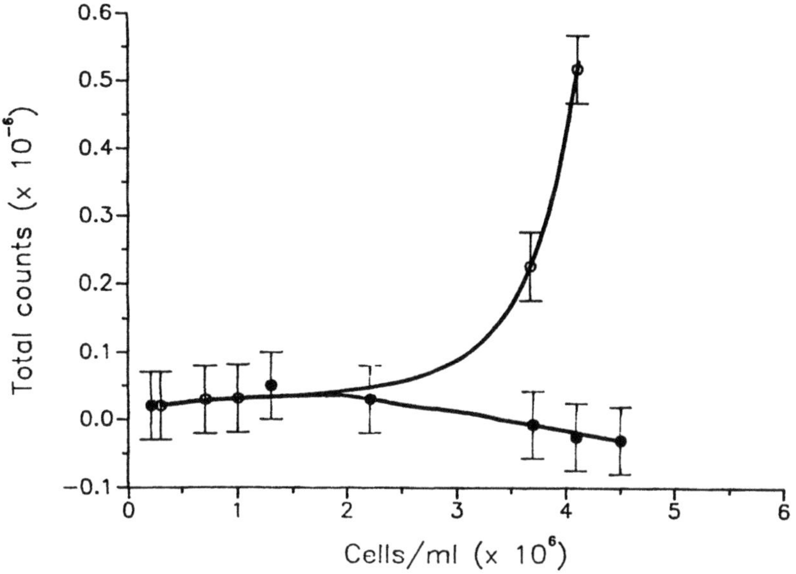

Figure 15.7: *"Delayed Luminescence" from tumour cells (data on upper curve) and normal cells (data on lower curve), as measured by Schamhart* [8]. *The different curves can be approximated by a non-linear (cubic) dependence of intensity on cell-number n, see text and formula (15.11).*

based on mutual triggering of internal clocks by biophoton emission, as is evident from the circadian rhythm which is triggered by sunlight.

Once we agree with the general validity of this destructive interference for any living system, similar results have to be expected for cellular systems. If n gets very high, the resolution of the equipment does not allow measurements of the interference pattern.

Instead, one register an "overall"-effect which will change quasi-continously with the number n of individuals. The term $\frac{\partial k(n)}{\partial n}$ determines the "social" behaviour of the system under study. A striking example is the behaviour of normal cells compared to that of tumour cells.

Fig. 15.7 demonstrates the typical results of such a measurement.

While normal cells exhibit decreasing photon intensity with increasing cell density, tumour cells display just the opposite behaviour: With increasing cell density the light emission increases in a non-linear way.

A mathematical analysis of the results of Fig. 15.7 shows qualitatively the following expression

$$i(n) = an\left(\gamma - bn + \frac{b^3}{3!}n^3\right), \tag{15.11}$$

where $a > 0$, $b > 0$, and $\gamma \approx 1$.

Fig. 15.7 display the results (continous curve), where $a = 0.078$, $b = 0.632$ and $\gamma = 1.136$ for tumour cells and $a = 0.059$, $b = 0.292$ and $\gamma = 0.831$ for normal cells. The deviation of γ from 1 can be explained in terms of a linearization of ordinary self-absorption or rescattering effects within the system. γ is smaller than 1 for normal cells and higher than 1 for tumour cells, indicating thus that the linear optical properties (transparency etc.) are different for normal and tumour cells. This agrees with the results of van Wijk (Chapter 7). The nonlinear behaviour can be interpreted in terms of an expansion of the sinus-function $\sin k(n)n$, where $b \propto \frac{\partial k(n)}{\partial n}$. Thus, the model describes tumour development as a loss of resonator quality, that is

$$\frac{d}{d(\Delta k)}\left(\frac{|\Delta k|}{k}\right) > 0, \tag{15.12}$$

as has been proposed already [6].

In most general terms, this reflects a process where, for tumour cells, the degree of coherence of biophoton emission decreases with increasing cell number in the population, but in a normal tissue this effect is exactly opposite. Hence, one can again trace back this effect to Dicke's theory or to Cavity Quantum Electrodynamics.

In order to elucidate the connection between this loss of coherence and the malignancy of cell population, let us describe the cell growth rate $\dot{n}$ in terms of $\tau(n, \cdots)$,

$$\langle \dot{n} \rangle = \frac{\langle n \rangle}{\langle \tau(n, \cdots) \rangle}, \tag{15.13}$$

where $\tau(n, \cdots)$ is a characteristic time constant which depends on the cell number n and further on a variety of other parameters including – besides of the degree of differentiation – all external influences, as, for instance, food supply, pH-value etc. The task of tumour research is the determination of τ. The brackets $\langle \cdots \rangle$ in (15.13) have been used in order to point out that (15.13) represents some average values over definite time periods of, say, one day.

From (15.13) it follows that $\langle \tau(n, \cdots) \rangle$ has to increase stronger than $\langle n \rangle$ in order to avoid unlimited growth. It is obvious that $\frac{\partial \langle \tau \rangle}{\partial \langle n \rangle}$ and $\frac{\partial k(n)}{\partial n}$ in our model of Eq. (15.11) are interlinked quantities. $\langle \tau \rangle$ represents some coherence time of the system under study, assigned to a resonator value of the cell population. k is expected to be inversely proportional to τ, since the connection to biological rhythms invites the relation

$$k \propto \omega \propto \frac{1}{\tau}\,, \tag{15.14}$$

where ω describes the long-wave limit of ELF-modes of DNA-lattice vibrations which serve for communication of cells (see Chapter 1).

Expansion of τ around an average n_0:

$$\tau(n) = \tau(n_0) + \left(\frac{\partial \tau}{\partial n}\right)_{n_0} (n - n_0) + \cdots$$

results in

$$\langle \dot{n} \rangle = \frac{\langle n \rangle}{\langle \tau(n_0, \cdots) \rangle} \left(1 - \left(\frac{\partial \tau}{\partial n}\right)_{n_0} \frac{\langle n - n_0 \rangle}{\langle \tau(n_0, \cdots) \rangle}\right). \tag{15.15}$$

This differential equation can be solved exactly: $n = \frac{1}{B+(1-B)\exp(-Ct)}$, where $B = \frac{\left(\frac{\partial \tau}{\partial n}\right)}{\tau}$ and $C = \frac{1}{\tau}$. It yields actually the well-known proliferation curves of cell populations and tissues (see Fig. 15.8).

From (15.15) we get an estimate of the total number $\langle n_g \rangle$ of the adult system:

$$\langle n_g \rangle \simeq \left(\frac{\langle \tau \rangle}{\langle (\frac{\partial \tau}{\partial n}) \rangle}\right)_n . \tag{15.16}$$

For the human body reasonable values of $\langle \tau \rangle$ are between 10^5 to 10^7s, while $\langle \frac{\partial \tau}{\partial n} \rangle$ is about 10^{-8} to 10^{-6}s.

A sufficient condition of limited proliferation is obviously

$$\langle \ddot{n} \rangle_{n > n_0} < 0\,, \tag{15.17}$$

where n_0 is a cell number which is smaller than that of the outgrown population.

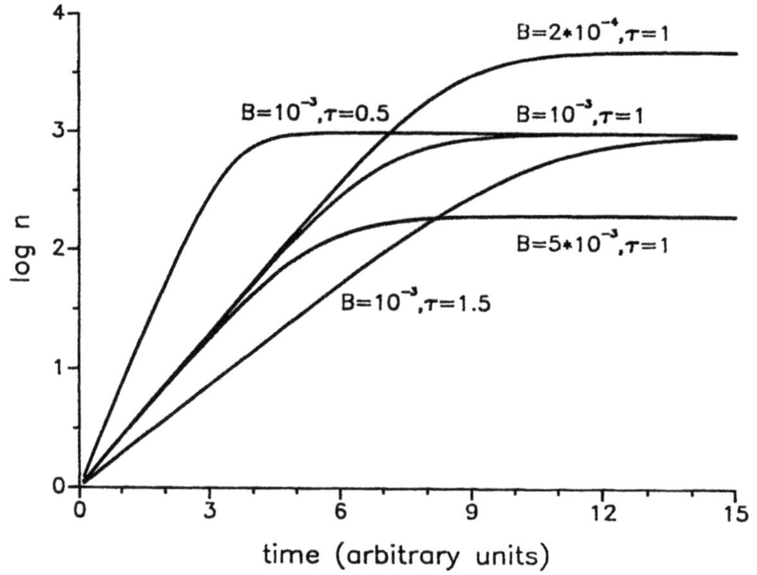

Figure 15.8: *The growth curve* $n(t) = \frac{1}{B+(1-B)\exp\left(-\frac{t}{\tau}\right)}$ *displays different limiting cell numbers* n_g *of the outgrown population for different values of* B*, where* $n_g = \frac{1}{B}$*. The relaxation time* τ*, on the other hand, determines the time of growing up to the limiting size.*

Since $\ddot{n} = \frac{\dot{n}}{\tau} - \frac{n}{\tau^2}\dot{\tau}$, we get after insertion of $\dot{n}$ according to (15.15):

$$\langle \ddot{n} \rangle = \frac{\langle n \rangle}{\langle \tau(n, \cdots) \rangle^2}(1 - \langle \dot{\tau}(n, \cdots) \rangle). \tag{15.18}$$

From (15.18), sufficient conditions of normal and tumour growth are obtained immediately

$$\langle \dot{\tau}(n, \cdots) \rangle \geq 1 \text{ for normal growth;} \tag{15.19}$$

$$\langle \dot{\tau}(n, \cdots) \rangle \leq 1 \text{ for tumour growth.} \tag{15.20}$$

This sensitive dependence on the critical value 1 is shown in Fig. 15.9.

Normal growth can then be characterized by a permanent gain of coherence time which reflects probably a very fundamental property of evolution (see Chapter 21).

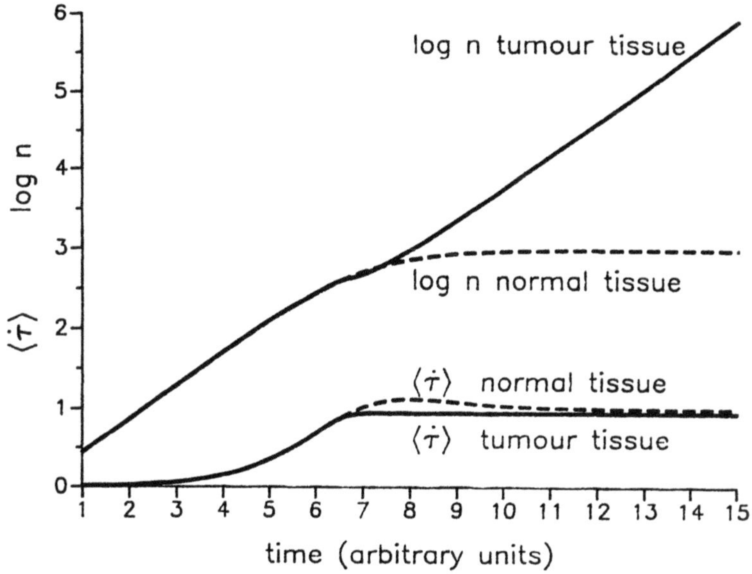

Figure 15.9: *Necessary and sufficient for normal growth is that $\langle \dot{\tau} \rangle$, the increase of coherence time with increasing time, exceeds the critical value 1. $\langle \dot{\tau} \rangle$ determines whether further growth extends the informational base in a tissue or not. For $\langle \dot{\tau} \rangle < 1$ the growth becomes malignant. The examples display normal growth according to figure 8 and malignant growth according to the same formula, but, as soon as $\langle \dot{\tau} \rangle$ reaches the value 0.95 it is kept constant. As a result, normal growth turns into malignant one.*

For $\langle \dot{\tau}(n, \cdots) \rangle = \left\langle \frac{\partial \tau}{\partial n} \right\rangle \langle \dot{n} \rangle$, (15.19) can be rewritten as

$$\langle \dot{n} \rangle \left\langle \frac{\partial \tau}{\partial n} \right\rangle \geq 1, \tag{15.21}$$

which means that cell proliferation is controlled by two antagonistic properties [9], e.g. (1) the growth rate $\langle \dot{n} \rangle$ on the one hand, corresponding to a loss of coherence below the threshold ($f < 1$ in the DNA system), and (2) the gain of coherence $d\tau$ with increasing cell number, reflecting a state above $f = 1$ ($f > 1$ in the DNA system). Roughly speaking, this model reduces the evolutionary development to the constraint that "socialization" should always result in an improvement of "communication". The integration of new members requires phases of localization on account of breaking down the overall coherence, while the delocalization, which improves the coherence, works against the production of new members.

In terms of biophoton emission, this effect can be traced back to the f-value through the introduction of a resonator value Q_ν [9]. The final result

$$\tau \propto \frac{1}{f} g \tag{15.22}$$

combines the coherence time τ with the f-value $\left(f = exp\left(-\frac{h\nu}{k\Theta(\nu)}\right)\right)$ and the "geometry" factor g of the resonators within the biological system. g describes some coherence-time of the elementary photon stores of the system. Since f is subject to measurements

$$i(\nu) \propto f(\nu)\,, \tag{15.23}$$

the conditions (15.21) and (15.22) can be translated into the terms of biophoton emission. From (15.21) we then have

$$\frac{\partial \tau}{\partial n} = \tau \left(\frac{1}{g}\frac{\partial g}{\partial n} - \frac{1}{f}\frac{\partial f}{\partial n}\right) = \tau \frac{\partial}{\partial n} \ln(\frac{g}{f}) \geq \frac{1}{\langle \dot{n} \rangle}. \tag{15.24}$$

Unless there is no change in the geometry of the cells, which means that $\frac{\partial g}{\partial n} = 0$, normal cell proliferation should always be assigned to a *decrease* of biophoton intensity $\left(\frac{\partial f}{\partial n} < 0\right)$. (15.24) explains all the observations on tumour and normal cells (see, for instance, Chapters 7, 8, 12) which show evidence of $\frac{\partial f}{\partial n} < 0$ for normal cells and $\frac{\partial f}{\partial n} > 0$ for tumour cells.

References

1. E.A. Power, *Introductory Quantum Electrodynamics* (Longmans, London, 1964).

2. W. Nagl and F.A. Popp, A physical (electromagnetic) model of differentiation. *Cytobios* **37** (1983) p. 45, 71.

3. F.A. Popp and W. Nagl, DNA conformations. Towards and understanding of stacked base interactions: Non-equilibrium phase transitions as a probable model. *Polymer bull.* **15** (1986) p. 89-91.

4. F.A. Popp, *Biologie des Lichts* (Paul Parey Verlag, Berlin, 1984).

5. J. Perina, *Coherence of Light. 2. ed.* (D. Reidel P.C., Dordrecht, 1985) p. 127.

6. F.A. Popp, Coherent Photon Storage of Biological Systems, in *Electromagnetic Bio-Information. 2. ed.*, eds. F.A. Popp, U. Warnke, H.L. Koenig and W. Peschka (Urban & Schwarzenberg, Muenchen, 1989) p. 144-167.

7. M. Galle, R. Neurohr, G. Altmann, F.A. Popp and W. Nagl, Biophoton emission from Daphnia magna: A possible factor in the self-regulation of swarming. *Experientia* **47** (1991) p. 457-460.

8. D.H.J. Schamhart and R. van Wijk, Photon Emission and the Degree of Differentiation, in *Photon Emission from Biological Systems*, eds. B. Jezowska-Trzebiatowska, B. Kochel, J. Slawinski, and W. Strek (World Scientific, Singapore, 1987) p. 137-152.

9. F.A. Popp, *Biophotonen – Ein neuer Weg zur Loesung des Krebsproblems* (Verlag fuer Medizin, Dr. E. Fischer, Heidelberg, 1976).

Chapter 16

Non-linear Optical Properties of Delayed Luminescence from Cress Seeds

Ralf Neurohr

Abstract

The present investigations on the properties of delayed luminescence (DL) from air dry cress seeds give indications for the existence as well as for the biological significance of optical couplings in living systems.

A nonlinear increase of DL with respect to the mass (emitting surface) of seed samples is observed. This observation cannot be explained in terms geometrical parameters because the effect is independent of the sample surface that is exposed to the detector.

Long term studies of DL uncover a complex rhythmical pattern of reemission intensities which do not completely confirm a former proposed mechanism for the explanation of temporal changes of delayed luminescence from the seeds. Correlations between temporal changes of DL and other rhythmical processes observed in seeds indicate a close connection between the biological clock and light emission in seeds.

16.1 Introduction

Delayed luminescence (DL) from air dry seeds is known for some years. Most of the work in this field has been done by Veselova & Veselovskii et al.[1–4]. They reported that there is a close relationship between the age of seeds and the intensity of DL, thus making it possible to estimate seed viability and storage capacity. Oxygen seems to have little influence on the DL, whereas moisture content of the seeds strongly affects the emission intensity (1% increase in water content leads to a 2-3 times weakened intensity of DL). A similar behaviour as well as relaxation characteristics can be observed from dry powders of proteins, starch, sugars and amino acids that are constituents of the seeds. Although the source(s) of emission is not yet identified, it can be assumed that excited states of macromolecules are mainly responsible for delayed luminescence. The hypothetical mechanism, underlying these observations, is considered to be a disturbance of biopolymers during aging processes, leading to a lower water binding capacity.

Aging processes cause irreversible "transitions" of biological systems and so the decrease of water binding capacity due to aging has to be a monotonic process as well. Therefore a monotonic increase of delayed luminescence may be expected if one follows the assumption that the time course of DL is mainly governed by aging processes which lead to lower water contents of the seeds.

Within the framework of biophoton research this interpretation may be considered as incomplete. According to the coherence of photon emission from biological systems, light signals from the seeds may show some properties that cannot be explained by the proposed mechanism.

We decided to study DL from air dried cress seeds (*Lepidium sativum*) in more detail. The experiments started with some preliminary investigations, to examine whether the relationship between seed age and intensity of delayed luminescence also holds for cress seeds. The dependence between seed age and DL intensity was verified for naturally as well as for artificially aged seeds. The results also indicate some nonlinear absorption/emission effects depending on the mass of seed samples.

16.2 Methods

The measurement equipment used for photon detection is described in Chapter 6. The measurement of DL intensity for each weighed seed sample (error ±0.01g) was

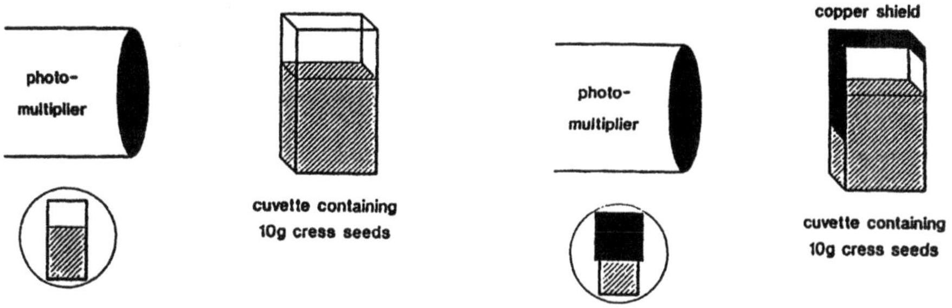

Figure 16.1: *Geometrical arangement during studies of seed DL. Left: The surface of seed samples exposed to the detector increases almost linearly with the amount of seeds. Right: The results discribed in 16.3.2 were obtained by shielding the upper part of the cuvette leaving only a constant area "visible" for the detector.*

performed three times to obtain a mean value for the initial counting rate for the first 500ms after excitation (excitation time=1s; white light/tungsten lamp).

For investigating the relationship between seed age and DL during natural aging, several measurements of reemission intensities were performed on four different seed lots harvested in 1989 and one seed lot harvested in 1988. Artificial aged seeds were obtained by incubating the seeds at 40°C and relative humidity of 100%[5]. To obtain two different stages of artificially aged seeds, one sample was incubated for 24h while a second one was incubated for 88h.

Changes of DL intensity as a function of sample mass (emitting surface) had been investigated under two different geometrical conditions. In the first measurement the whole area of the sample (increasing almost linearly with the amount of seeds) was exposed to the detector. The measurement is repeated with the upper part of the cuvette covered by a piece of copper, leaving only a small constant area of the sample "visible" for the detector (Fig. 16.1).

The seeds (*Lepidium sativum*; lot no. 1033-89) used in the long term studies were stored under room conditions. Temperature varied between 20°C and 24°C and air moisture fluctuations showed the same characteristics as measured outside the laboratory (meteorological observation station). The goal of these storage conditions, which allow the seeds to change their water content according to air moisture fluctuations, is to keep the seeds in equilibrium with laboratory surroundings, thus avoiding

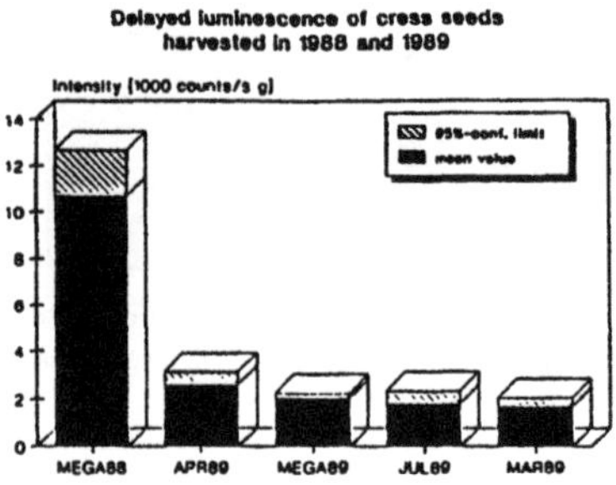

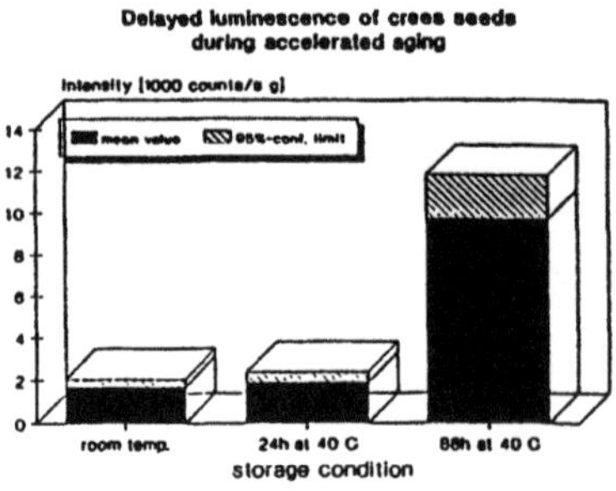

Figure 16.2: *DL intensity of naturally and artificially aged cress seeds. Left: DL of a sample stored for one year under room conditions (MEGA88) in comparison with the DL of 4 seed lots harvested in the following year. Right: DL from 3 samples of one seed lot that were stored under different conditions to obtain different stages of accelerated aged seeds.*

uncontrolled water uptake or loss during the meassurement procedure. The latter may lead to some difficulties in interpreting the results, while equilibrium changes in seed water content due to air moisture fluctuations can be considered via correlation analysis.

16.3 Results

16.3.1 Delayed luminescence (DL) of natural and artificial aged cress seeds

The reemission intensities of cress seeds exhibit the same behaviour in respect to seed age, as reported for various kinds of other species. Older seeds emit significantly more light after external light stimulation than younger seeds. This holds for naturally aged seeds, stored under ordinary room conditions (20°C, 40-70% relative humidity), as well as for artificial aged seeds, stored at 40°C and relative humidity of 100%. Fig. 16.2 displays the results (initial DL) for natural (left) and artificially (accelerated) aged seeds (right).

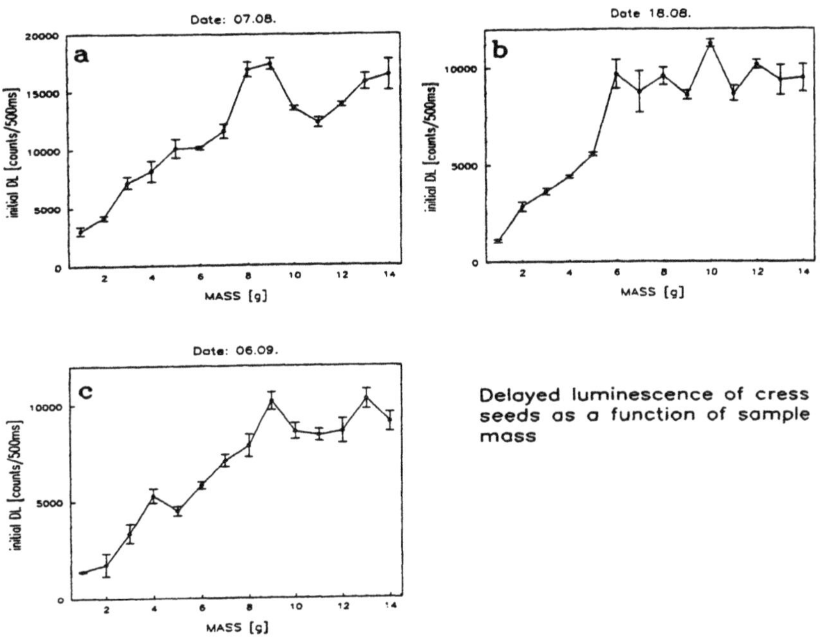

Figure 16.3: *The DL of cress seeds shows a nonlinear increase of intensity as a function of increasing amount of seeds. Error bars represent triple standard deviations of the mean value obtaind from 3 excitations. Because the emitting surface increases almost linearly with the mass of samples, the observed local maxima can not be explained by the linear superposition of stationary and independent emission from single grains.*

16.3.2 Nonlinear changes of DL intensity as a function of increasing mass (surface) of seed samples

Figure 16.3 shows DL intensities for increasing sample mass from 1 to 14g in three different experiments. There is a clear local maximum at 8-9g in two of three experiments, which cannot be derived from geometrical considerations. Increasing mass also increases the surface of the sample that emits photons towards the detector system. Even if one takes into account that self absorption will lead to some decrease in DL intensity, no local maximum can be expected. So this observation cannot be explained by assuming an independent, temporal constant contribution of every

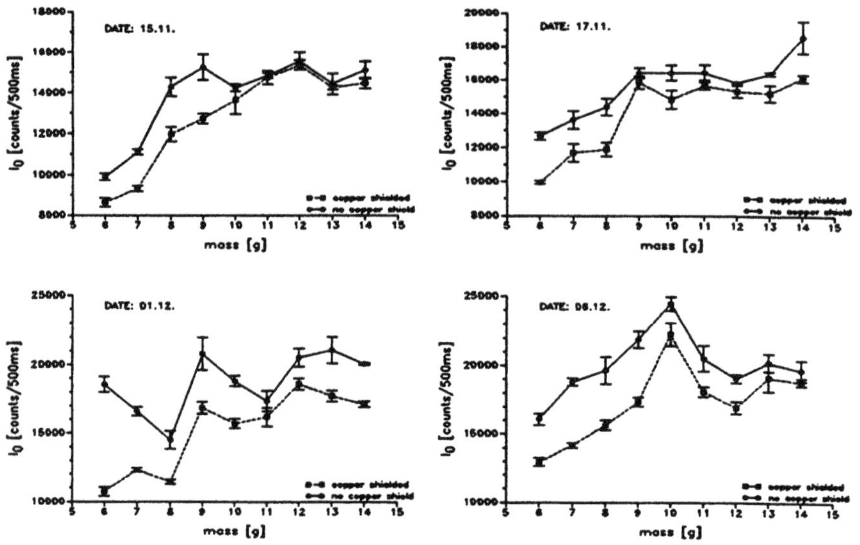

Figure 16.4: *DL from cress seeds of different masses. The graphs show mean values and triple standard deviations for constant (lower curves) and variable (upper curves) sample surfaces exposed to the photomultiplier. The characteristic relationship obtained in different experiments does not depend on the area that is "seen" by the detector.*

single grain to the DL of the whole sample. Optical couplings as well as temporal fluctuations of optical properties have to be considered.

Repeated measurements on identical samples confirm these findings. The plot of DL intensity versus sample mass displays a characteristic shape, which is independant of the area that is "seen" by the detector. The emission of a constant area from the lower part of the cuvette shows almost the same nonlinear behaviour (local maxima and minima) that is observed when the emitting surface of the sample increases linearly with the mass of the sample (Fig. 16.4).

16.3.3 Time course of DL intensity from cress seeds

Figure 16.5 shows the time series of DL intensities for 4 different sample masses (7, 9, 11 and 13g). The whole period is about one year (342 days). Measurements

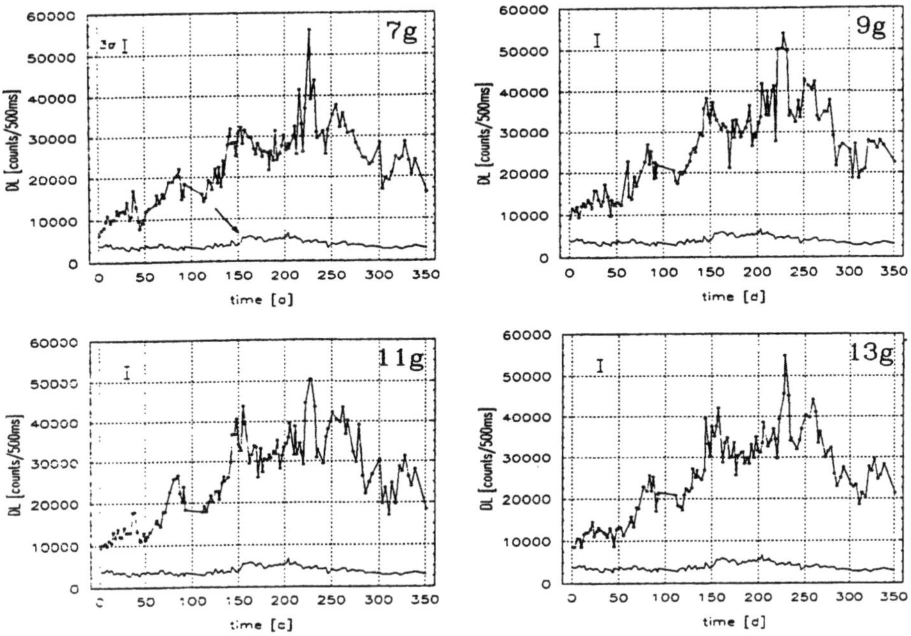

Figure 16.5: *Long-time evolution of the DL from cress seeds for samples of 7, 9, 11 and 13 g. Average values for the triple standard deviation of the single data points are shown in the upper left corner of each graph. The curve at the bottom of the 4 graphs shows the time course of noise (signal after excitation of the empty cuvette). The marked increase of the noise (arrow) can be assigned to the exchange of the cuvette used for the measurements.*

took place every 2 (respectively 3) days. Error bars in the upper left corner of each graph display the mean value of triple standard deviations for the single data points of the shown series (each data point is the mean value from 3 excitations). The noise term (signal after excitation of the empty cuvette) is shown in the lower part of the graphs. There is a small but clear increase in the noise term (arrow in Fig. 16.5), which can be assigned to the exchange of the cuvette that was used. So this discontinuity does not reflect a change in detector sensitivity but only a neglectible increase of noise compared to signal intensities obtained from the seeds.

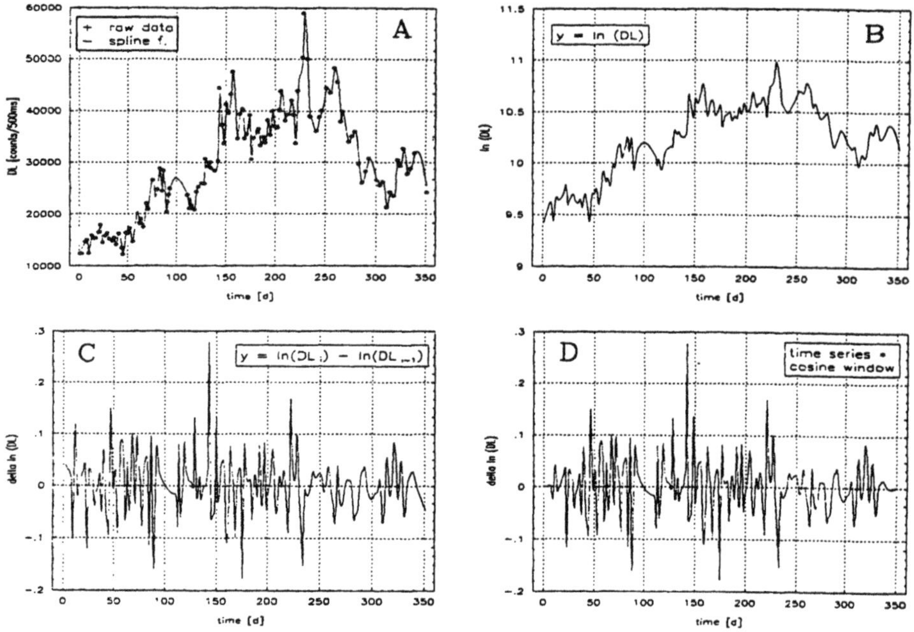

Figure 16.6: *Signal processing of the time series before entering into time series analysis. A) Interpolating Akima spline. B) Stabilizing variance by taking the logarithm of the sreies. C) Numerical differentiation removes the first order trend. D) Tapering with a bell shaped cosine window to prevent the Fourier spectrum from leakage.*

Spectral analysis

Obvious correlations between the 4 time series indicate a close connection between the signals, which can be examined by time series analysis. Prior to analysis[6,7], equally spaced intermediate data points ($\Delta t = 1\text{d}$) were generated by spline interpolation (Akima spline)[8]. Then the data had to be transformed ($x'(t) = \ln x(t)$) in order to stabilize variance. Finally, the trend was removed by numerical differentiation ($x''(t) = x'(t-1) - x'(t)$). In order to prevent the Fourier spectrum from leakage phenomena, data preparation was completed by the application of a cosine window that affected 7.5% of data at each end of the series (Fig. 16.6).

A typical example for the spectral power density functions of the 4 time series is shown in Fig. 16.7. The raw spectra had been smoothed by a 5-pass Daniell filter

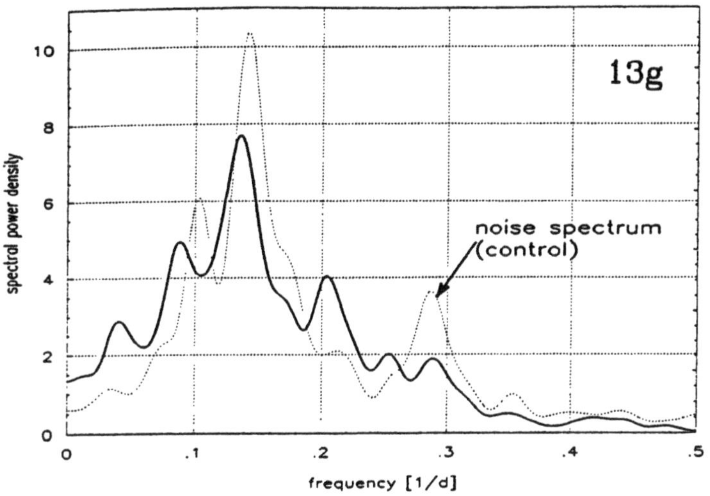

Figure 16.7: *Daniell smoothed spectral power density of the time series obtained from the 13g sample. The solid line displays the spectrum of the seeds. The dotted line displays the noise spectrum (empty cuvette). The spectra of the remaining 3 time series of the seeds are similar to the shown example.*

(filterwidths: 5, 7, 11, 7, 5). All spectra include 3-4 broad bands in the low frequency range ($f < 0.25d^{-1}$; $T > 4d$). There are no clear peaks from which a common single mode oscillator can be deduced. But the overlapping of the broad spectral bands indicates the connection between the 4 series. Because there is also a low frequency oscillation (T=7d) in the noise term (empty cuvette), seed spectra might be influenced or disturbed by this rhythm. It has to be shown, that the correlations between the DL from the samples are not caused by this external influence.

Coherence spectra

Based on the calculation of bivariate coherence spectra (Fig. 16.8)[7,9,10] between each of the 4 time series and the noise term, a correlation between noise and DL from the seeds can not be verified. There are no significant coherence estimates in the low frequency range, where the spectral bands of seeds coincide with the peak in the noise spectrum. Single coherence estimates exceeding the 5% threshold in the whole frequency range are less than 5%, thus indicating a complete independence of the

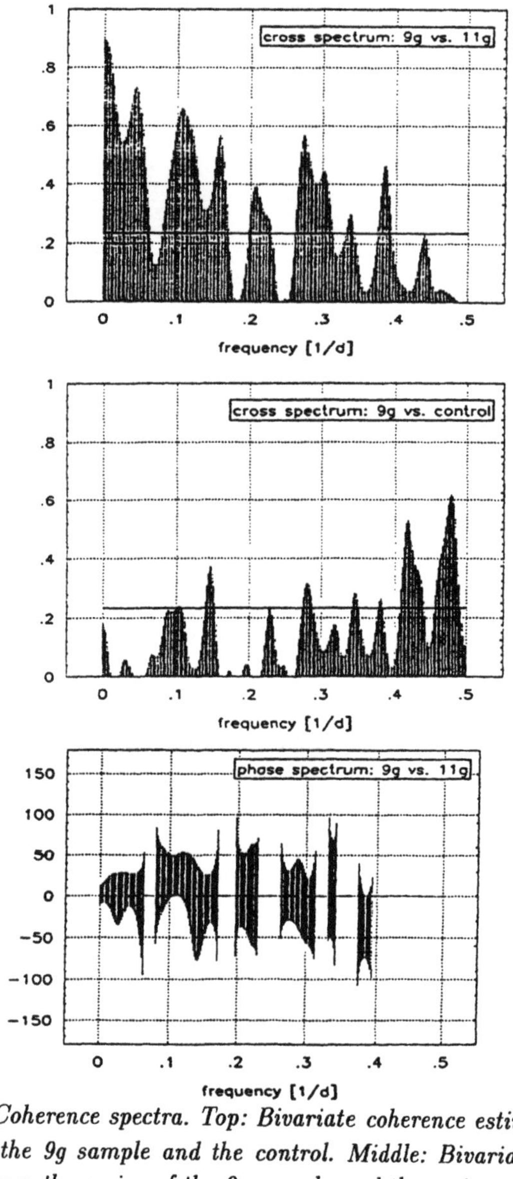

Figure 16.8: *Coherence spectra. Top: Bivariate coherence estimate as obtained from the series of the 9g sample and the control. Middle: Bivariate coherence estimate as obtained from the series of the 9g sample and the series of 11g sample. Bottom: Phase spectrum (95%-confidence limit) as obtained from the series of the 9g sample and the series of the 11g sample.*

series. In comparison, significant and high coherence estimates (up to 0.9) can be found in any of the 6 possible cross spectra for the time series of the seeds. Coherence estimates are significant over a wide frequency range and phase shift confidence limits include zero. Figure 16.8 displays typical examples for the coherence spectra of noise vs. seed DL, seed DL vs. seed DL and cross spectra of the phase angle in seed DL.

Bivariate cross correlations

Although coherence spectra indicate correlations between the time series as well as the absence of correlations between noise and each of the time series, it is possible, that correlations originate from autocorrelations which are still present in the series. Thus cross correlations may be explained in terms of interfering autocorrelations and not in terms of real correlations. To remove this uncertainty, all autocorrelations of the single series have to be eliminated before estimating bivariate cross correlations of the remaining residuals.

In order to remove all autocorrelations, stochastic models (Box-Jenkins-ARIMA models)[11] are fitted to the series. The fitting procedure led to an ARMA(4,3) model in case of noise and to ARMA(6,3) or ARMA(7,3) models in case of the DL series, respectivly. The square sum of errors (residuals) is in minimum and autocorrelations completly vanish. No difference between the integrated residual spectra and the white noise spektrum is found (Test according to Jenkins & Watts[12]). Cross correlations between the residuals of seed series show a clear peak for time lag zero, wheras no correlation can be found between the residuals of noise and the residuals of the seed series (Fig. 16.9).

These results indicate the close relationship between the 4 DL signals from the seeds and their independance from the noise. The peak shaped correlation estimate implies phase locked coherence of the signals in the whole frequency range, thus confirming the results of cross spectral analysis.

16.3.4 Conclusion

The results of time series analysis show that the periodic behaviour of DL is highly correlated in the 4 series and completely different from the oscillations of the noise. Since there is some overlapping between the noise spectrum and one of the spectral bands of DL, it still has to be considered that seed spectra are disturbed by noise. Therefore it seems to be more reliable to calculate the mean spectrum from the 4 single DL spectra. This mean spectrum is characterized by a main peak at $f=0.135d^{-1}$

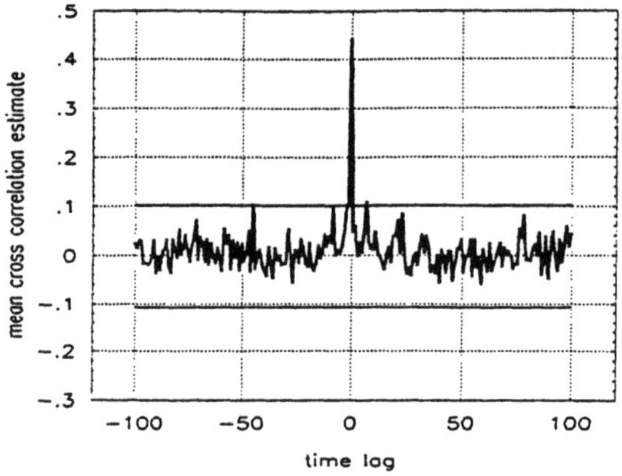

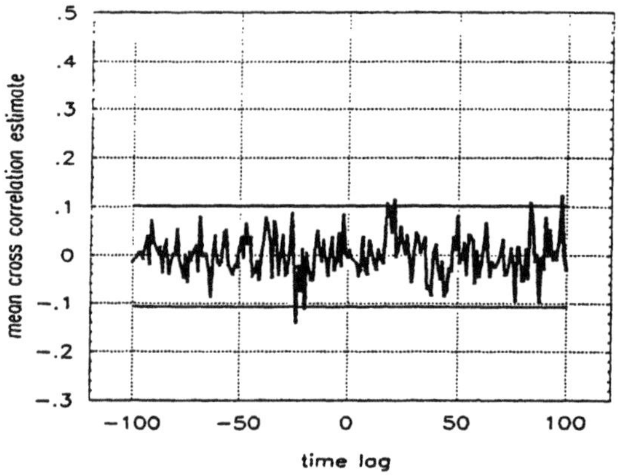

Figure 16.9: *Mean cross correlation estimate for the bivarite cross correlation between the residuals of the noise and the residuals of each of the 4 time series obtained from the seeds (left). Mean cross correlation estimate for the bivarite cross correlation between the residuals of the 4 time series obtained from the seeds (right).*

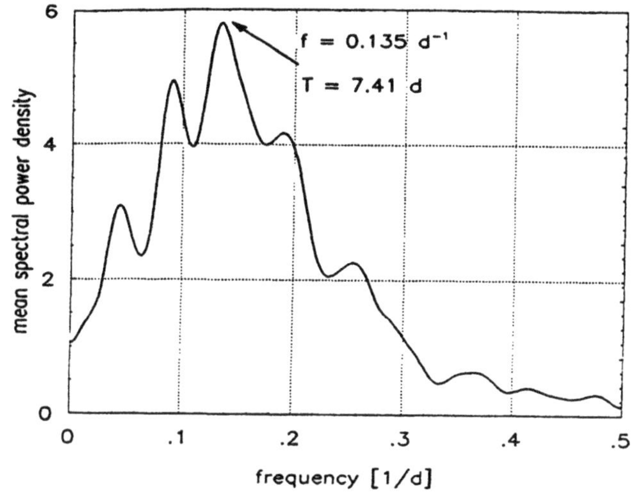

Figure 16.10: *Mean spectral power density as calculated from the single spectra of the 4 time series of seed DL.*

(T=7.4d) leaking out in 4 more or less symmetrical side lobes (Fig. 16.10). The reliability of these side lobes is still uncertain because of the possible disturbance by noise oscillations.

16.4 Discussion

The results presented in Fig. 16.3 and Fig. 16.4, showing the nonlinear increase of DL intensity over increasing mass/surface of the invastigated samples, disprove the assumption that delayed luminescence from seeds only depends on seed water content. If this were the case, a more or less linear relationship between the surface of the sample and the intensity of DL would be expected.

Differences in the results (shape of data plots) from repeated experiments may be caused by complex seasonal rhythms of DL discussed below.

Analysis of long time changes in DL again contradicts Veselova & Veselovskii's asumptions on the mechanisms governing the light emission. If irreversible aging processes are responsible for a loss in water binding capacity and enhancement of DL, the clear decrease of DL, after reaching the maximum intensity in May, cannot

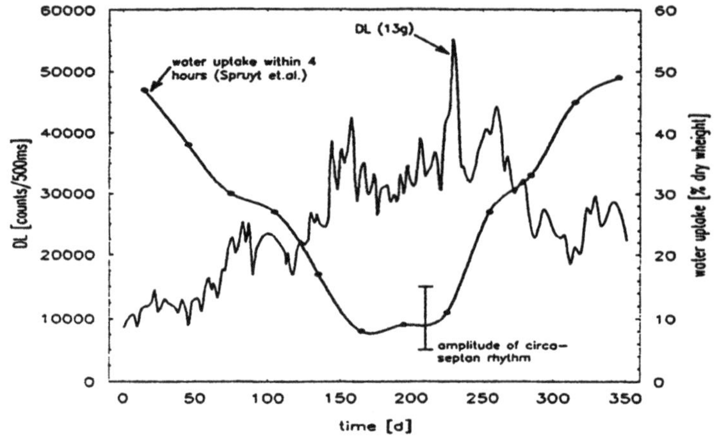

Figure 16.11: *The time course of water uptake within 4 hours in bean seeds shows a similar seasonal pattern as DL from cress seeds. Additionally there is a corresponding circaseptan rhythmicity in both signals whose amplitude is in the same order of magnitude compared to the seasonal pattern.*

be explained. Aging processes at best can be responsible for the slight increasing trend of DL between the beginnig and the end of the series. The seasonal pattern as well as the highly correlated shorter oscillations of DL must be caused by other factors. According to the close relationship between seed moisture content and DL, the observed patterns of DL intensity as well as their correlations may be explained by fluctuations of seed moisture content due to climatic fluctuations of air moisture. Cross spectral and cross correlational analysis of DL intensity and meteorological data do not confirm this assumption (see Apendix). There is no significant correlation between air moisture and delayed luminescence from the seeds.

One factor which remains to be considered is biological rhythms, which are ubiquitous to all living organisms.[13] Spruyt et al.[14] report a circannual rhythmicity in the water uptake of bean seeds which is superposed by a circaseptan rhythm showing a period length of T = 7.4 days. The time plot of DL and rate of water uptake shows a clear negative correlation as shown in Fig. 16.11. In addition, the superposed circaseptan rhythm of water uptake is in the same order of magnitude as the high frequency DL oscillations which reach their maximum spectral power density also at a period length of 7.4d. This indicates a connection between delayed luminescence and biological rhythms. At present it only has to be speculated whether DL is triggered by the biological clock (being just a hand) or whether the coupling of excited states is a constitutive element in the generation of biological rhythms. Nevertheless, the present state of knowlege about the biological clock is very incomplete and biophysical considerations involving the concept of coherent excitations and photon emission may contribute to our understanding of the nature of biological clock.

16.5 References

1. T.V. Veselova, V.A.Veselovsky, A.B.Rubin, P.Z.Bochvarov, Delayed luminescence of air-dry soybean seeds as a measure of their viability. *Physiol Plant* **65** (1985) p. 493-497.

2. E.R. Kartashova, T.V.Veselova, V.A.Veselovskii, V.D.Nadykta, The delayed luminescence of sunflower seeds during prolonged storage under conditions of a regulated gas medium and in the air. *Biol. Nauki* (1988) p. 31-35.

3. T.V. Veselova, V.A.Veselovskii, V.I.Kozar, A.B.Rubin, Delayed luminescence of soybean seeds during swelling and accelerated ageing. *Seed Science and Technology* **16** (1988) p. 105-114.

4. T.V. Veselova, V.A.Veselovskii, A.A.Krasnovskii, J.K.Lichshteld, Delayed luminescence of seeds. *Biophysics* **30** (1985) p. 775-777.

5. J.C. Delouche, C.C.Baskin, Accelerated aging techniques for predicting the relative storability of seed lots. *Seed Science and Technology* **3** (1973) p. 427-452.

6. P.J. Diggle, *Time Series: A Biostatistical Introduction* (Clarendon, Press, Oxford, 1990).

7. C. Chatfield, *Analyse von Zeitreihen* (Carl Hanser, München-Wien, 1982).

8. H. Akima, A new method of interpolation and smooth curve fitting based on local procedures. *J. Assoc. Comp. Math.* **17** (1970) p. 589-602.

9. J.S. Bendat, A.G.Piersol, *Random Data: Analysis and Measurement Procedures* (Wiley & Sons, New York, 1971).

10. R.K. Otnes, L.Enochson, *Applied Time Series Analysis. Volume 1: Basic Techniques* (Wiley & Sons, New York, 1983).

11. G.E.P. Box, G.M.Jenkins, *Time Series Analysis: Forcasting and Control* (Holden Day, San Francisco, 1976).

12. G.M. Jenkins, D.G.Watts, *Spectral Analysis and its Applications* (Holden-Day, Oakland, 1968).

13. E. Bünning, *Die physiologische Uhr* (Springer, Berlin, Heidelberg, New York, 1977).

14. E. Spruyt, J-P.Verbelen, J.A.DeGreef, Expression of circaseptan and circannual rhythmicity in the imbibition of dry stored bean seeds. *Plant Physiology* **84** (1987) p. 707-710.

16.6 Appendix

16.6.1 Relationship between seed DL and air moisture fluctuations

If the water content the seeds is mainly responsible for the intensity of DL signals, there should be a correlation between the time course of DL and air moisture fluctations, since the seeds were allowed to change their equilibrium water content according to external water potential[1]. As mentioned already in Sec. 16.4, no correlations between climatic air moisture fluctuations and the fluctuations of delayed luminescence were found. This finding is demonstrated in Fig. 16.12 which shows the estimates for bivariate cross correlation and coherence between the time series of seed DL (13g) and the time series of the mean daily air moisture as measured by a meteorological observation station not far from the laboratory.

16.6.2 Water uptake of the dry seeds

Because of the lack of correlation between DL and air moisture fluctuations, on the one hand, and the strong dependence of DL on seed water content reported by Veselova & Veselovskii on the other hand, an unexplained contradiction arises. In order to solve this problem, it is proposed to assume that changes of seed water content due to external water potential fluctuations are not high (quick) enough to affect the delayed luminescence significantly. Figure 16.13 displays the water uptake

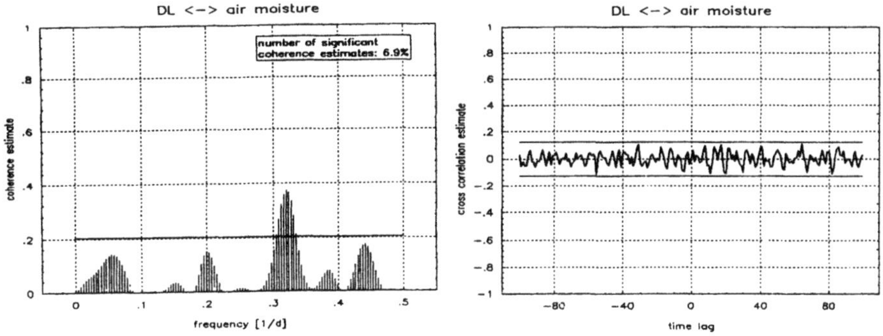

Figure 16.12: *Left: Bivariate coherence estimate as obtainend from time series for DL (13g) and time series of the mean daily air moisture. Right: Bivariate cross correlation estimate as obtained from time series for DL (13g) and time series of the mean daily air moisture. Horizontal lines indicate the 5% significance levels.*

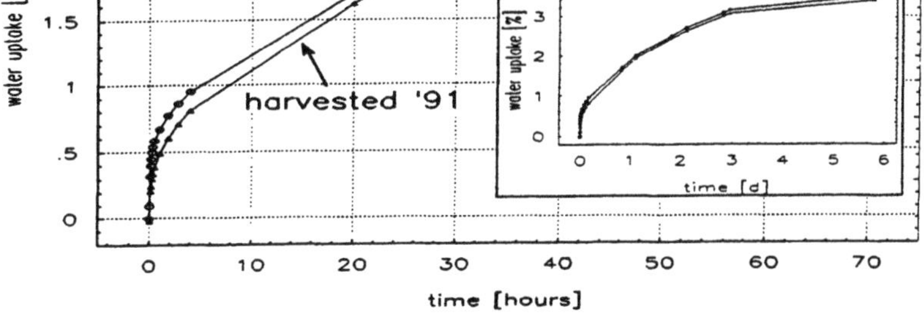

Figure 16.13: *Water uptake of dry cress seeds. The seeds (4g) were dried according to the ISTA rules[2] (1h at 130°C). Subsequent water uptake was measured at 20°C and 50-60% humidity. Equilibrium water content of 6-7% was not reached within 14 days.*

of completely dried (drying procedure according to ISTA rules[2]) cress seeds during 3 (6) days at 20°C and 50-65% humidity. The rate of water uptake rapidly decreases and initial water content (before drying) of 6-7% was not reached within 14 days, thus confirming the assumption that changes of water equilibrium content are too slow to affect DL in the way that can explain the observed DL fluctuations.

16.6.3 References

1. E.H. Roberts, R.H.Ellis, Water and seed survival. *Annals of Botany* **63** (1989) p. 39-52.

2. International Seed Testing Association (ISTA), International rules for seed testing. *Seed Science and Technology* **13** (1985) p. 299-355.

Chapter 17

Biophoton Measurement as a Supplement to the Conventional Consideration of Food Quality

K. Lambing

Abstract

The biophoton measurement can be used as a method complementary to chemical analysis:

- Heating influences causing internal changes of the milk are detectable, especially after illumination with red or green light. The milk can be regarded as a non-thermal resonator whose quality is dependent of the resonator effectiveness Q_ν. The higher the light storage capacity of the resonator (milk) the lower is the intensity of the emission after illumination. The more the resonator is destroyed by heating, the higher is the intensity of the emission after illumination or the higher is the leaking stream i_ν.
- The origin of hen's eggs is detectable by the biophoton measurement. Free range eggs exhibit a higher emission rate after white light illumination because they have a higher amount of stored energy U_ν than eggs from soil or cages. Light illumination leads to a kind of induced emission so that the stored energy U_ν is released.

The ChiHyp-value, as a measure of non-agreement with a hyperbolic relaxation kinetics after light illumination, is influenced by the hens' exposure to the sun. Free range eggs exhibit a nearly hyperbolic relaxation dynamic, but eggs from soil or cages exhibit a nearly exponential decay. The agreement with a hyperbolic relaxation can be regarded as a measure of internal order.

- The quality of vegetable oils can be characterized by the biophoton parameters, e.g. the self-emission, the emission rate after white or red light illumination. A high emission rate after red light illumination indicates decomposition procedures due to aging, poor storage etc. An objective classification of the oils can be made by a factorial analysis using the first three factors.
- The irradiation of food induces a breaking of chemical bonds and the formation of radicals. Some of these radicals can react with other food components leading to a varity of products and some radicals can remain in a long living stable state. By the illumination with light of different wavelengths the excited radicals fall down to their ground state and emit the energy difference as light. So, irradiated foodstuffs exhibit a much higher emission rate after light illumination than non-irradiated ones.

17.1 Introduction

In general food is regarded as a mixture of chemical compounds, supplying the organism with material that maintains reactions necessary for life. Dependent on the functional and physiological significance the food components can be divided up into 3 groups:

1. energy substrates (caloric value),
2. building material (synthetic value), and
3. catalytic substrates (e.g. enzymes, minerals,trace elements, water).

This point of view determines the biological or healthy value of food, while the food quality can be described as a function of contents regarded as positive. Through the drastically increased ubiquitous environmental pollution and the resulting contamination of food, the consumer expects from qualitatively good food not only a high amount of positive and healthy components but also a low amount of substances regarded as negative and harmful (e.g. residues of organic halid compounds, animal medicines, heavy metals, nitrate etc.).

This fact led to the formation of a new industrial scope, that produces in "controlled cultivation" or "controlled animal breed" biological healthy food to win the sympathy (and money) of environmentally and health-conscious consumers. However, examinations have shown that those biological products are not always as healthy as expected or promised, because the air and water as the main carriers of pollutants can not be excluded.

The existing food-law prescriptions and laws are not sufficient, because the empirically fixed and tolerated amounts of residues include only single substances. They do not take into account the possibility of cumulatory harmful effects caused by interaction of the different residues and substances. A remedy could be a total chemical analysis of all foods coming into trade. But, because of the high costs and time necessary for a total analysis only random sample examinations are carried out, while the choice for the parameters analyzed is more or less accidental. An alternative to the total chemical analysis is the application of biophoton measurement, which is being used with increasing frequency and not only for food quality analysis (see other Chapters).

In literature several designations for the phenomenon of photon emission occur and can be used as synonyms, e.g. low-level-luminescence, ultraweak luminescence, weak luminescence, ultraweak photon emission, delayed luminescence, ultraweak bioluminescence, biophoton emission etc..

17.2 The Principle of Biophoton Measurement and the Correlation to Food Quality

In chemical analysis the amount of single, isolated components is determined and quantified by specific detection and indication methods.

In contrast, the biophoton measurement, as a kind of holistic analysis, examines the food in a non-invasive way as a unit. The food is illuminated with light of different wavelengths. By means of optical parameters, such as the light storage capacity

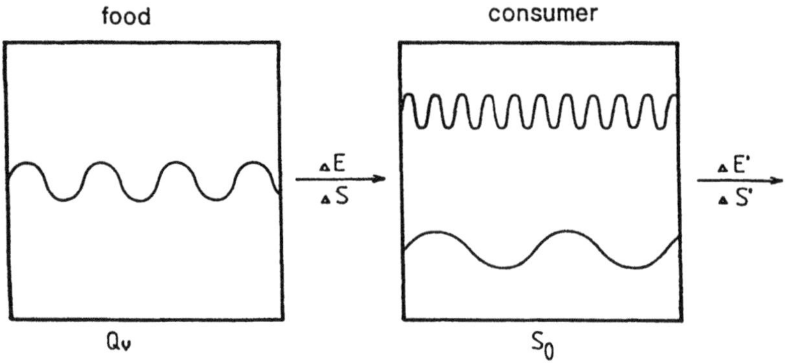

Figure 17.1:

and the kinetics of the light reemission, a quality declaration of the food can be made[7–10,13].

This approach can be traced back to the quantum physicist E. Schroedinger, who connected the quality of food with the expression "negentropy" as measure of order. The quality of food increases as it is able to keep the entropy or entropy production of the consumer at a minimum, with a minimum of energy display. By the stabilisation of the entropy the living system is able to stabilize itself against external influences and can remain in a state S_0 far away from thermal equilibrium. The attainment of the thermal equilibrium is synonymous with death, because all potentials and gradients are leveled out and no further gain of energy and no further life is possible.

Technically the negentropic capacity of food cannot be measured. The food can be regarded as a non-thermal resonator far away from thermal equilibrium which empties its reservoir ΔE and ΔS (energy- and negentropy content as a regulative information) inside the consumer (Fig. 17.1). The negentropy content of the resonator (food) can be expressed by its resonator effectiveness Q_ν, in the form

$$Q_\nu = 2\pi\nu\left(\frac{U_\nu}{i_\nu}\right)\left(\frac{V}{F}\right) = 2\pi\nu\tau_\nu,$$

where ν is the frequency, U_ν and i_ν the stored energy or the leaking stream, V and F are the volume or the surface area of the object. τ_ν is the coherence time, which can be measured as decay time after external (light-)stimulation.

All parameters describing the resonator effectiveness are detectable by the biophoton measurement.

Further parameters to describe the quality of food are for example:

- the total and spectral intensity i or $[(\frac{\partial i}{\partial \nu})\, \Delta\nu]$ of the food in counts/s or the photon countrate/time interval,
- the number of reemitted quanta $[< i >]$ in a defined time T after defined light illumination (defined by wavelength,polarisation, intensity and irradiation time),
- the relaxation kinetics after excitation, that can alternatively be described by the best adaptable exponential functions $Ae^{(-\beta t)}$ and the hyperbolic functions $Bt^{-\mu}$ (with t as time after excitation, A, B, β and μ as parameters to be fitted),
- the ChiHyp-value that describes the non-agreement of the observed relaxation kinetics with a hyperbolic one,
- the aHyp-value that describes the (negative) slope of the relaxation kinetics after light illumination.

While every change in the quality of food influences directly the optical parameters of the biophoton measurement, not every change in the optical parameters can be assigned to quality-changes of the food (e.g. temperature-, pressure-, mass dependence). The qualitative holistic aspect of the biophoton measurement can be understood as a complementary supplement to the partial aspect of the chemical analysis, where the two methods are complementary to each other. In the following the applicability of the biophoton measurement for the quality description and determination of food is described.

17.3 Biophoton Measurement of Milk

The milk is usually treated by different heating methods to improve its durability. It is known that the heating leads to a change of the milk compounds and diminishes the quality of the milk. Even digestive problems can occur in little children. The internal changes of the milk quality caused by heating consist in general of:

- protein denaturation (association, coagulation),
- dephosphorylation,

- nitrogen release in the form of low-molecular fragments,
- changes in the casein micelles,
- thiol-disulphide exchange,
- maillard-reactions between lactose and aminogroups,
- formation of δ-lactones and methyl ketones,
- loss of vitamines (especially C, B_{12}, B_6, B_1),
- lysin degradation,
- enzyme inactivation, etc..

In view of the manifold of possible influences on the milk composition and quality, it can be expected that these changes are detectable by the biophoton measurement. In the following some milks from the trade are described by the optical parameter mean value 100 (MW100, see below) received by red-, green-, blue-, white-light illumination.

The different milks investigated were:

- 31 samples of pasteurized, homogenized milk with 3.5%-fat (VM),
- 28 samples of ultrahigh temperature, homogenized milk with 3.5%-fat (HVM),
- 29 samples of ultrahigh temperature, homogenized milk with 1.5%-fat (HMM),
- 16 samples of pasteurized, homogenized milk with 1.5%-fat (MM),
- 7 samples of pasteurized, homogenized milk with 3.8%-fat (LL).

17.3.1 Performance of measurement

Two hours before the measurement the milk was taken out of the refrigerator and warmed up by room temperature ($\sim 25°C$). Then, 12.5 ml of the milk was placed in a quartz cuvette and put into the dark measuring chamber for 5 minutes before the measurement cycle was carried out 3 times for every illumination with white-, red-, blue- or green-light.

Measurement cycle

- 30s of light illumination with a 150 W tungsten lamp (white light =without filter and red-, green-, blue-light = with colour filter),
- registration of 101 biophoton emission data (time interval for the registration = 500 ms, this means that the photons emitted and registered in the photomultiplier tube were registered every 500 ms),
- 1 minute break, and
- start again with 30s illumination etc..

From 100 registrations in the photomultiplier tube the mean values of the emission rate were calculated for each illumination, then combined to one mean value (MW100). Similarly the empty cuvette was treated as reference and its measurements subtracted from those of the milk samples.

17.3.2 Results from the biophoton measurement of milk

The optical parameter MW100 (Fig. 17.2, 17.3) was compared in a variance analysis (95% level of significance) to find significant differences between the different milks. In the MW100-value there is a tendency increase from LL, VM, MM, HVM to HMM for all colour illuminations, while the ultrahigh temperature milks show in general higher values than the pasteurized milks. The ultrahigh temperature milks as well as the pasteurized milks with 1.5%-fat exhibit a higher emission rate than those with 3.5/3.8%-fat. Apparently two parameters influence the optical properties of milk:

1. the kind of heating (pasteurized or ultrahigh temperature) and
2. the fat content (3.5/3.8%- or 1.5%-fat).

In the MW100-value after green light illumination nearly all samples are significantly different from each other, while no significant difference could be found between MM and HVM (Fig. 17.2).

In the MW100 after red light illumination there is no significant difference within the pasteurized milks (LL, VM, MM) and the ultrahigh temperature (HVM, HMM) milks (Fig. 17.3). It seems that the kind of heating and not the fat content strongly

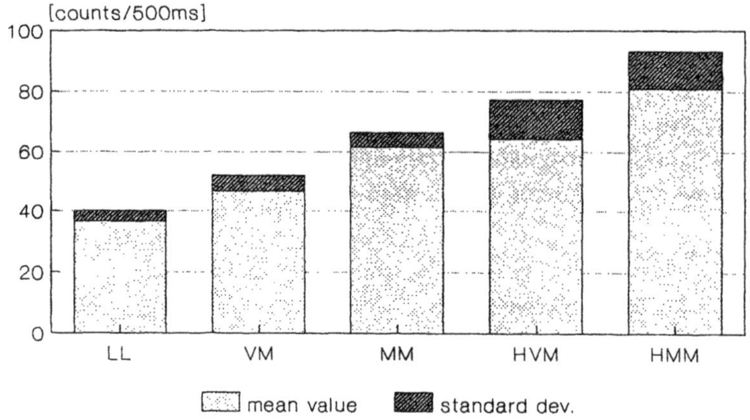

Figure 17.2: *Green light illumination of milk. The mean values of the first 100 values after illumination are displayed.*

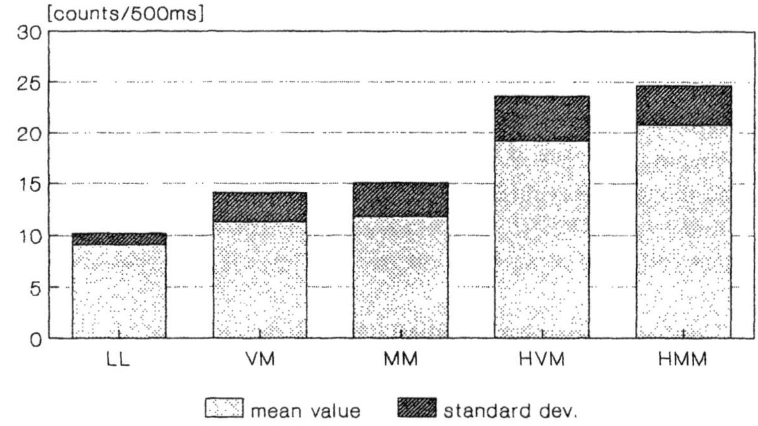

Figure 17.3: *Red light illumination of milk. The mean value of the first 100 values after illumination are displayed.*

influences this parameter of red light illumination, because significant differences are evident between the two different heated groups.

The MW100-values after white light- and blue light-illumination resemble each other and show, in general, the same pattern as the MW100-values after green light illumination, although not all samples differ significantly from each other.

The difference in the MW100-values between the different kinds of milk can be interpreted by changes in the light storage capacity induced by internal changes of the milk compounds (see Chapter 3). Heating procedures, leading to conformational changes in the albumin fraction, reduce the light storage capacity of the milk by lowering the resonator effectiveness Q_ν. This fact is detectable by an increased photon emission after light illumination, while the leaking stream i_ν (equation (1)) is increased. A reduced fat content of the milk (1.5%-fat) also decreases the light storage capacity, whereas the milks with a natural fat content (3.5/3.8%) exhibit a higher resonator effectiveness Q_ν.

17.4 Biophoton Measurement of Hen's Eggs

The majority of the available eggs originate from cages where the hens are kept in a small cage without sunlight and no freedom of movement. To increase their wholesomeness, the hens are constantly treated with chemicals (antibiotics, coccodiostatics, psychopharmaka etc.), which accumulate in the yolk.

Most of the consumers associate a bright yolk with a high quality, so synthetic dyes are added to the feed to intensify the colour of the yolk. The consumers prefer eggs produced by free range methods without addition of chemical substances and would pay a higher price. In the past, eggs were often declared to be from free range or soil keeping although they originated from cages. This was possible because there was no suitable detector method for the determination of the origin of eggs. With the help of the biophoton parameters the origin of the hen's eggs can be determined, as will be shown in the following.

To check whether the optical parameters can be correlated with chemical components or whether there are chemical differences between eggs of different origin, chemical parameters of the eggs were determined.

17.4.1 Performance of the experiments

In cooperation with the "Institut fuer Kleintierzucht Celle" a project was undertaken to investigate the influence of breeding conditions on the biophoton parameters.

325 brown ISA-Warren hens (one year old, same origin and cage as breeding condition) were separated into 4 different enclosures (groups of keeping):

- cage,
- soil,
- soil with free range on sand,
- soil with free range on vegetation.

The date when the hens were separated is marked as (-21d), because there followed a period of 21 days to accustom, before the two free range groups were allowed to go out at the day 0 (0d). All the hens were fed with the same grain mixture and only the free range group on vegetation was able to pick grass, flies, beetles etc..

About always 20 eggs from each enclosure were sampled and coded in the "Institut fuer Kleintierzucht" before they were sent for the biophoton measurement to Kaiserslautern. So the origin of the eggs was not known until the end of the project. Once delivered, the eggs were stored in a refrigerator at 4^oC for 4-8 days before the chemical and biophoton parameters were investigated.

The biophoton measurement was carried out 4 times for every enclosure, while each time the yolks of 3 eggs were separated by a syringe (about 6 ml from each egg) and mixed as one sample. From this mixture 9 ml of yolk was used for the biophoton measurement and the rest for chemical analysis. The chemical parameters investigated were lecithin (%), cholesterol (mg/g fat), fat (%) and the total amount of carotenoids (mg/kg).

The biophoton measurement was carried out as described in the Sec. 17.3.1., with the exception that only a white light illumination was carried out. The time interval for the registration of the biophotons was 1 second. The MW90- and the ChiHyp-values were used to compare the eggs from the different origins.

The ChiHyp-value describes the deviation of the observed relaxation kinetics from a hyperbolic function $Bt^{-\mu}$. While the relaxation kinetics after excitation exhibits a deviation from the theoretical hyperbolic function, the ChiHyp-value corresponds to the sum of squares.

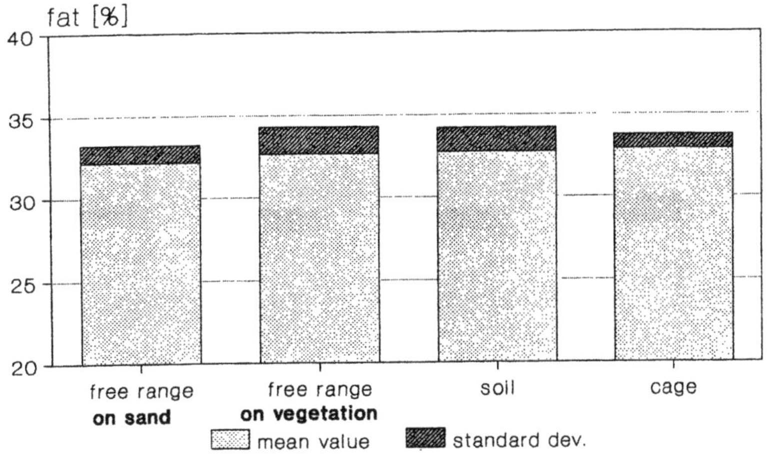

Figure 17.4: *Fat content of eggs from different breeding conditions.*

For the calculation of the mean value of the emission rate the first 5 values after illumination were rejected because the counts of the empty cuvette were higher than those of the cuvette with the yolk. Hence, the calculated mean value (MW90) is a measure of the total registrated counts in the time interval from 5 to 95 seconds after light illumination.

17.4.2 Results from the biophoton measurement of hen's eggs

No significant differences could be detected in the chemical parameters between the eggs from the different origins (Fig. 17.4, 17.5, 17.6, 17.7).

The biophoton parameters MW90 and ChiHyp from the eggs originating from the soil and the cages remained constant during the whole experiment, indicating that no "internal" change of the eggs occurred.

The free range groups exhibited changes in the optical parameters as they are displayed in the figures 17.8 and 17.9.

The MW90-values remained nearly constant for one week. The intensity of the eggs from free range on vegetation remained on that higher level, before increasing again after 12 weeks, whereas the eggs from free range on sand exhibited a slightly decreasing tendency after 3 weeks in free range (Fig. 17.8).

Concerning the ChiHyp-value (Fig. 17.9), there was a decreasing tendency in both

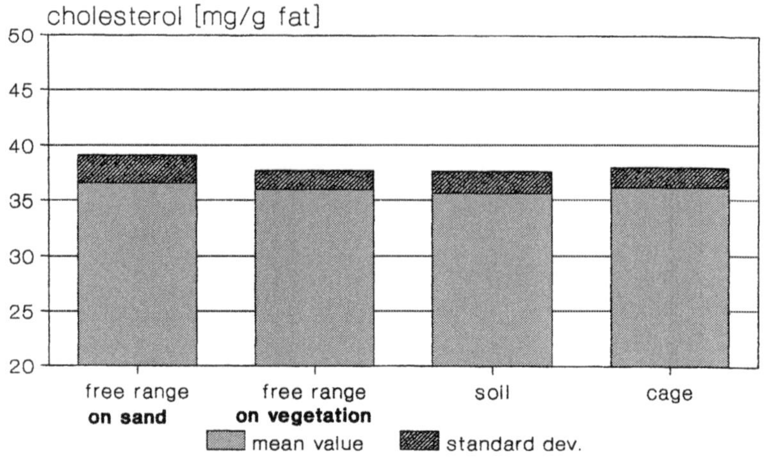

Figure 17.5: *Cholesterol content of eggs from different breeding conditions.*

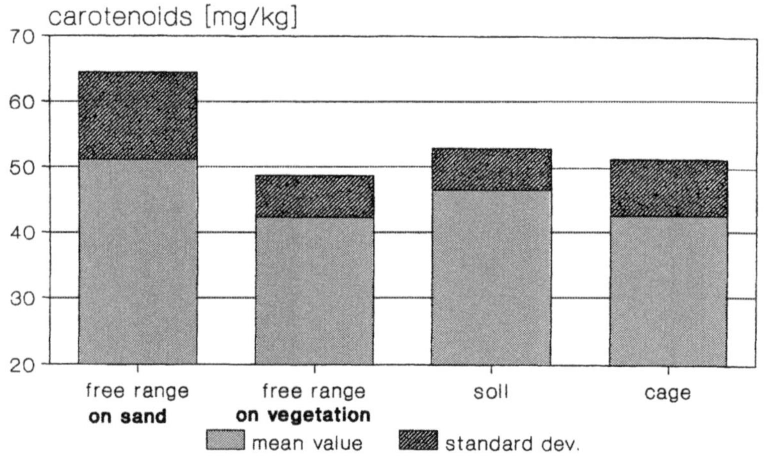

Figure 17.6: *Carotenoid content of eggs from different breeding conditions.*

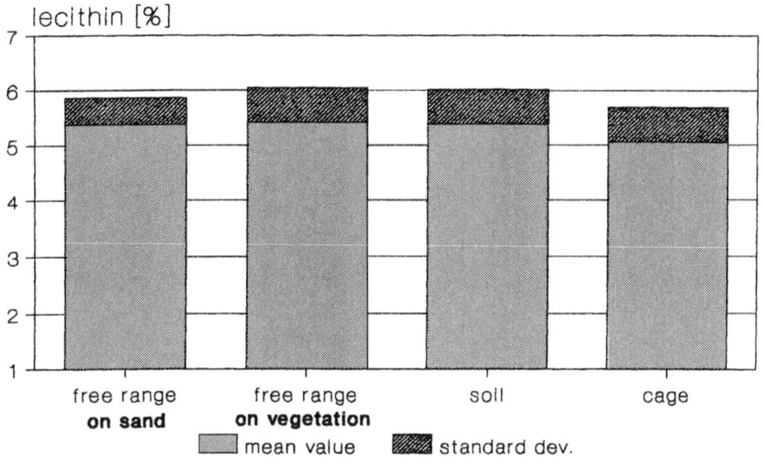

Figure 17.7: ***Lecithin content of eggs from different breeding conditions.***

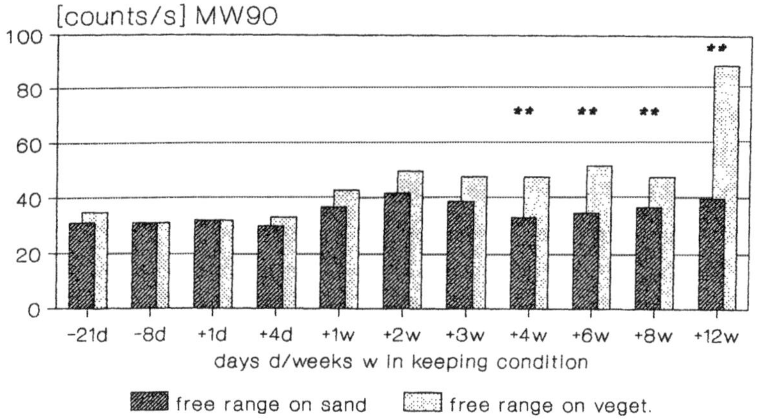

Figure 17.8: ***Intensity of biophoton emission free range on sand versus free range on vegetation. (*p = 0.05, **p = 0.01).***

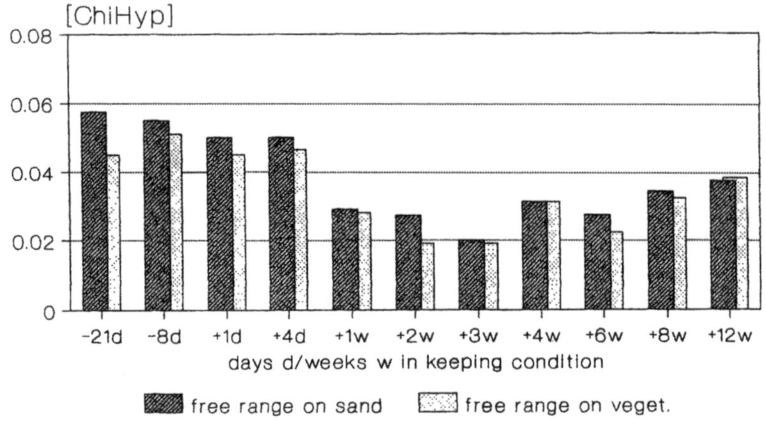

Figure 17.9: ***Agreement with a hyperbolic relaxation kinetic free range on sand versus free range on vegetation. (*p = 0.05, **p = 0.01).***

groups during 3 weeks in free range before the ChiHyp-values increased again. It could be shown that the ChiHyp-value of the eggs is influenced by the sun exposure of the hens. This means that sun exposure leads to a minimal ChiHyp-value. For the ChiHyp-value, which is a measure of the non-agreement of the observed relaxation kinetics with the hyperbolic relaxation kinetics after light illumination, it may be concluded that the sun exposure increases the order of the eggs.

Figures 17.10 and 17.11 show the comparison of the free range on vegetation with the soil eggs (significant differences are marked by * or **).

The MW90 of the eggs from soil remained at a constant level while the values for the eggs from free range increased with time. The eggs from soil exhibit relatively constant ChiHyp-values, whereas the eggs from free range exhibit a drastic decrease after one week in free range.

The chemical parameters of the eggs could not be correlated to the biophoton parameters. As the eggs from the free range breeding exhibit a higher amount of carotenoids than the eggs from other breeding (see Fig. 17.6), an additional study was made to check whether the carotenoids are responsible for the differences in the biophoton parameters. However, no correlation between the level of carotenoids and the biophoton parameters could be found. This fact indicates that the source of biophotons is not a single chemical substance, but internal properties of the sample as a unit, e.g. inter- and intramolecular interactions behaving as a resonating light-store.

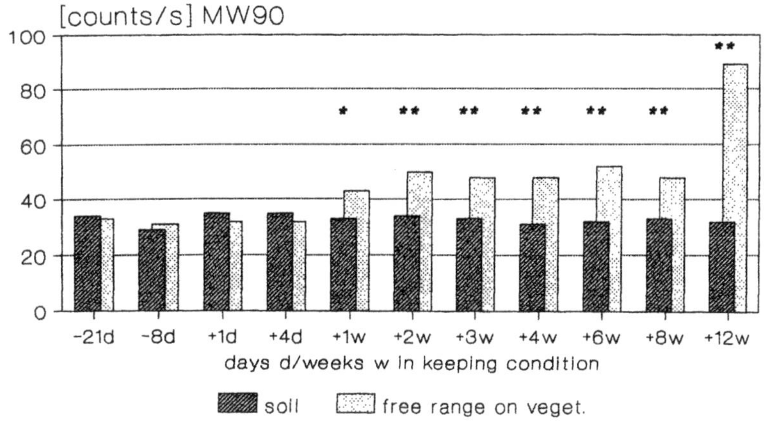

Figure 17.10: ***Intensity of biophoton emission keeping condition soil versus free range on vegetation. (*p = 0.05, **p = 0.01).***

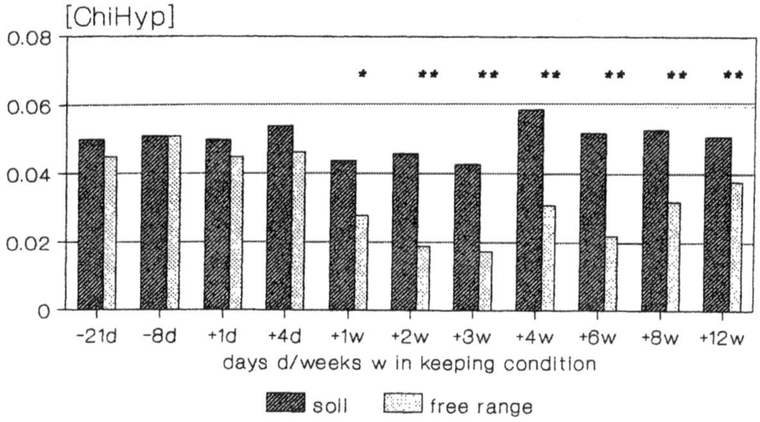

Figure 17.11: ***Agreement with a hyperbolic relaxation kinetic soil versus free range on vegetation. (*p = 0.05, **p = 0.01).***

Table 17.1: Factor loadings.

	factor 1	factor 2	factor 3
DA	0.17	0.78	0.36
NB_w	0.95	-0.03	0.10
NB_r	0.37	0.73	0.38
aH	-0.94	0.25	0.20
ChiH	0.45	0.35	-0.80
ChiEH	0.46	-0.66	0.51

17.5 Quality Analysis of Sunflower Oil

24 different kinds of sunflower oils from the trade were examined in the biophoton measurement (10 ml each). The measured optical parameters were:

- self-emission (D.A. in counts/500 ms) without external illumination,
- average emission after white light illumination (NB_w) in counts/500 ms, within a time interval of 10s after exposure,
- average emission after red light illumination (NB_r) in counts/500 ms, within a time interval of 10s after exposure.

Also parameters describing the non-agreement with a hyperbolic decay (ChiHyp), the relationship of the non-agreement of an exponential to a hyperbolic relaxation (ChiEH) and the relaxation kinetics (aHyp) were examined. The aHyp-value as a parameter describing the (negative) slope of the relaxation kinetics after light illumination corresponds to the light storage ability of the sample.

To ensure an objective, relative classification of the oils, a factorial analysis with all measured data was carried out. The primary aim here is to record the factors (= linear combination of data) describing the properties of the oils relatively independently of each other. It is sufficient to use only the first three factors, because the others determine only 5% of all the properties. The analysis exhibited the following "factor loadings" of the first three factors (Tab. 17.1).

The factor 1 describes a process, correlated to the stimulative ability of white light ($NB_w = 0.95$), while the aHyp-value decreases (becoming more negative). This is a good indication, because through this effect the light-storage capacity of the oil increases (aHyp = −0.94). A high positive value of the factor 1 indicates, in general, a high light-storage capacity of the oil.

The factor 2 describes, in general, the self-emission, leading to a drastic increase of the values in the red part of the spectrum (sector of oxidation processes). Simultaneously, the agreement with an exponential relaxation kinetics increases. It seems that the factor 2 describes decomposition processes. A high positive value of factor 2 is unfavourable.

The factor 3 exhibits its highest contribution in the ChiHyp-value, and with increasing influence of the factor 3, the "state of order" increases in the sense of a physiological value.

Summarized, the three factors could be entitled as followed:

factor 1: light storage capacity (characteristic of the sunflower oil),

factor 2: decomposition procedures (e.g. aging, poor storage etc.),

factor 3: value of order in the sense of a physiological, nutritional value.

The quality of the oils can be characterised by these three factor values. A better qualitative comparison of all samples is possible when the values are standardised first by the following equation:

$$Q_F = F_1 x_1 - F_2 x_2 + F_3 x_3$$

with F_1 = eigenvalue of the first factor (2.33); F_2 = eigenvalue of the second factor (1.77); F_3 = eigenvalue of the third factor (1.20). The minus of the factor 2 is due to its unfavourable influence on the quality of the oils. x_1, x_2, x_3 represent the normalised factor values of the oils for the factors 1, 2, 3.

17.6 The Irradiation of Food

The increasing number of approved food irradiations in the 1980s requires control procedures to ensure safety, health standards and proper trade regulations with regard to irradiated products. Postirradiation identification methods are of general interest to ascertain that legally imposed food laws are not violated, prevent irradiation from becoming a compensation treatment for low level hygienic products,

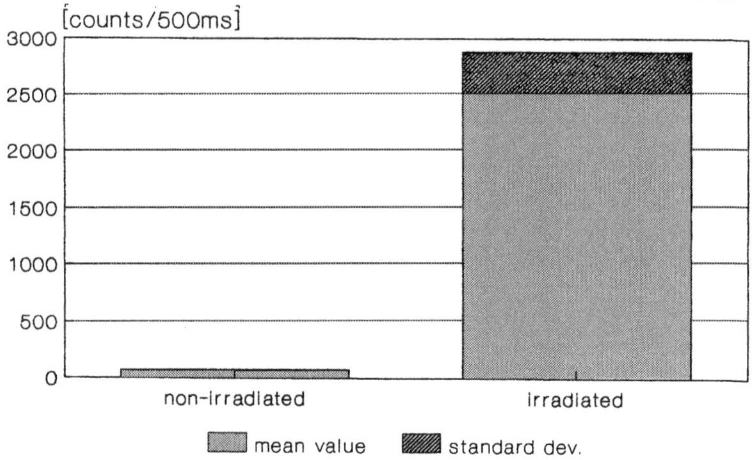

Figure 17.12: *Self-emission of rice (10g) 5 minutes after irradiation with 1 Gy.*

supervise correct labeling, and control the absorbed dose. The biophoton emission can be used for the detection of irradiated food.

Even low doses (1 Gy) of X-rays induce a breaking of chemical bonds and the formation of radicals. The figure 17.12 shows the effect of 1 Gy-irradiation on rice. The non-irradiated sample exhibits a low level of self-emission, whereas the irradiated sample exhibits a high count rate (factor 50 higher) due to chain reactions and recombinations of the generated radicals. After several hours (10-20 hours) the self-emission is again at the initial level. To detect irradiation of food which was irradiated a longer time ago, the food can be illuminated with light of different wavelengths. The irradiated food then exhibits a higher emission rate than the non-irradiated ones (Fig. 17.13).

17.7 Further Examples for the Application of the Biophoton Measurement

In 1965[1] the photon emission of plant seeds was used to investigate their heat and cold resistance. It was found a lower temperature point, corresponding to the cold death, and an upper heat point, corresponding to the heat death of the plants. During heating or cooling of the seeds there was a dramatic change in the photon emission upon reaching the lower or upper temperature point.

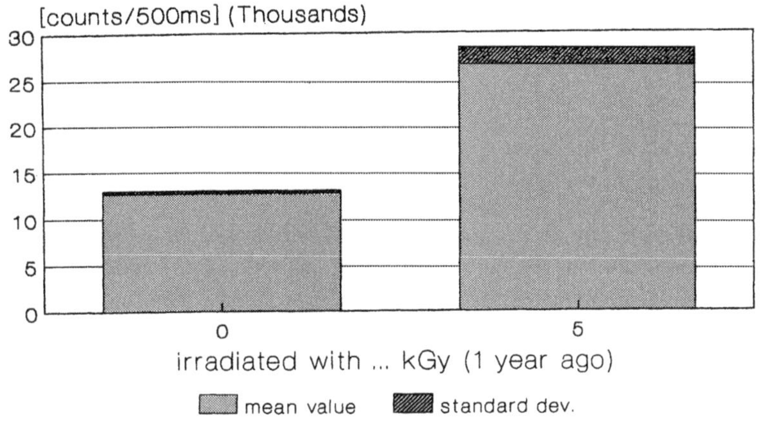

Figure 17.13: *White light illumination of irradiated and non-irradiated rice (10g, MW20 = mean value 20).*

Mendenhall[5], Mendenhall and Agarwal[4], and Mendenhall[6] characterized the habit and age of polymers by their photon emission.

Loeliger[3] correlated the photon emission of food with oxidative chemiluminescent processes.

Veselova et al.[8] detected a negative correlation between the photon emission and the viability of soybean seeds.

Teubner et al.[9], Teubner[7], and Popp[10] investigated different cultivated plants and fruits with the help of the biophoton emission.

Schamhart and van Wijk[11] differentiated tumor cells and normal cells with the help of the photon emission. With an increase in cell number the tumor cells show an increase in the photon emission whereas the normal cells exhibit a decrease.

Kaneda et al.[2] investigated the photon emission of beer during oxidation processes.

Albrecht et al.[12] described the application of chemiluminescence for immunoassays, clinical chemistry, tumor therapy and the detection of inorganic ions.

17.8 References

1. A.S. Agaverdiyev, Y.Y. Doskoch and B.N. Tarusov, Effect of low temperatures on the ultraweak luminescence of plants. *Biofizika* (1965) p. 832-836.

2. H. Kaneda, Y.Kano, T. Osawa, S. Kawakishi and M. Kamimura, Role of free radicals in chemiluminescence generation during the beer oxidation process. *Agric. Biol. Chem.* **54** **(8)** (1990) p. 2165-2166.

3. J. Loeliger, Food oxidation as measured by chemiluminescence. *J. of Luminescence* **31+32** (1984) p. 908-910.

4. G.D. Mendenhall and H.K. Agarwal, Slow luminescence emission from polymers with a hyperbolic decay law. A survey of commercial materials with an apparatus of wide aperture. *J. of Applied Polymer Science* **33** (1987) p. 1259-1274.

5. G.D. Mendenhall, Analytische Anwendungen der Chemilumineszenz. *Angew. Chem.* **89** (1977) p. 220-228.

6. G.D. Mendenhall, Charakterisierung von Materialien durch Chemilumineszenz-Messungen. *Angew. Chem.* **102** (1990) p. 380-391.

7. R. Teubner, Zur Qualitaetsbestimmung von Nutzpflanzen, insbesondere Medizinalpflanzen, mit Hilfe der ultraschwachen Photonenemission. *Dissertation*, Fachbereich Agrarwissenschaften (Goettingen, 1983).

8. T.V. Veselova, V.A. Veselovsky, A.B. Rubin and P.Z. Bochvarov, Delayed luminescence of air-dry soybean seeds as a measure of their viability. *Physiol. Plant.* **65** (1985) p. 493-497.

9. R. Teubner, M. Rattemeyer and W. Mehlhardt, Eine neue Methode zur Untersuchung der Qualitaet von Pflanzen und Fruechten. *Aerztezeitschrift fuer Naturheilverfahren* **4** (1981) p. 204-205.

10. F.A. Popp, Qualitaetsuntersuchungen an Gemuese aus konventionellem und alternativem Anbau mit Hilfe der Biophotonen-Analytik in Qualitaetsuntersuchungen an Gemuese aus konventionellem und alternativem Anbau. Deutsche Gesellschaft fuer Qualitaetsforschung e.V., *Berichtsband XXIV, Vortragstagung* (1989) p. 102-185.

11. D.H.J. Schamhart and R. van Wijk, Photon emission and the degree of differentiation in photon emission from biological systems, in *Photon Emission from Biological Systems*, eds. B. Jezowska-Trzebiatowska, B. Kochel, J. Slawinski and W. Strek (World Scientific, Singapore, 1986) p. 137-152.

12. S. Albrecht, H. Brandl and W. Adam, Chemilumineszenz-Reaktionen. *Chemie in unserer Zeit* **5** (1990) p. 227-238.

13. B. Koehler, K. Lambing, R. Neurohr, W. Nagl, F.A. Popp and J. Wahler, Photonenemission – Eine neue Methode zur Erfassung der Qualitaet von Lebensmitteln. *Deutsche Lebensmittel-Rundschau* **3** (1991) 78-83.

Chapter 18

Biophoton Emission as an Indicator of "Biological Quality"

F. Rohner

Abstract

Two examples, one concerning the bio-indication of natural and synthetic plant extracts, the other quality of water, show evidence that biophoton emission provides a powerful tool of investigating the efficacy of external influences. This methods is of high accuracy and reliability.

18.1 Introduction

At present it is well known that there exist synergetic effects in biological systems which cannot be explained in terms of common biochemistry [1].

A method of revealing "biophysical" interactions of living tissues in a non-invasive way is based on biophoton emission. In order to demonstrate the sensitivity and reproducibillty of this tool, this paper is confined to bio-indication of natural and synthetic compounds of plant extracts as well as of water by use of the "delayed luminescence" from *Acetabularia acetabulum.*

Actually, the difference in the response to a natural and to the corresponding synthetic compound can be hardly traced back to different biochemical pathways.

In addition, concerning water, at present there is no scientific explanation of the specific and individual efficacy of medical spring water. This is why a differential balneotherapy could not be developped [2–4]. However, since balneologists agree in that the substantial components of water are probably necessary but not sufficient for explaining the biological effects [2–6], biological tools like biophoton-indicators may be powerful enough to examine the influence of water on living systems.

18.2 Materials and Method

Flowers of orchid (Dendrobium Nobile) and seeds of lotus (Nelumbo Nucifera) have been used as subjects under investigation. By blending 50/50 these extracts, the so-called natural hydrovegetal extract (N) has been prepared. From the composition (aminoacids, oses, vitamines, minerals and trace elements) the corresponding synthetic orchid-like and lotus-like extract with the same blending 50/50, resulting in the so-called synthetic hydrovegetal extract (S), have been produced, too.

Acetabularia acetabulum (a green, unicellular algae (Chlorophycea)) has been used as a bio-indicator, by measuring the biophoton emission as well as the "delayed luminescence" under the influence of the natural and the synthetic compound, respectively. Since stress, as vitamine reduction, induces a significant lowering of photon emission, the different biological effects of these extracts can be registered by measuring the change of the suppression of excitability by a definite stress under the influence of the synthetic and the natural compound, alternatively.

In second series of experiments, instead of different extracts simply different qualities of water have been used as external influence i.e. tap water, spring water (Zurzach) and the corresponding synthetic water, composed of the same chemicals as the spring water.

A variety of investigations has been performed, where the age of the acetabularia, the stress conditions, the duration of the exposure to the agents and the method of registering "delayed luminescence" were altered.

18.3 Results and Discussion

All the experiments result in significant differences of biophoton emission from acetabularia under the influences of different external agents, with the exception that no significant difference between synthetic and natural spring water has been found.

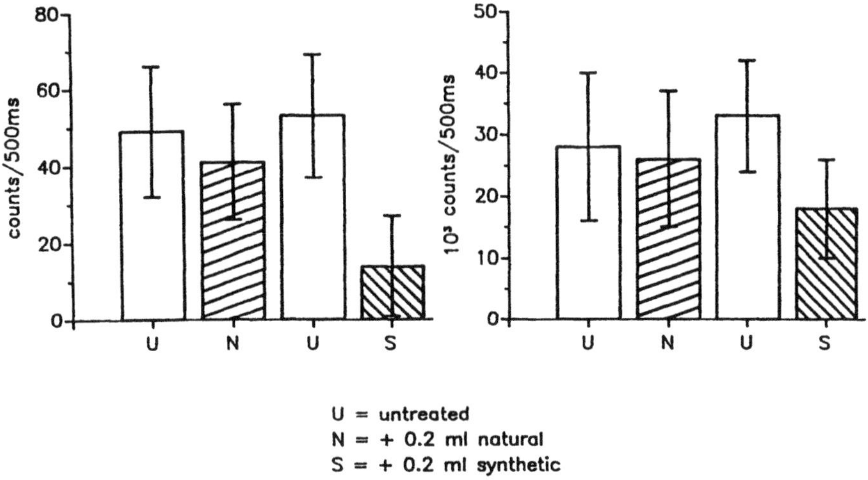

Figure 18.1: *Always 5 acetabularia have been treated by 90 % vitamine-reduction and addition of a natural (N) and synthetic compound (S, see text). The biophoton emission (left side) as well as the "delayed luminescence" (right side) significant differences between (1) controls (U) and S-treated ones, and (2) N- and S-treated ones. (Courtesy of ROCHAS, Paris).*

Fig. 18.1 displays an example, where the biophoton emission of always 5 adult acetabularia has been registered before treatment and after three days vitamine reduction of 90 % and exposure to 0.2 ml N-extract and, alternatively, S-extract in always 10 ml medium (salt water with only 10 % vitamines).

It demonstrates that

- there is a significant difference in the self-emission values as well as in "delayed luminescence",
- the removal of stress in N-treated acetabularia is significantly stronger than that in S-treated ones.

Fig. 18.2 exhibits the response of "delayed luminescence" to monochromatic light illumination of always 18 acetabularia immediately after exposing them to spring water, the corresponding synthetic one and tap water.

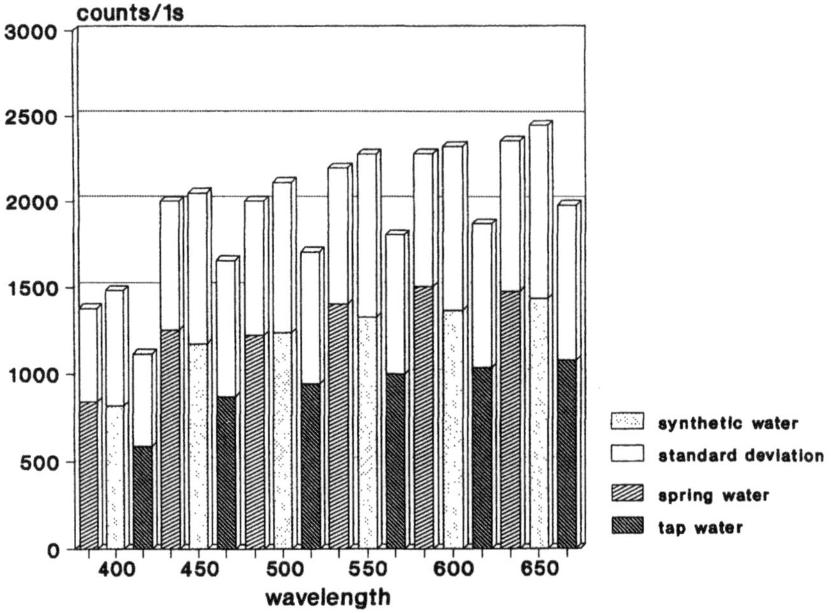

Figure 18.2: *For different wavelengths of exciting light acetabularia show almost the same depression of "delayed luminescence" immediately after addition of (1) spring water, (2) the corresponding synthetic one and (3) tap water. However, over all the different wavelengths there is - besides of some tendency no significant difference between spring water and synthetic one, - a significant difference between spring water, synthetic one one the one hand and tap water on the other hand.*

The results show evidence that this test provides a powerful and sensitive tool to examine biological effects of external influences, like pollutants, remedies and other agents, demonstrating significant differences in cases where they can be expected, and rejecting differences, if there are none.

18.4 References

1. F.A. Popp, Some elements of homeopathy. *Br. Hom. J.* **79** (1990) p. 161-166.

2. O. Knuesel, Badekuren: Historische Entwicklung und Stellenwert in der modernen Medizin. *Der informierte Arzt (Gazette Medicale)* **2** (1990) p. 93-100.

3. O. Knuesel, Die Grundlagen und Wirkungsmechanismen der Balneologie, in *Thermen der Schweiz: Die Schweizer Badekurorte* (Offizin, Zuerich 1991) p. 161-169.

4. E. Senn, Die Frage nach der Zukunft der ortsgebundenen natuerlichen Heilmittel. *Therapiewoche* **37** (1987) p. 3413-3414.

5. H.G. Pratzel und W. Schnizer, Medizinisch orientierte Grenzwerte für Badeheilwaesser auf der Grundlage von Gesetzmaessigkeiten der Hautpermeation. *Z. Phys. Med. Baln. Med. Klim.* **16** (1987) p. 367-373.

6. E. Senn, Zukunftsperspektiven der Kurortmedizin. *Heilbad und Kurort, Zeitschrift des Deutschen Bäder Verbandes e.V.* **40** (1988) p. 156-158.

Chapter 19

Coherence and Some Quantum Paradoxes

K.H. Li and F.A. Popp

Abstract

We re-analyse the so called de Broglie's paradox and the EPR phenomenon, taking into account the coherence properties of the Schrödinger wave functions. We find that the wave function describing de Broglie's paradox, and therefore also the interference pattern, would be destroyed by the process of separating the box confining the electron. For the EPR paradox the wave function used by the original authors represents not two particles but a single entity. This resolves the EPR paradox phenomenon.

Due to the fact that properties of ultra weak photon emission compatible with non-classical light from biological tissues (biophotons) have been experimentally observed in our laboratory[1], we are encouraged to investigate more deeply some fundamental problems of quantum theory for understanding the essence of eletromagnetic information in biological systems. In that case we can clearly see that some, perhaps the majority, of biological processes only occur at the microscopic level, i.e. at the level of excited states of atoms and molecules (including very complicated biochemical reactions), while all biological phenomena are manifested macroscopically.

We are thus obliged to plead that the universal character of the uncertainty relations of Heisenberg i.e. the coherence properties of matter waves, must be taken into

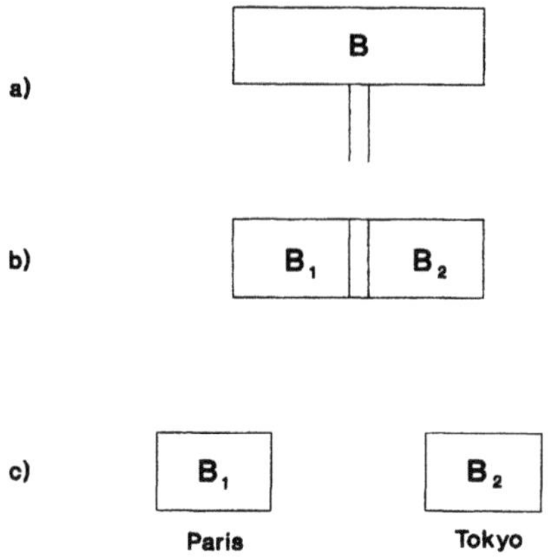

Figure 19.1: *An electron is initially contained in a Box B a) which is then divided into the compartments B_1 and B_2 by a double sliding partition b). Part B_1 is taken to Paris and part B_2 to Tokyo c).*

account during the investigations of biological processes. Hence, from the point of view of consistency, it is also necessary to reconsider some quantum paradoxes in the light of communications between biological cells by means of coherent electromagnetic fields.

The second part of this chapter: "Coherence and the EPR paradox" was completed in September 1990. Copies of it were circulated widely amongst our interested friends, reported on at some meetings and is incorporated in this book, without any alterations.

19.1 Coherence and de Broglie's Paradox[2]

F. Selleri[3] has described de Broglie's paradox in his new book as follows: "Consider a box B, with perfectly reflecting walls, which can be divided into two compartments B_1 and B_2 by a double sliding partition (Fig. 19.1).

Suppose that B contains initially an electron, whose wave function ϕ(x,y,z;t) is defined in the volume V of B. The probability density for observing the electron at the point x,y,z at the time t is then given by the squared modulus of ϕ(x,y,z;t).

Next, B is divided into the two compartments B_1 and B_2. Part B_1 is then taken to, say, Paris and part B_2 to Tokyo. The new situation is described quantum mechanically by two wave functions: ϕ_1(x,y,z;t), defined in the volume V_1 of B_1 , and ϕ_2(x,y,z;t), defined in the volume V_2 of B_2.

These two wave functions could simply be equal to the 'old' ϕ when the latter is calculated for the two volumes V_1 and V_2, respectively. Or they could be different from ϕ, in the case where the process of separating the two boxes itself modifies the state of the electron. Which alternative occurs is however irrelevant; the important thing is that there are now two probabilities P_1 and P_2, both different from zero and adding up to unity, such that P_1 is the probability for finding the electron in box B_1, and P_2 the probability for finding it in box B_2.

Also important is the fact that P_1 can be calculated as the volume integral over V_1 of the squared modulus of ϕ_1 and that P_2 can similarly be calculated as the volume integral over V_2 of the squared modulus of ϕ_2. *Note that the completeness of the theory forces us to shun statements like: "The electron is actually either in B_1 or in B_2". For, such a statement implies a more precise spatial localization than that provided by the wave functions ϕ_1 and ϕ_2. The only assertion that is compatible with the completeness of quantum mechanic will be: "The particle is in B_1 and in B_2"* (stress ours)

Suppose an observation was performed in Paris at time t_0 and the electron found in B_1, then one is certain that for all $t \geq t_0$ the probability P_2 of observing the electron in B_2 in Tokyo is zero. Since P_2 is given by a volume integral of the squared modulus of ϕ_2 over V_2, one must necessarily conclude that the vanishing of P_2 for $t \geq t_0$ implies the venishing of ϕ_2(x,y,z;t) for $t \geq t_0$. Thus the observation of the electron in Paris changes the wave function in Tokyo, reducing it to zero !"

In order to solve the paradox contained in this case let us first revise some statements about probability interference by Heisenberg[4] and Feynman[5]. Heisenberg stated:

> "A great deal of quantum paradoxes may be caused by insufficient consideration of the influence of measurements."

As a typical example the following simple Gedankenexperiment ought to be discussed:

A Stern-Gerlach's beam of atoms assumed primarily to be in the stationary state n, would be sent into a strongly inhomogeneous field F_1 (Fig. 19.2) which can cause some transitions of atoms from state n to state m. The probability amptitude after F_1 to find the atoms in the state m is given by $S^{(1)}_{nm}$. The corresponding probability

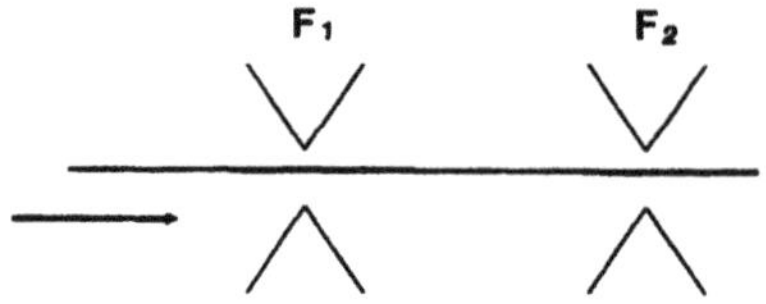

Figure 19.2: *A Stern-Gerlach's beam of atoms would be sent into an inhomogeneous field F_1 which can cause some transitions of atoms. After a definite distance the beam enters the second field F_2, where it can once more be excited. How the transition probability in different cases ought to be caculated would be answered in the text.*

is $| S^{(1)}_{nm} |^2$. After a definite distance the beam enters the second field F_2, where it can once more be excited to a state l. For the atoms which were in the state m after F_1, the probability of changing to state l in field F_2 is surely $| S^{(2)}_{ml} |^2$. According to this argument the total number of atoms in the state l after F_2 must be given by

$$\sum_m | S^{(1)}_{nm} |^2 | S^{(2)}_{ml} |^2 . \tag{19.1}$$

Otherwise the total transition amplitude of atoms in the fields as a whole ought to be given by

$$S^{(12)}_{nl} = \sum_m S^{(1)}_{nm} S^{(2)}_{ml} \tag{19.2}$$

and the number of atoms in the state l after F_2 is

$$| S^{(12)}_{nl} |^2 = | \sum_m S^{(1)}_{nm} S^{(2)}_{ml} |^2, \tag{19.3}$$

disagreeing with the Eq. (19.1).

The contradiction can soon be solved if we note that the mathematical expressions (19.1) and (19.3) really belong to two different experiments: The mathematical deductions of Eq. (19.1) can only be correct when an experiment has actually been performed between F_1 and F_2 to establish the stationary state of atoms. Such an experiment cannot avoid changing the phase of the Schrödinger wave function belonging to the state m. If we use the Eq. (19.3) to describe this experiment, we must multiply each term $S^{(1)}_{nm} S^{(2)}_{ml}$ within the summation symbol by a phase factor $e^{i\varphi_m}$,

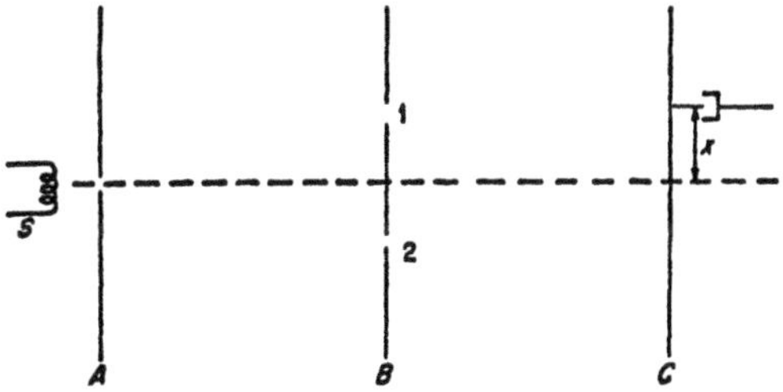

Figure 19.3: *The experimental arrangement. Electrons emitted at A make their way to the detector at screen C, but a screen B with two holes is interposed. The detector registers a count for each electron which arrives; the fraction which arrives when the detector is placed at a distance x from the center of the screen is measured and plotted against x, as in Fig. 19.4*

then the average of phases on the right hand side of Eq. (19.3) must be taken. After this change we find that Eq. (19.3) is identical to Eq. (19.1).

If no experiment is made between F_1 and F_2, then we do not know whether the atom has stayed in a stationary state. In this case the interference field of the Schrödinger wave would not be influenced. The principlal difference of the case described by Eq. (19.3) from that described by Eq. (19.1) indicates the essential point of quantum theory.

Feynman has used another type of experiment to illustrate the change in calculation of probability in quantum theory. It is not the concept of probability that is altered, when we switch to quantum from classical mechanics. What is changed and changed radically, is the method of calculating probabilities.

His imaginary experiment is illustrated in Fig. 19.3. At A we have a source of electrons S. The electrons at S all have the same energy but (emit) in all directions to impinge on a screen B.

The screen B has two holes, 1 and 2 through which the electron may pass. Finally, behind the screen B at a place C we have a detector of electrons which may be placed at various distances x from the center of the screen. The same experiment with photons is well known as Young's interference experiment.

What we shall measure, for various positions x of the detector, is the mean number of pulses per second. In other words, we shall determine experimentally the (relative) probability P that the electron passes from S to x, as a function of x. The graph of

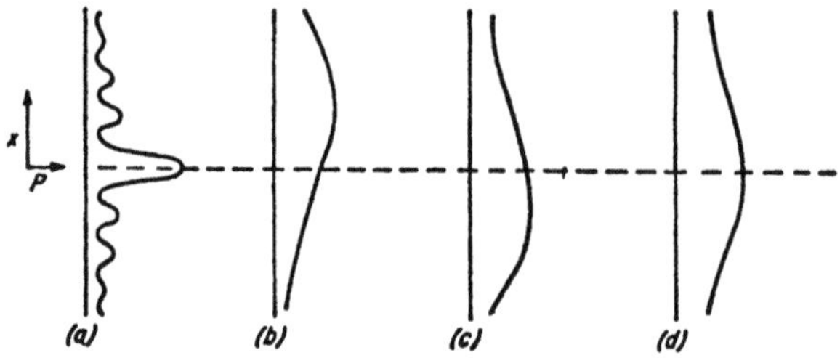

Figure 19.4: *Results of the expeiment. Probability of arrival of electrons at x plotted against the position x of the detector. The result of the experiment of Fig. 19.3 is plotted here at (a). If only one hole is open, so the electrons can go through just hole 1, the result is (b). For just hole 2 open, it is (c). If we imagine each electron just goes through one hole or the other, we expect the curve (d) = (b) + (c) when both holes are open. This is considerably different from what we actually get, (a).*

this probability as a function of x is the complicated interference curve illustrated qualitatively in Fig 19.4(a).

It has several maxima and minima and there are locations near the center of the screen at which electrons hardly ever arrive.

We might at first suppose (since the electrons behave as particles) that:

1. Each electron which passes from S to x must go through either hole 1 or hole 2. As a consequence we expect that:

2. The chance of arrival at x should be the sum of two parts, P_1, the chance of arrival through hole 1 (Fig. 19.4(b)), plus P_2, the chance of arrival through hole 2, (Fig. 19.4(c)).The sum of these (Fig. 19.4(d)) clearly does not agree with the curve (Fig. 19.4(a)). Hence, experiment tells us definitely that $P \neq P_1 + P_2$.

However, we can easily design an experiment to test our expectations directly. We have merely to place a source of light behind the holes and observe through which hole the electron passes. As electrons scatter light, if the light is scattered behind hole 1, we may conclude that an electron passed through hole 1; and if it is scattered in the neighbourhood of hole 2, the electron has passed through hole 2. The result of this experiment is to show unequivocally that the electron does pass through either hole 1 or hole 2 ! What has been changed?

When we watch the electrons to see through which hole they pass, we obtain the result $P = P_1 + P_2$. When we do not watch, we get a different result $P \neq P_1 + P_2$. Just by watching the electrons, we have changed the chance that they arrive at x . How is this possible? The answer is that , to watch them, we used light and the light in colliding with the electron may be expected to alter its motion, or, more precisely, to alter its chance of arrival at x.

Thus, for example , the overall probability amplitude for arrival at x is given as

$$\phi = \phi_1 + \phi_2, \tag{19.4}$$

then the probability that an electron reaches the detector is

$$P =\mid \phi_1 + \phi_2 \mid^2 \tag{19.5}$$

but not

$$P =\mid \phi_1 \mid^2 + \mid \phi_2 \mid^2 \tag{19.6}$$

with $P_1 =\mid \phi_1 \mid^2$, $P_2 =\mid \phi_2 \mid^2$.

Now we come to analyse de Broglie's paradox. For box (Fig. 19.1) B with perfectly reflecting walls we obtain from the Schrödinger equation

$$\nabla^2\Psi + \varepsilon\Psi = 0, \tag{19.7}$$

the general solutions

$$\begin{aligned} \Psi_n &= A_n \cos(\frac{n\pi}{L}x) \qquad if\ n\ is\ odd \\ \Psi_n &= B_n \sin(\frac{n\pi}{L}x) \qquad if\ n\ is\ even, \end{aligned} \tag{19.8}$$

with discrete values of ε and the length of box L:

$$\varepsilon_n = \frac{n^2\pi^2}{L^2} \qquad (n = 1, 2, 3, ...\infty),$$

thereby we have solved only a one-dimensional problem without loss of generality. As we know that the solutions can be multiplied by an arbitrary phase factor $e^{\pm i\omega t}$ because Eq. (19.7) is a modified form of the time-dependent Schrödinger Equation.

However, each component of this standing wave can be identified by the interference of two oppositely travelling waves. This means that a standing wave in the box B (here one-dimensional box) can still be considered as superposition of two travelling waves reflected by the two oppositely reflecting walls, which is completely understandable within the uncertainty space-time of electron waves. For example, we have two travelling waves

$$\begin{aligned} y_1 &= A_1 \cos(\omega t - kx) \\ y_2 &= A_2 \cos(\omega t + k(x - L)) \end{aligned} \tag{19.9}$$

coming from two opposite reflecting box-walls.

For simplicity we assume

$$A_1 = A_2, \tag{19.10}$$

and rather than carry out the trigonometric addition indicated, we will perform the summation using the equivalent but more elegant exponential rendition. If

$$\begin{aligned} y_1 &= Ae^{i(\omega t - kx)} = Ae^{i\omega t}e^{-ikx} \\ y_2 &= Ae^{i(\omega t + k(x-L))} = Ae^{i\omega t}e^{+ik(x-L)} \end{aligned} \tag{19.11}$$

represent the two oppositely moving waves, their sum is

$$\begin{aligned} y = y_1 + y_2 &= \bar{A}e^{i\omega t}(e^{ik\bar{x}} + e^{-ik\bar{x}}) \\ &= 2\bar{A}e^{i\omega t}\cos k\bar{x}, \end{aligned} \tag{19.12}$$

where $\bar{A} = Ae^{-\frac{i}{2}kL}$, $\bar{x} = (x - \frac{1}{2}L)$. The real part of which is the trigonometric sum

$$Re\ y = 2A(\alpha \cos \omega t + \beta \sin \omega t) \cos k(x - \frac{1}{2}L) \tag{19.13}$$

with constants $\alpha = \cos \frac{1}{2}kL$, and $\beta = \sin \frac{1}{2}kL$.

This can be interpreted as a position-dependent amplitude $2A \cos k\bar{x}$, every point of which oscillates in simple harmonic motion according to the function $\cos \omega t$ and

$\sin \omega t$. The sum of the opposite travelling waves produces a stationary standing wave form changing only in size, and having fixed points of zero displacements, determined by the condition

$$\cos k\bar{x} = o, \tag{19.14}$$

making

$$k\bar{x}_n = (2n+1)\frac{\pi}{2} \qquad n = 0, 1, 2, ... \tag{19.15}$$

or

$$\bar{x}_n = \frac{(2n+1)\pi}{2k} = (2n+1)\frac{\lambda}{4}. \tag{19.16}$$

These permanent zero displacements or nodal points occur at half wave length intervals giving credence to the standing wave. In a sense there are no longer any travelling waves, only the varying standing wave form.

If we take the distance between two successive nodes to be L, condition (19.14) to (19.16) leads to

$$kL = k\bar{x}_2 - k\bar{x}_1 = \pi \qquad (or\ n\pi), \tag{19.17}$$

making

$$k = \frac{\pi}{L} \qquad or \qquad \lambda = 2L,$$

or

$$k = \frac{n\pi}{L} \tag{19.18}$$

in accordance the Eq. (19.8).

Now we see how the interference pattern has to display in the box

$$|\ y\ |^2 = |\ y_1 + y_2\ |^2$$

by using the two oppositely travelling waves of Eq. (19.9)

$$\begin{aligned} | y |^2 &= | A_1 \cos(\omega t - kx) + A_2 \cos(\omega t + k\bar{x}) |^2 \\ &= A_1^2 \cos^2(\omega t - kx) + A_2^2 \cos^2(\omega t + k\bar{x}) \\ &\quad +2A_1 A_2 \cos(\omega t - kx) \cos(\omega t + k\bar{x}). \end{aligned} \tag{19.19}$$

If, however, we divide the box B into the two compartments B_1 and B_2 by inserting a double sliding partition (Fig 19.1(b)), which is also assumed to be perfectly reflecting, then at the same time we destroy the interference pattern Eq. (19.19) due to the fact described by the " thought experiment " of Feynman.

The destruction of the interference pattern only means the disappearance of the interference term of Eq. (19.19) as discussed above. That is like saying that either $A_1 = 0$, the disappearance of the component of the left coming travelling wave, or $A_2 = 0$, the disappearance of the component of the right coming travelling wave. As a result of the destruction of the interference pattern of the standing wave field within box B, we have learnt that we can only say: *the particle either comes from the left reflecting end wall, or from the right reflecting end wall of the box B.* With other words: *From the moment of inserting the double sliding partition the particle will be either in the left half-box B_1 or in the right half-box B_2.*

This statement is rejected by the Selleri formulation and is very different from the one formulated by de Broglie's paradox: "The particle is in B_1 and in B_2". The case $A_1 = A_2 = 0$ must be eliminated due to particle conservation, i.e. the electron ought not to be completely destroyed during the separation process.

Hence, if the particle has been observed in the Box B_1 in Paris, we can on the basis of logic conclude that the box B_2 in Tokyo has never been occupied by the particle. This conclusion has nothing to do with the collapse of the wave function. If we say about the collapse of the wave function, that the collapse had already taken place at the moment when the whole box B was divided into two parts B_1 and B_2, this collapse leads to the fact that

$$\int_B | A_1 \cos(\omega t - kx) + A_2 \cos(\omega t + k\bar{x}) |^2 \, dx = 1 \tag{19.20}$$

shrinks to

$$\int_{B_1} | \bar{A}_1 \cos(\omega t - kx) |^2 \, dx = 1 \tag{19.21}$$

$$\int_{B_2} \mid \bar{A}_1 \cos(\omega t - kx) \mid^2 dx = 0 \tag{19.22}$$

if $A_2 = 0$, or vice versa, where $\bar{A}_1$ is also not identical to the original A_1. Because, the process of dividing box B into the two compartments B_1 and B_2 by a double sliding partition is corresponding to keep watch the electrons in the Feynmans imaginary experiment. The interference field would be influenced and wave becomes particle.

Hence, we see that the so-called de Broglie paradox is actually non-existent. In formulating that paradox, the destructive effect on the coherent wave structure, caused by the separation process, was inappropriately neglected.

19.2 Coherence and the EPR Paradox

The Einstein-Podolsky-Rosen paradox[3,6] is concerned with the problem of deciding whether a measurement carried out on one system can affect a second system which previously interacted strongly with it but is now far away. If this is possible, it implies the existence of superluminal communication between systems. If it is not possible, we are forced to conclude that the wave function of quantum mechanics does not give a complete description of physical reality. The majority of scientists find both alternatives unpalatable.

By an effect on the second system, we mean some physical interaction, some constraints or interference between the two systems. The stronger the disturbance, the less the individuality of the two systems and the higher the collectivity. Two systems which exchange information in the form of energy or particles must to some extent be considered as a single system, not two.

In quantum mechanics, a system, for example a particle, can be described by a wave function obeying Schrödinger's equation with the appropriate Hamiltonian. Within the region covered by the wave function, the particle is completely delocalized and therefore loses its particle character in the classical sense. Instead, we can only assign a probability to a position, using the wave function Ψ. For example, the probability of finding two identical particles for a wave interference experiment is according to Feynman[3]:

$$P = \mid \Psi_1 + \Psi_2 \mid^2 \neq P_1 + P_2, \tag{19.23}$$

where

$$P_1 = \mid \Psi_1 \mid^2, \qquad P_2 = \mid \Psi_2 \mid^2 \tag{19.24}$$

and Ψ_1, Ψ_2 are the wave functions of the two particles.

The deviation from the classical result stems from the coherence property of particle wave functions, i.e. from the correlation functions $< \Psi_1\Psi_2 >$ and $< \Psi_2\Psi_1 >$. These have the same meaning as in optics and are responsible for the interference between wave packets. When the correlation functions vanish, the two wave functions are independent and the probabilities add as in the classical case. This happens when the space-time separation of the centres of the wave packets is greater than the coherence length and the coherence time.

No signal can travel between the two systems at a speed greater than that of light. Within the coherence length and the uncertainty length, however, the particles are completely delocalized and we cannot define the velocity of a signal between them. The system of two particles must be considered as a whole.

To continue our discussion of the meaning of coherence in this problem, let us first recall the important connection between Heisenberg's uncertainty principle and coherence length[4]. We have for the one-dimensional case

$$\Delta x \Delta p \geq \hbar. \tag{19.25}$$

Here for coherent states, the state of minimun uncertainty, only the equals sign is valid. Where

$$\Delta p = \hbar \Delta k = \hbar \Delta \frac{2\pi}{\lambda} = h \frac{\Delta\lambda}{\lambda_o^2} \tag{19.26}$$

according to the de Broglie relation. Here λ_o denotes the mean value of the wave length.

From Eq. (19.25) and Eq. (19.26) we obtain

$$\Delta x = \frac{1}{2\pi}\frac{\lambda_o^2}{\Delta\lambda} = L, \tag{19.27}$$

where L is the coherence length of the wave under consideration[7].

At this point we remind the reader of Feynman's formulation of the uncertainty principle: Any determination of the alternative taken by a process capable of following more than one destroys the interference between the alternatives[5].

In classical optics, the coherence length L is equivalent to the width of the wave packet[8]. Similarly, the uncertainty Δx in the position of a particle in quantum mechanics corresponds to the width of the wave function within which the matter is statistically distributed in the sense of Born[9]. This may also be shown from the general time-dependent Schrödinger equation:

$$i\hbar\frac{\partial}{\partial t}\Psi = (-\frac{\hbar^2}{2m}\nabla^2 + V)\Psi, \tag{19.28}$$

which can be rewritten in the form

$$\hbar^2\nabla^2\Psi = 2m\{V - i\hbar\frac{\partial}{\partial t}(\ln\Psi)\}\Psi \tag{19.29}$$

with the momentum operator

$$\vec{p} = \frac{\hbar}{i}\nabla \qquad and \qquad p^2 = -\hbar^2\nabla^2. \tag{19.30}$$

Since Eq. (19.29) constitutes an eigenvalue problem, we obtain the appropriate eigenvalue of the momentum operator

$$P = \sqrt{(-2m)\{V - i\hbar\frac{\partial}{\partial t}(\ln\Psi)\}}. \tag{19.31}$$

If the solution Ψ of the Schrödinger equation factorizes into space- and time- parts, i.e. $\Psi(x,t) = \Psi(x)f(t)$, the momentum eigenvalue Eq. (19.31) has the simple form[10]

$$|\, p \,| = \sqrt{2m(E - V)}, \tag{19.32}$$

where E is the (well-defined) energy of the whole system in a certain state.

On account of the relation

$$\Delta\,|\, p \,| = \sqrt{<|\, p \,|^2> - <|\, p \,|>^2} \tag{19.33}$$

and the condition $|\,\frac{<V>}{E}\,| < 1$, we obtain for Eq. (19.26)

$$\Delta x = \hbar\sqrt{\frac{2E}{m}\frac{1}{\Delta V}} = L. \tag{19.34}$$

Taking for V a simple Coulomb potential of a hydrogen atom, we have[9]

$$\begin{aligned} < V > &= < \frac{-ke^2}{r} > = -\frac{ke^2}{n^2 a_o}, \\ < V^2 > &= \frac{2k^2e^4}{(2l+1)n^3a_o^2}. \end{aligned} \tag{19.35}$$

Here n is the principal quantum number and l is the angular momentum quantum number, a_o the radius of the first Bohr orbit. If $\mid E \mid = \frac{mk^2e^4}{2\hbar^2}$ is constant, then Eq. (19.34) becomes

$$\Delta x = L = \frac{n^2 a_o}{\sqrt{\frac{2n}{2l+1} - 1}}. \tag{19.36}$$

Thus in this case the coherence length is no smaller than the Bohr radius of hydrogen atom.

When the wave function $\Psi(x,t)$ cannot be factorized, the problem is very complicated, but the fundamental correspondence between the coherence length and the width of the wave packet must remain.

We now turn to the EPR paradox. Einstein, Podolsky and Rosen wrote in their original paper: "We have two systems, I and II, which we permit to interact from time t=0 to time t=T, after which time we suppose that there is no longer any interaction between the two parts. We suppose further that the states of the two systems before $t = 0$ were known." With these assumptions we understand that for $t < 0$ there are two definite independent systems I and II, which are well specified and can be described by the complete sets of wave functions $\{\mid \Phi_i >\}$ and $\{\mid \phi_j >\}$, respectively.

At time $t = 0$ the two systems begin to interact, and in principle combine to form an integrated system (I+II) at time $t = T$. While this is happening, the time development of the state vector of the whole system, which we denote by $\mid \Psi_{in} >$ to indicate the effect of the interaction, is determined by the relevant time evolution operator, which is a linear operator in the composite Hilbert space. Since a linear operator transforms a single state vector into another single state vector, the initial pure state $\mid \Psi_{in} >$ must evolve in such a way that the final state of the combined system can also be represented by a single state vector, which we will denote by

$$\mid \Psi(\alpha,\beta)_{out} >= \sum_{i,j} a(\alpha\beta, ij) \mid \Phi_i > \mid \phi_j > . \tag{19.37}$$

In general, the summation in Eq. (19.37) contains more than one term and the coefficient $a(\alpha\beta, ij)$ cannot be factorized as $a(\alpha,i)a(\beta,j)$[11]. In fact, the factorization is only possible for two-component systems with no interaction.

After the interaction has ceased, i.e. for $t \geq T$, the combined system (I+II) can either persist in its combined state or can separate into two or more parts (subsystems) which no longer interact. For simplicity, we shall assume that it separates into two systems. These will be different in principle from the original systems I and II, because in their interaction the original systems exchanged information in the form of energy or particles. Thus if two sub-systems have interacted in the past, it is generally not possible to assign a single state vector to either of them. We denote the new subsystems by I_α (at the point X_α) and II_β (at the point X_β). They can be described by two different complete sets of wave functions $\{\mid \Phi_\alpha >\}$ and $\{\mid \phi_\beta >\}$, respectively.

Wave packets in either system will be of the form

$$\mid \Phi(\alpha) >= \sum_\alpha a_\alpha \mid \Phi_\alpha >, \qquad \mid \phi(\beta) >= \sum_\beta a_\beta \mid \phi_\beta >, \tag{19.38}$$

respectively. Equation (19.37) can then be written as

$$\mid \Psi(\alpha,\beta)_{out} >= \sum_{i,j} a(\alpha\beta, ij) \mid \Phi_i >\mid \phi_j >=\mid \Phi(\alpha) >\mid \phi(\beta) > . \tag{19.39}$$

A problem such as the scattering of the two newly produced systems (particles) I_α and II_β can be described by the wave functions $\mid \Phi(\alpha) >$ and $\mid \phi(\beta) >$, which may be built up from a combination of the original bases $\mid \Phi_i >$ and $\mid \phi_j >$.

We have now reached the heart of the EPR paradox. Can the second wave function of system II_β be collapsed by a measurement carried out on the first system I_α? To answer this question, we have to consider the wave properties of indistinguishable particles. In the two-body problem of wave mechanics, the general (time-dependent) combined wave function in the six-dimensional configuration space must be either symmetric or antisymmetric[12]:

$$\mid \Psi_{total} >= \frac{1}{\sqrt{2}}(\mid \Phi(x_\alpha) >\mid \phi(x_\beta) > \pm \mid \Phi(x_\beta) >\mid \phi(x_\alpha) >). \tag{19.40}$$

The probability density is therefore

$$\begin{aligned} < \Psi_{total} \mid \Psi_{total} >= \frac{1}{2}\{&\mid \Phi(x_\alpha) \mid^2\mid \phi(x_\beta) \mid^2 + \mid \Phi(x_\beta) \mid^2\mid \phi(x_\alpha) \mid^2 \\ &\pm < \phi(x_\beta) \mid< \Phi(x_\alpha) \mid \Phi(x_\beta) >\mid \phi(x_\alpha) > \\ &\pm < \phi(x_\alpha) \mid< \Phi(x_\beta) \mid \Phi(x_\alpha) > \mid \mid (x_\beta) >\}. \end{aligned} \tag{19.41}$$

The last two correlation functions can be evaluated[11]:

$$\begin{aligned} &< \Phi(x_\beta) \mid \Phi(x_\alpha) >=\mid \Phi(x_\alpha) \mid^2 exp\{(i\vec{k} - \frac{1}{L}) \mid x_\beta - x_\alpha \mid\}, \\ &< \phi(x_\alpha) \mid \phi(x_\beta) >=\mid \phi(x_\beta) \mid^2 exp\{(i\vec{k'} - \frac{1}{L'}) \mid x_\alpha - x_\beta \mid\}, \end{aligned} \tag{19.42}$$

where L, L' are the coherence lengths for the two systems (i.e. particles) I_α, II_β, in the directions of $\vec{k}$ and $\vec{k'}$ respectively.

The answer we require is now clear. The particle wave functions $| \Phi(\alpha) >$ and $| \phi(\beta) >$ can be considered to be independent only if the correlation functions are very small, i.e. if

$$\frac{| x_\beta - x_\alpha |}{L} \gg 1 \qquad and \qquad \frac{| x_\alpha - x_\beta |}{L'} \gg 1 \tag{19.43}$$

If this is not the case, then they are correlated through the functions Eq. (19.42) and the measurement of either system does collapse the wave function of the other. For if two identical particles are within a coherence region, they are completely delocalized and constitute a single system in accordance with the uncertainty principle. In that case we may change the notation, and write the wave functions as

$$\begin{aligned} | \Phi(x_\beta) > &= | \phi(x_\beta) > \\ | \phi(x_\alpha) > &= | \Phi(x_\alpha) >, \end{aligned} \tag{19.44}$$

as in the original EPR paper. Equation (19.41) then reduces to the simpler form

$$< \Psi_{total} | \Psi_{total} > = | \Phi(x_\alpha) |^2 | \phi(x_\beta) |^2 \{1 \pm exp[2(i\vec{k} - \frac{1}{L}) | x_\beta - x_\alpha |]\}, \tag{19.45}$$

where

$$| \Phi(x_\alpha)\phi(x_\beta) |^2 = | \phi(x_\beta) |^2 | \Phi(x_\alpha) |^2 . \tag{19.46}$$

Within the coherence length the commutation relation

$$[X_\alpha - X_\beta, P_\alpha + P_\beta] = 0$$

cannot be valid, although a number of authors have asserted that it is. This is the principal flaw in the EPR paper: they constructed a wave function for a system of two particles using the constraints

$$P_\beta = -P_\alpha \qquad and \qquad X_\beta = X_\alpha + X_o, \tag{19.47}$$

but then interpreted the particles as independent. Their wave function

$$\Psi(x_1, x_2) = \int_{-\infty}^{\infty} e^{i\frac{2\pi}{h}(x_1 - x_2 + x_o)p} dp = h\delta(x_1 - x_2 + x_o) \tag{19.48}$$

with $x_o = const.$, represents the two particles as one unified system and actually implies a very strong connection between their positions. For instance, if after a measurement the first particle is left in the state given by the wave function

$$e^{i\frac{1}{\hbar}x_1 p}$$

with eigenvalue p, as given by EPR in Eq. (19.31) of their paper, then the second particle can be found in the state

$$e^{-i\frac{1}{\hbar}(x_2-x_o)p},$$

with eigenvalue $-p$.

These two states clearly represent one-dimensional plane waves with infinite coherence lengths, $\Delta x = L \rightarrow \infty$. The two particles are therefore completely delocalized in space. In such a case we cannot say that carrying out a measurement on one particle does not destroy the other. The two cannot be distinguished, and the expression "a measurement carried out on one particle" is meaningless. The source of the EPR paradox is thus the following. The two particles are represented by a single wave function, which implies that they are a single system, but the results are interpreted as though they were independent entities.

While the EPR paradox suggests that the wave function in quantum mechanics is incomplete, no concrete proof of this has ever been given.

In quantum mechanics, the uncertainty relation is a necessary consequence of the non-commutability of conjugate operators. Within the region of uncertainty, which corresponds to the coherence volume, we can speak only of a probability distribution described by a wave function. The form of this wave function exhibits a spatio-temporal structure determined by the interference of wave functions. Thus within the uncertainty region there exists a definite order.

Within the coherence zone, the particles constitute a single entity. In this zone we have to give up the conservation law of particle number, since particles cannot be observed at all. For example, an atom within the zone does not radiate independently. Instead, it radiates as though it had reached some form of agreement with its neighbours. This is precisely the source of Dicke's regimes of superradiance and subradiance[14,15].

Beyond the coherence volume, two systems can be considered independent. A measurement on one does not disturb the other. But then we cannot derive any result like the EPR paradox. The two systems behave exactly as we would expect from two systems described by different wave functions.

19.3 References

1. F.A. Popp, K.H. Li and W.P. Mei, *Experientia* **44** (1988) p. 576.

2. L. de Broglie, *J. Phys. Radium* **20** (1959) p. 963.

3. F. Selleri, *Quantum Paradoxes and Physical Reality* (Kluwer Acad. Publ. Dordrecht-Boston-London, 1990) p. 37-38.

4. W. Heisenberg, *Physikalische Prinzipien der Quantentheorie* (Bibliographisches Institut AG, Mannheim, 1958) p. 10.

5. R.P. Feynman and A.R. Hibbs, *Quantum Mechanics and Path Integrals* (McGraw-Hill Book Company, 1965).

6. A. Einstein, B. Podolsky and N. Rosen, *Phys. Rev.* **47** (1935) p. 777.

7. M. Garbuny, *Optical Physics* (Academic Press New York and London, 1965) p. 303.

8. E. Goldin, Waves and Photons, *An Introduction to Quantum Optics* (John Wiley & Sons, Inc., 1982) p. 78-83.

9. A. Messiah, *Quantum Mechanics* **Vol.I** (North-Holland Publishing Company, 1965) p. 129-133.

10. R.T. Weidner and R.L. Sells, *Elementare moderne Physik* (Friedr. Vieweg & Sohn, Braunschweig-Wiesbaden, 1982) p. 182.

11. Karl Blum, *Density Matrix Theory and Applications* (Plenum Press. New York and London, 1981) p. 64.

12. H.J. Treder and H.-H. von Borzeskowski, *Foundations of Physics* **18** (1988) p. 77.

13. W.H. Louisell, Quantum Theory of Noise, in *Quantum Optics, Proc. Inter. School of Phys.* **"Enrico Fermi" Course XLII** (1969) p. 715.

14. R.H. Dicke, *Phys. Rev.* **93** (1954) p. 99.

15. H. Paul, *Rev. Mod. Phys.* **58** (1986) p. 209.

Chapter 20

Bioluminescence as a Signature for Dark Matter Reactions

Konstantin Zioutas

Abstract

The photon emission of all living matter can be utilized as a signature for monitoring the weakest external impact. The postulated dark matter in the Universe follows from astrophysical-cosmological considerations and must also interact with living matter, giving rise to biochemical changes, which are accompanied by the emission of "biophotons". Then "biophotons" may be the signal of a dark matter reaction with the living state.

20.1 Introduction

Astrophysical observations ask for a non-luminous mass (dark matter (DM)) in the cosmos, which is estimated to exceed ordinary matter by a factor of ~ 10 [1]. This DM is considered to consist of different types of weakly interacting particles. There are many candidates following from theoretical considerations. Their flux is taken to be as much as

$$\Phi_{DM} \approx 1\ mW/cm^2,$$

whilst their galactic velocity is expected to be

$$v_{DM} \approx 300 \; km/s.$$

However, much smaller velocities relative to an observer on Earth come also into question, if gravitationally clustered DM around the Sun, the Earth, the Moon, etc. is taken into account. All conventional DM detectors failed so far to provide a DM signature. We must keep in mind, that the seasonal modulation of a potential signal for DM reactions with ordinary matter is taken as the best criterion for its significance in those experiments; their detection principles are based almost exclusively on physical reactions between DM and ordinary matter. In other words, we have to do with a "wind" of DM particles and their effective flux on Earth varies with the season as well as with the time of the day/night, as the Earth revolves and rotates.

It was previously suggested[2] that living matter (LM) might well be able to "see" this enormous flux of weakly interacting DM particles. Their flux is estimated to be $\sim 10^{20}$ per second and per human (for $m_{DM} \approx 1 \; eV/c^2$). One of the most celebrated DM candidates is the invented *axion* with spin-parity 0 (see reference 3). In a simplified picture the axion is a weakly interacting photon; in fact it can replace the photon involved in a (magnetic) atomic, molecular or nuclear transition, with an estimated yield being as small as $10^{-10} - 10^{-20}$ or even below, depending on the coupling strength.

It is interesting to note that biorhythms follow in fact seasonal, circadian and other astrophysical frequencies. Despite the fact that models in this field are plentiful in experimental and theoretical attempts to elucidate the clock mechanism(s), it is certainly remarkable that organisms measure astronomical time in some manner. Indeed, the very fact that so many models have been put forward should serve as an indication that the clock's working principle is different than it was earlier envisioned. However, time modulations are expected in accordance with the generally accepted properties of a quasi DM wind.

That is to say, DM-induced biochemical reactions at a variable rate are the alternative explanation for biorhythms as well as for a plethora of other biological phenomena. A collective mode of interaction (coherent scattering) of the DM field with quasi point-like LM-constituents as are proteins, DNAs, cells or even whole organisms might well be the reason for the enhanced reaction rate (N^2-coherence factor) [3]. In addition, we know that lowest threshold energies [3] are registered by

LM; both can be the principal reasons for distinguishing LM from all known conventional DM detectors made of non–living matter [3]. Electroreception is mentioned in this context too [4].

It arises then the question, why not to use LM as the new active detector material? The 'only' problem behind this approach is that we will use biological material (preferentially simplest microorganisms like bacteria), whose utmost complex and finely tuned biomachinery we do not know in all its details. Not to mention of course basic phenomena as is the unknown DNA–replication trigger, which is one of the major unsolved problems in biology and might well be linked to the postulated DM; so, the number of the not-replicated cells shows usually a very characteristic exponential time distribution [5].

We must note that, if $\sim 1 - 10$ eV photons can initiate biological cascade reactions as the one that leads to cell division, the maximum performance of LM as a detector system becomes obvious. In fact such an energy is less than the binding energy of the hydrogen, which is the simplest atom we have. At the end such an energy is *materialized* to macroscopic masses (e.g. $ng - \mu g$), which is the mass of the replicated cell. Certainly a unique feature encountered only in the living world!

20.2 Why "Biophotons" ?

The observed hyperbolic instead of an exponential 'biophoton'-decay intensity after exposure of LM to white-light illumination (see Chapter 1), is the most characteristic experimental result for coherent rescattering. Coherent interaction within a DNA-molecule or even a whole cell [3] can give rise to enormous enhancement factors of as much as $10^{10} - 10^{20}$. However, biophotons have an additional feature of not minor importance for utilizing LM as a detector for weakly interacting DM-candidates: their spatiotemporal features reflect LM's dynamics; their connection to the outside world takes place via the emission of a neutral particle, which is the "biophoton". Therefore, any kind of feedback associated with the emission of the photon is practically excluded as should be the case for a charged particle/ion emission.

It is interesting to recall that the same argumentation holds in Astrophysics : with neutral and (almost) massless particles (e.g. photons and neutrinos) they can trace back in time and space astrophysical reactions taking place even deep inside remote stars.

Thus, the intensity of spontaneously emitted "biophotons" is characteristic for the underlying biorhythm. Furthermore, in certain cases – like the light-emitting

Gonyaulax single cells – the ***spontaneous*** burst-like photon emission on top of a longer time component may reflect, in an alternative scenario, DM-involvement with corresponding spatiotemporal fluctuations [3]. Like other conventional cosmic radiations, the postulated DM-flux can be variable in time and/or in space. The observed correlation (see Chapters 1, 14, 16) of the spontaneously emitted "biophotons" in the ~sec range is of potential importance for this work too.

In addition, with the stimulated fluorescence we might well have available another method of studying LM response to weak external stimuli-changes, whatever their origin; its sensitivity, compared to the spontaneous bioluminescence, under certain conditions might even be much higher. The reason for this is the fact that we have the detector material not in its (stable) ground state as is usually the case, but in a transient excited state (thermodynamically unstable), being more reactive according to chemical as well as physical reasoning. Such a detector configuration should also be taken in DM research using non-living matter as active material. Therefore the observed diurnal [3] as well as seasonal variations of the induced fluorescence of *cress* seedlings by Popp's group [6] is of interest also for this approach to biological phenomena.

20.3 Conclusion

The "biophoton" emission – whether spontaneous or induced – might well be related to still unknown mechanisms underlying diverse biological phenomena of fundamental importance. DM-initiated biochemical reactions could be studied by bioluminescence, provided that they give rise to the emission of electromagnetic radiation we can detect. Many known features of "biophotons" fit remarkably also to a DM-scenario [3]. We specified axions or axion-like particles as DM candidates for the sake of definiteness. After all, bioluminescence seems to be an effect unique for the understanding of the connection of biological with physical phenomena of cosmic origin, whilst moving into unexplored land.

Acknowledgements

I wish to thank F. A. Popp for introducing me into the field of bioluminescence with many interesting and stimulating discussions. His interest as well as that of his group in this work was very encouraging. I thank all of them for the very friendly working atmosphere. The support I had from C. Daskaloyannis in preparing this work is gratefully acknowledged.

References

1. P.F. Smith and J.D. Lewin, *Phys. Rep.* **187** (1990) p. 203;

   J.R. Primack et al., *Annu. Rev. Nucl. Part. Sci.* **38** (1988) p. 751.

2. K. Zioutas, *Phys. Lett. B* **242** (1990) p. 257 and CERN / LAA / PC - 12 / 17.9.1990.

3. A. Zichichi and K. Zioutas, to be published.

4. F. A. Popp, U. Warnke, H.L. König and W. Peschka (eds.), *Electromagnetic Bio-information*, 2nd ed. (Urban and Schwarzenberg, München 1989).

5. R. Shields, *Nature* **267** (1977) p. 704.

6. R. Neurohr, *Dissertation* (University of Saarbrücken, 1991).

Chapter 21

Evolution as the Expansion of Coherent States

F.A. Popp

Abstract

Biological evolution is discussed in term of the optimization of the signal/noise-ratio of living systems. By integrating longer and longer coherence lengths into this procedure, biological systems "conquer" always larger areas of the spatio-temporal reality. This expansion of coherent states provides a guide line of evolution where the basis of communication provides a deterministic development, while the contents remain completely undetermined.

21.1 Introduction

Ernst Mayr, a renowned evolutionary biologist at Harvard University and author of numerous monographs on evolution, has recently claimed that the history and interactions of living systems are well understood in terms of the generally accepted theory of the origin of life, i.e., the theory of evolution by natural selection[1]. Evolution theory was first enunciated more than a hundred years ago by Charles Darwin and has been modified and developed recently through genetics. This theory is accepted today as the most important organizational principle of biology.

According to this theory, evolution is a two-step process, where (1) by recombination, mutation or similar random events a genetic variant is produced, and (2) a new order arises by selection.

Following Manfred Eigen's theory[2], this principle can be traced back to the molecular level, where the "alternate play between cooperation and rivalry among the molecules stimulated the fashioning and information".

This theory states that self-replicating DNA-strands of different sequences associated with proteins, competed for energy-rich monomers which served as food for them. The durable production of new mutant sequences led to a permanently renewed assessment of most of the fit species. Later on, as the RNA-dependent protein synthesis became significant, a new kind of dynamic interaction between the molecules came into play. This induced "hypercycles", e.g. the development of self-organized associates. In a certain way the role of chance is now partially taken over by natural laws, based on the binding forces between atomic and molecular units.

One may ask whether these postulates can bear any risk at all of being wrong. Of course, the mechanisms are described so generally that any strategy of evolution, even a completely random or a fully teleological one, fits into this concept. This comes from the fact that any kind of interaction can be divided into a repulsive and an attractive one. The first one may be assigned to "competition" and the second one to "cooperation". The advantage of such a theory is that it can never fail. The disadvantage is its lack of any predictive power.

All the other modifications and versions of evolution theory, such as neo-Lamarckism[3], neo-Darwinism[4], mutation theory (H. de Vries[1]), synthetic theory[5,6], saltationism[7], punctualism[8], neutralization[9], critical evolution theory[10], semantic theory[11] or system theory[12] have, by no means, changed this situation.

The principal problem is the avoidance of the essential question of evolution itself, e.g. the question of its purpose.

Wuketits[13] wrote that it is the aim of all the scientifically well established theories of evolution to avoid a teleological concept. This means that any predictable future in terms of a general determinism is disproved. The future development thus remains subject to the free play between randomness and necessity, based on natural laws. The term "Selforganization", modern, but not new in evolutionary science, implies that evolution originates from free, unpredictable decisions of the "self", and also from the organizing power of natural laws, in terms of the "Yin-Yang"-principle of the elementary "deterministic chaos" of the world.

At a first glance, one may believe that incorporating the question of a possible purpose of evolution into science and making it even a central topic can never lead to a "scientific" approach to the problem. However, bearing this question in mind instead of avoiding it may be helpful for interpreting valuable scientific results just as the idea of an ideal gas or a closed system has been very useful in understanding the structure of matter irrespective of the fact that an ideal gas, or a closed system does not exist in reality.

It should be noted that the exclusion of a possible aim of evolution will also exclude an essential part of possible scientific solutions, mainly the solution of leaving the question open, and thus always obtaining new exciting answers due to the simple fact that our consciousness is subject to evolutionary development, too. In other words, by avoiding the question of an aim of evolution, we cut off the development of our own mind, spirit, phantasy and last, but not least, of our intellectual power.

It will now be shown that there are ample indications of a deterministic feature in evolutionary development, based on the f=constant-rule. This rule does not contradict the conventional approach, but helps to explain it more profoundly than before. It has the advantage that it divides the complete uncertainty about future events into two parts: a deterministic one and a "free" part.

The deterministic part of the f=constant-rule describes the development of the basis of information-platform, and the "free" part its contents. This theory essentially claims that the guide-line of evolution is the expansion of coherent states with increasing spatio-temporal extension, while the "modulations" of this special kind of expansion remain unpredictable. Providing that we can assign an aim to this process, it would be the one which retains the capacity for asking the question about the aim of evolution. In scientific terms, it means that the balance of actual and potential information is the only guide-line of evolution. This balance determines the direction of time.

21.2 Molecular Aspects

Today, it is well known that the Einstein coefficient A of spontaneous emission depends on the interaction of the excited atom or molecule with the vacuum. As soon as the mode structure of the vacuum field is altered, e.g. in a cavity whose size is smaller or comparable to the wavelength, A changes dramatically.

Depending on the boundary conditions, this coefficient A will either significantly increase or decrease, corresponding to Dicke's regimes of superradiance or subradiance, respectively.

Provided that the DNA is subject to biophoton emission, (we refer to the experimental results[14]), the essential question is that of the creation of this stable molecule and of its extraordinary properties which could be one of the starting points of biological evolution.

From the viewpoint of quantum theory, there are different tools available for obtaining a mechanistic description of the molecular events, e.g.

- Thermodynamics, including Bose-condensation, applied on a molecular level[15],
- the exciplex-model (see Chapters 3, 4),
- Casimir forces[16],
- coherent state-description[16],
- the Dicke theory[17],
- chaos theory[18].

All these models arrive at essentially the same results:

1. Around f=1, where f is the probability of occupying the phase space cells, there is a region where the interaction of photons and matter becomes conditionally stabilized.
2. The system exists between an ordered regime tending towards a crystalline-like state, and a disordered one tending towards thermal equilibrium.
3. The system represents a coherent state.
4. It is an ideal open system, where the pumping source always has enough energy available. The pumping power can drop to zero when the Einstein coefficient A becomes very small.
5. The signal/noise-ratio is optimized with respect to the frequency distribution.
6. The entropy of the system takes the highest possible relative value and can achieve, at the same time, the lowest possible absolute value by minimizing the number of degrees of freedom.
7. The system has a tendency towards:

(a) autocatalytic self-reproduction,

(b) integrating foreign systems as subsystems,

(c) communication by means of resonance-like interactions.

In order to explain these features, let us discuss some essential ones by looking at the results of the different papers on this topic. For a more detailed evaluation, the reader is referred to the papers mentioned in this volume or elsewhere[19].

As an open system always relates to an environment, the most important influence on evolutionary development at a molecular level is the temperature gradient between sun and earth. The sun's rays provide the elementary food for living systems, as is evident from photosynthesis.

According to the van Cittert-Zernike theorem[20] the coherence area of the sun's rays is $10^{-6}cm^2$ which corresponds to about the surface area of a living cell. Consequently, on the level of molecular evolution the coherent excitation of molecules within at least the volume of a cell has to be considered as a "self-organizing" source.

By applying the Brillouin theorem[21] to an "average" coherent photon absorption process, a photon with two possible directions of polarization may induce an entropy change $\delta S = k \ln 2$. As the f-value can be described as

$$f(\lambda) = \frac{1}{\exp(\frac{\delta S}{k}) - 1} \tag{21.1}$$

(see Chapter 1), the average molecular interaction of a coherent photon may produce an f-value which is just

$$f(\lambda) = \frac{1}{\exp(\ln 2) - 1} = 1, \tag{21.2}$$

independent of the wavelength λ.

So if a coherent excitation of matter is stabilized against thermal dissipation, an optimum value of $f(\lambda)$ around 1 can be expected.

Similar results of f=1 are obtained by an optimization of the entropy due to the interaction of a radiation field with matter (see Chapter 1) or by a thermodynamic description of Casimir forces within the lattice system of a polymer. The latter argument will be repeated: Take $U(V, T, g, \lambda)$ as the internal energy of the molecular system, where the distance d between absorbing units (e.g. exciplex states) is small compared to the wavelength λ. V, T and g are the volume, temperature and number of available phase space cells, respectively. An equilibrium $\delta U = 0$ is obtained as soon as the volume work $(\frac{\partial U}{\partial V})_{\lambda,T}\,\delta V$ of the radiation pressure of exciplex decay between the molecular units (e.g. base-pairs) is compensated by the change of the mode structure of the vacuum δg, where g is the number of degenerated states within the phase space:

$$\left(\frac{\partial U}{\partial V}\right)_{\lambda,T} \delta V + \left(\frac{\partial U}{\partial g}\right)_{\lambda,T} \delta g = 0. \tag{21.3}$$

Consequently, from $\delta U = 0$ it follows that

$$\left(\frac{\partial U}{\partial T}\right)_{g,V} \delta T + \left(\frac{\partial U}{\partial \lambda}\right)_{g,V} \delta \lambda = 0. \tag{21.4}$$

Since $U \propto gf = g\exp(-\frac{hc}{kT\lambda})$, where g does not depend on λ in view of the Casimir effect, Eq. (21.4) results in the solution $T \propto \frac{1}{\lambda}$, which again gives f=constant.

It is worthwhile to note that f=1 (or f=constant) enables the temperature dependence of biophoton emissions to be described in a surprisingly good agreement with the experimental results[22].

We have already shown that f=constant and the hyperbolic relaxation behaviour are closely connected (Chapter 1). Consequently, because the hyperbolic decay law under ergodic conditions is sufficient for a fully coherent field, it represents a coherent state. This state may turn into thermal equilibrium or into crystalline-like structures, as has been shown elsewhere[15,23].

It corresponds to an "ideal" open system since the optimization procedure $\delta S = 0$ without any constraint leads to the result f=constant. Of course, as

$$S \propto \sum_{\lambda} f(\lambda)\ln f(\lambda), \tag{21.5}$$

$\delta S = 0$ without any constraint leads again to $f(\lambda)$=constant. Compared to a thermal equilibrium system this means that there is always enough energy available. S from (21.5) is higher than that of a closed system. However, by mode-coupling it can take any value between 0 and $k\ln N$, where N is the number of independent modes. The restriction of the degrees of freedom leads to a decrease in entropy.

Hence, the tricky strategy of evolution by using the f=constant-rule is the maximization of entropy, in order to achieve an absolute minimum by freezing the degrees of freedom without considerable supply of energy. At the same time the signal/noise ratio which is proportional to f, also becomes optimized[19]. Roughly speaking, in this way biological systems become the most sensitive broad band detector systems and, at the same time, multimode lasers (-masers, - rasers) of the highest possible coherence.

In conclusion, the coherent excitation of matter by the sun's rays may well have built a living cell as the product of the f=1-stabilization of the DNA, and then initially extending the f=constant-behaviour over the coherence volume (about $10^{-9}\ cm^3$). This tendency subsequently has been extented to bigger and bigger coherence volumina in which biological evolution takes place.

21.3 From Molecules to Consciousness

The hypothesis $f(\lambda)$=constant is only based on measurements of biophotons, including their Fourier coefficients in the ELF-range. Naturally, the infrared-, microwave and radiowave regions must be examined for general validity.

However, up to now, there are no experimental results which contradict this rule. In the infrared range, for instance, $f(\lambda) \approx 10^{-20}$ approaches the experimental results of common "thermography".

Provided that $f(\lambda)$=constant holds over almost the whole electromagnetic spectrum, it invites its use as a general guideline of biological evolution in the following way:

Consider the case that n photons hit successively, but completely randomly a biopolymer-system (or, more general, antenna system) with $\langle n \rangle$ consecutively arranged exciplexes $1, 2, \cdots, \langle n \rangle$. After absorption, each photon is delocalized over the whole system, which returns then to its original absorbing capacity before a new photon falls in. Thus, the sequence of events can be expressed in term of the sequence $a_1, a_2, \cdots a_j \cdots a_n$ of exciplex units, in which the absorption takes place; a_j can take any value $1, 2, \cdots \langle n \rangle$.

The number of possibilities equals then the number of ways in which the $1, 2, \cdots, \langle n \rangle$ exciplexes can be randomly arranged. Thereby it is only necessary that at least one exciplex is represented in a series of n consecutive events. For example, the rare event $1, 1, 1, \cdots, 1$ (n-times) is as probable as any other sequence $a_1, a_2, \cdots a_j \cdots a_n$. The number M of ways of arrangements $a_1, a_2, \cdots, a_n$ of $\langle n \rangle$ elements is then

$$M(\langle n \rangle, n) = \langle n \rangle^n,$$

corresponding to the number of combinations of nth order for $\langle n \rangle$ different, but unlimited reproducible elements with regard to the arrangement. After absorption, the photons are delocalized, loosing then their identity. This leads to a reduction of distinguishable arrangements by a factor which describes the number of combinations of n distinguishable elements, e.g. $n!$.

Consequently, the probability $p(n)$ of registering n photons originating from such a source is then

$$p(n) = N \frac{\langle n \rangle^n}{n!},$$

where the normalization constant $N = \exp(-\langle n \rangle)$.

This means that the antenna systems of biological evolution can fairly well display the properties of coherent states.

The coherent photon field of the DNA has consequently the natural tendency of extending its coherence volume by incorporation of increasingly longer wavelengths. The reason for this can be found in resonance-like biophoton interactions of DNA between cells, organs (within an organism) and even between organisms. This would lead automatically to an integration of translational, rotational and vibrational modes, serving as a basis for development of further communication. It is evident that from the Darwinistic point of view as well as from the optimization procedure for ideal open systems that the $f(\lambda)$=constant-rule is the most likely of all the possible $f(\lambda)$-distributions. This means that in an autocatalytic way the biological evolution extends the coherent states of DNA to larger and larger coherence volumes, incorporating with increasing evolutionary development longer and longer wavelengths. This process can be written as

1. $f(\lambda)$=constant, and
2. $0 < \lambda \rightarrow \infty$.

The development state of a system is defined by the longest wavelengths which are incorporated into the communication basis. At the same time they are a measure of the spatio-temporal extension of its informational level. As well as the longest cut-off wavelength, the deviations around $f(\lambda)$=constant are to be expected species-specific. They may change with time, but have $f(\lambda)$=constant as an average. The way in which the systems oscillates around $f(\lambda)$=constant describes species specific

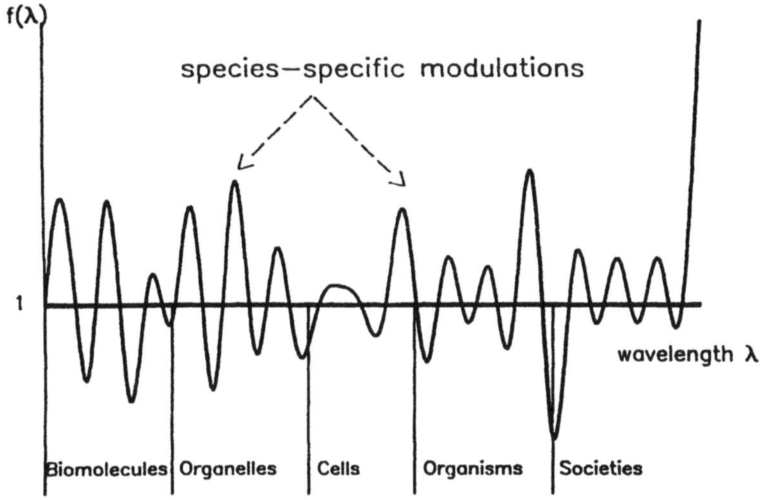

Figure 21.1: *The $f(\lambda)$=constant-rule provides that evolution expands around f=1 of the DNA-System of the living state. The more this development is extented to the long-wavelength-range, the higher the evolutionary development has to be looked upon. With extending wavelength the size of coherence volume and communication increases as well as the range of social interactions. Since f=1 is confined to an average value, time-dependent and species-specific modulations around f=1 determine the species of the living system under investigation.*

properties. Consequently, $f(\lambda)$=constant provides the basis of communication and the deviation $\delta f(\lambda, t)$ describes the informational contents.

Fig. 21.1 displays this picture of evolution which is a new approach to the problem.

We have to conclude that Darwin's interpretation together with all its modifications cannot truthfully describe the evolutionary development, because even after the best fit of the models a teleology is still disputed. In contrast to that, we hypothesize an optimization principle that cannot be traced back to a "survival of the fittest". The actual origin of it is the expansion of a spatio-temporal basis of communication with well-defined optimum properties.

Living systems are then neither mere subjects, nor objects, but subjects and objects at the same time.

In contrast to the neo-Darwinistic point of view, their capacity of evolutionary development does not originally depend on their rivalry and power in the fight for existence. Rather, it depends mainly on their capacity of communication. They can be looked upon as expanding antennae systems, as, for instance, has been postulated by P.S. Callahan[24] for many years.

The entity of all living systems (which can be considered as a more or less fully interlinked unit), rather than the individuals, is always developing.

One may ask about the advantages of such a process. From the point of view of natural sciences the satisfication of elementary curiosity can be discovered. It displays the property of insatiability. Of course, the more questions are answered, the more new questions can be asked. By conquering the world of "actual information" the other world of potential information – that is, the world of possibilities – is expanded. This is a permanently progressing process, which at the same time autocatalytically amplifies the capacity to look forward to the future as well as to remember the past by increasing the power of the memory.

By this process one becomes more aware of the present, simply by squeezing it in between future and past. According to the uncertainty relation, such a process is unlimited. This uncertainty is the essence of individuality, "freedom" and consciousness. We cannot exclude that the Planck constant itself is a subjective quantity, possibly based on the limited structure of our DNA. However, in that case again we had reason to reconsider the subjectivity of our view of the world.

21.4 References

1. E. Mayr, Evolution, *Spektrum der Wissenschaft* (Verstaendl. Forschung, Heidelberg, 1986).

2. M. Eigen, Darwin und die Molekularbiologie, *Angewandte Chemie 3, Int. ed. engl.* **20** (1981) p. 233.

3. E.J. Steele, *Somatic Selection and Adaptive Evolution* (University of Chicago Press, Toronto, 1979).

4. A. Weismann, *The Germ Plasm: A Theory of Heredity* (Scott Publishing, London, 1893).

5. T. Dobzhansky, E. Boesinger und D. Sperlich, *Beitraege zur Evolutionstheorie* (Fischer Verlag, Jena, 1980).

6. B. Rensch, *Biophilosophical Implications of Inorgamic and Organismic Evolution: Die blaue Eule* (Essen, 1985).

7. R. Goldschmidt, *Theoretische Genetik* (Akad. Verlag, Berlin, 1961).

8. N. Eldredge and S.J. Gould, in *Models in Paleobiology*, ed. T.J. M. Schopf (Freeman Cooper Comp., San Francisco, 1972).

9. M. Kimura, *Die Neutralitaetstheorie der molekularen Evolution* (Paul-Parey-Verlag, Berlin-Hamburg, 1987).

10. W.F. Gutmann und K. Bonik, *Kritische Evolutionstheorie* (Gerstenberg Verlag, Hildesheim, 1981).

11. M. Barbieri, *La teoria semantica cell 'evoluzione* (Harwood Acad. Publ., Turin, 1985).

12. R. Riedl, *Quart. Rev. Biol.* **52** (1977) p. 351.

13. F.M. Wuketits, Evolutionstheorien, in *Dimensionen der modernen Biologie* **vol. 7** (Wiss. Buchges., Darmstadt, 1988).

14. M. Rattemeyer, F.A. Popp and W. Nagl, Evidence of photon emission from DNA in living systems. *Naturwissenschaften* **68** (1981) p. 572-573.

15. F.A. Popp, K.H. Li, W.P. Mei, M. Galle, and R. Neurohr, Physical aspects of biophotons, *Experientia* **44** (1988) p. 576-585.

16. F.A. Popp, Coherent Photon Storage of Biological Systems, in *Electromagnetic Bio-Information, 2nd edition*, eds. F.A. Popp, U. Warnke, H.L. Koenig and W. Peschka (Urban & Schwarzenberg, Munich, 1989) p. 144-167.

17. R.H. Dicke, Coherence in spontaneous radiation processes *Phys. Rev.* **93** (1954) p. 99-110.

18. F.A. Popp and J. Deny, Biophotonen-Information und Chaostheorie, in *Ganzheitsmedizin*, ed. A. Stacher (Facultas Universitaetsverlag, Wien, 1991) p. 53-69.

19. F.A. Popp, Photon Storage in Biological Systems, in *Electromagnetic Bio-Information*, eds. F.A. Popp, G. Becker, H.L. Koenig and W. Peschka (Urban & Schwarzenberg, Munich, 1979).

20. M. Born and E. Wolf, *Principles of Optics* (Oxford Press, New York, 1975) p. 511.

21. F.J. Zucker, Information, Entropie, Komplementaritaet und Zeit, in *Offene Systeme* **I** ed. E. von Weizsaecker (Ernst Klett Verlag, Stuttgart, 1974).

22. J. Slawinski and F.A. Popp, Temperature hysteresis of low level luminescence from plants and its thermodynamical analysis. *J. Pl. Physiol.* **130** (1987), 111-123.

23. F.A. Popp, *Biologie des Lichts* (Paul Parey-Verlag, Berlin, 1984).

24. P.S. Callahan, Nonlinear Infrared Coherent Radiation as an Energy Coupling Mechanism in Living Systems, in *Molecular and Biological Physics of Living Systems*, ed. R.K. Mishra (Kluwer Academic Publishers, Dordrecht, 1990) p. 239-273.

Chapter 22

Mitogenetic Radiation as an Evidence of Nonequilibrium Properties of Living Matter

A.A. Gurvich

Abstract

The concept of the nonequilibrium molecular organization of living systems based upon the phenomenon of mitogenetic radiation (ultraweak ultraviolet luminescence of biological systems) and the theory of vectorial cellular fields elaborated by A.G. Gurvich in the 1940's are presented. The experimental data given in the second part of the paper are related to muscle radiation studied in *in vivo* conditions. They characterize some properties of the structural and energetic processes in the sarcoplasm of a rested muscle, which seem to be important and specific for maintaining the rested state of the muscle.

22.1 Nonequilibrium Molecular Organization: Biological Premises to the Theory of Cellular Fields

The ideas which led A. G. Gurvich to the study of nonequilibrium states of molecular substrate in biological systems came from the concept of the dominating role of form in biological processes.

"In the realm of very small dimensions the concepts of structure and of process are merged. The combination of material processes which is considered as metabolism only occurs regularly when the particles of a substrate capable of reactions are associated in definite configurations, that is when they are spatially distributed in a given way. Such a spatial distribution requires energy uptake and the existence of a regulative factor restricting the spatial degrees of freedom of reactions"[1,2].

Therefore the concept of nonequilibrium molecular states is in no way exhausted by nonequilibrium deformations of large molecules, for example proteins*.

It should be wider and include the concept of nonequilibrium organization of the molecular substrate regulated by the influences from cellular to supracellular levels. The components of nonequilibrium organization of molecular substrates may be exemplified by the highly labile temporal associations – "constellations" – of excited small and medium size molecules, such as peptides, aminoacids etc.[2,5,6,7,8].

Ascribing a functional importance to nonequilibrium molecular constellations (NMC) we must assume that a considerable part of the molecules of any biological system is capable of creating such ensembles. The necessary energetic level of molecules is supported, most probably, by oxidative reactions. The energy migrating along NMC is accumulated in such a way that it piles up to mitogenetic radiation. Therefore the excitation of NMC is of metastable character. During their destruction, NMC release energy partly in the form of radiation. Its detection during the destructive events (for example, cooling the object) permits the study of different variants of nonequilibrium states.

Here we shall consider only the principal idea of the repeatedly published experiments[5,6,9–11]. If a naked organ – liver, muscle of a homoeothermic animal – is

---

*The concept of nonequilibrium deformations of molecules was first proposed by E. Bauer who ascribed a decisive role to the maintenance of a "stable nonequilibrium state" of the living systems to a nonequilibratory deformed protein molecules[3]. These views have been later developed by L. A. Bluemenfeld[4].

rinsed by a cool physiological solution or if a fresh surviving organ is immersed in the cooled solution, a relatively intense outburst of mitogenetic radiation is registered. It has been suggested that the energy emission during the inhibitory influence is due to NMC disturbances. NMC should now be characterized as labile energy accumulators requiring at the same time energy inflow for their very existence.

This conclusion has been supported by centrifugation of kidney or liver, where the living tissue even at moderate accelaration displays outbursts of mitogenetic radiation. As a most probable explanation it was proposed that the relatively large intracellular inclusions shifted by centrifugal forces disturb NMC mechanically, thus leading to their energetical discharges. What is most important is that after the application of one of the influences and then of the second, whatever their succession (cooling after centrifugation or centrifugation after cooling), only the first one causes the radiation outburst.

Therefore, quite different factors disturb the same labile states of a substrate which should be considered as associations of molecules requiring energy inflow for their structural stability. The next problem was as follows. Evidently the formation and maintenance of nonequilibrium structural states, as well as the changes in their configurations cannot be reduced to the molecular interactions. In other words, a factor preventing their transition towards the equilibrium state should play a role.

This very role has been ascribed by A. G. Gurvich to the cellular fields. Here we present the main principles of the cellular field theory[5–8,2].

The cellular or cell nuclei field provides a definite degree of regular spatial arrangement of excited molecules. The work required for this effect is performed at the expense of molecular excitation energy which is transformed into the common directed movement of molecules or their coordinated orientations. In both cases NMC are created. The general field concept has been specified by the author by suggesting that the immediate sources of vectorial influences are located in the nuclei where some undiscovered chromatine structure exerts a directed repulsive action upon the different surrounding excited molecules. If the other excited molecules of nucleotides and histones are situated in the immediate vicinity of this influence, they also become capable of the repulsive vectorial influence, and so forth.

The temporal continuity of a common chromosomal (nuclear) field is considered as a statistical (fluctuating) phenomenon. The quantity of the simultaneously active field sources is a function of chromatine metabolism and depends upon the level of the splitting and synthesis of the general metabolism. In different periods of cell life the intensity of its field may also be different. It is also suggested that the field is of anisotropical character, its anisotrópy may be approximated by a rotational ellipsoid.

It is worth mentioning that only the combined, or integral –"actual" – fields should be of real biological significance. "Each cell creates a field spread out from its borders. That determines the vectorial cellular interactions which take place also in the intercellular spaces. At each point of a system and at each given moment the actual fields depend not only upon the numbers and the functional states of cells but also upon their spatial arrangement. In other words, they depend upon the shape and the configuration of the system".

Let us return now to the processes of molecular level directly connected with the concepts of cellular and actual fields. The common energetic levels of NMC can transform into more stable bonds. In other words, something like "traces" ("vestigia") of the successively replaced NMC states are created. These transformations are of importance because the vestigia, as the more stable states, may "inform" the actual fields about the nonequilibrium processes which have just occurred, thus causing the corresponding changes of the integral fields.

In A. G. Gurvich's words: "Such a continous interaction of causes and effects determines, without any additional assumptions, the continuity of the developmental course and the impossibility of a strictly stationary living state"[2]. Not only intercellular processes but also macroprocesses such as cell and nucleus migration, cell multiplication etc. are embraced by the notion of vestigia.

In connection with the above, A. G. Gurvich stresses: "The cellular field, be it a theoretically constant field of a single cell or a continously changing actual field, should not fulfil any complicated work. The task of each actual field (which is always short-lived) is in the regulation of an infinitesimal step of embryogenesis because in each next moment the field has been changed under the influence of the next stage of development. The actual fields and the results of their action are developing by continous interactions".

Along these lines we can see the idea of the coupled interactions of processes on different levels, which led A. G. Gurvich to the study of nonequilibrium organization of molecular substrate. The NMC concept which requires the spatial normalizing influence of the actual fields fits with these biological presumptions.

22.2 The Evolution of the Mitogenetic Radiation of Skeletal Muscle during Postnatal Development of the Organism

We describe here some parameters of the radiation of musculus gastrocenemius in the rabbit and frog which are correlated in the rabbit with the growth of muscle fibers and development of their regulative potentials. By analysing them from the point of view of the coupled dependence between the molecular processes and the spatial dynamics of actual fields we would like to demonstrate the application of the field concept to the functional states of the neuro-muscular system†.

From our point of view that means that the concept of physiological state is inseparable from the concept of shape in its general meaning, that is from the arrangement and configuration of the elements of different levels whose spatial parameters are dynamic to a different extent.

From this point of view it is necessary to compare the molecular processes with those of the cellular level. The works of R. Couteaux[12] which analyse the formation of the neuro-muscular connections during the latest stages of prenatal development give us a possibility to do that.

One may certainly doubt the value of a comparison of the dynamic molecular processes with the fixed microscopical material but in any case we compare the stages of development which are close to each other. The inclusion of some hypothetical links into the study of the dynamics of biological processes is inevitable and it seems to be justified if based upon a generalizing principle.

The mitogenetic radiation of m. gastrocnemius in frog and rabbit have been studied in *in vivo* experiments‡. It was then observed that in both species the radiation increased by 2-3 times during the cooling (10-12°C for rabbit and 1-2°C for frog) compared to the radiation at the normal temperature [11].

Therefore, the concept of a regulated nonequilibrium organization of a molecular substrate may also be related to a muscle. We suggest that at the expense of the

†The detailed analysis of the cerebral cortex processes from the point of view of the cellular field theory was given by A. G. Gurvich[2]

‡Yeast cell culture served as a detector; the detection was based upon the radiation stimulation of cell divisions. If the experimental conditions are strictly kept, the results are significant and reproducible. The method is highly sensitive, but the radiation intensity can only be estimated relatively, assuming the reverse relation between intensity and threshold exposure[10,11,13]

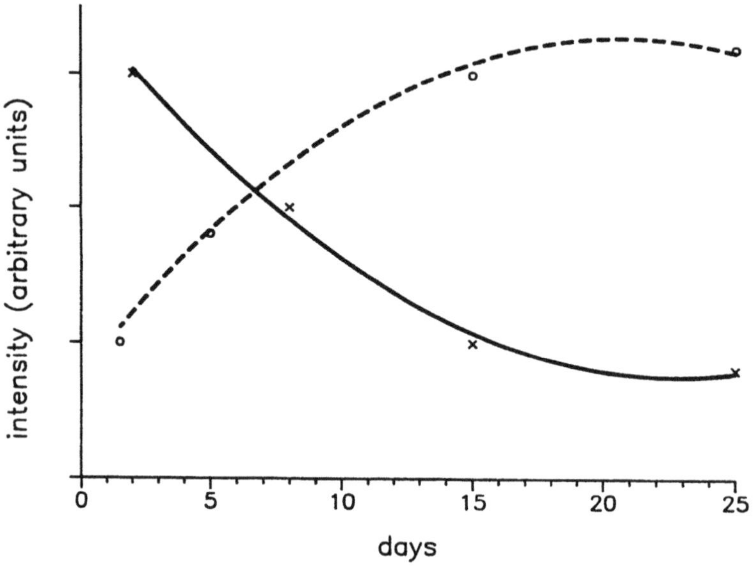

Figure 22.1: *Developmental age evolution of the muscle radiation intensity (I) at the normal temperature (×) and under cooling (∘). Abscisse: days of the postnatal development. Ordinate: conventional intensity.*

normal metabolic energy, that is by direct participation of excited molecules, NMC are created in the sarcoplasm. It is highly probable that the interactions between NMC and more stable structures – actomyosine chains, membrane components etc. – are realized by structural formations of a different degree of stability.

It was shown experimentally that the muscle innervation is necessary for maintaining its unequilibrium state[11]. The data presented led to the suggestion that the disturbances and existence of NMC are natural processes for a muscle, which occur somewhat more seldom under normal temperature and full-scale innervation than under experimentally caused degradation. The decisive arguments came from the spectral analysis of radiation (Fig. 22.1 and 22.2).

The results are presented in Table 22.1.

They show that under all the experimental conditions the spectral range of radiation is the same, showing one rather wide maximum in the short UV range (220-230 nm). This permits the conclusion that energetic discharges of similar level occur in the rested muscle at normal temperature as well, but are, as estimated by radiation intensity, much more rare than under cooling.

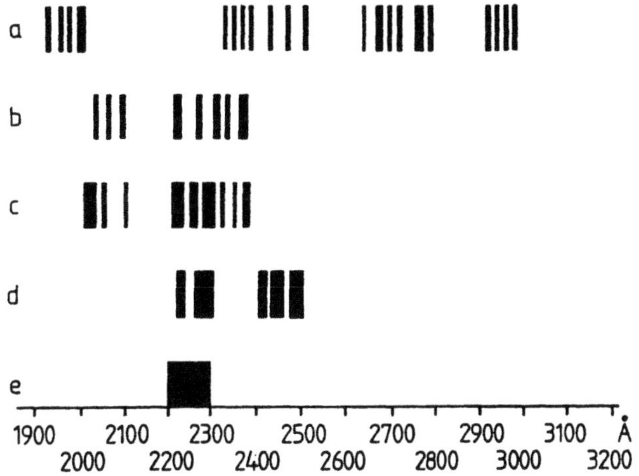

Figure 22.2: *Developmental dynamics of the radiation spectra under the normal temperature. Abscisse: wavelengths in A; Ordinate: the postnatal growth in days: (a) one, (b) three, (c) six, (d) nine, and (e) sixteen.*

The molecular system modelled in such a way seems to be a necessary prerequisite for maintaining the excitability of the muscular and particularly the neural system[14]. The spectral similarity of the different species of animals points to the functional specificity of the structural nonequilibrium organization of a muscle.

In the broadest sense the spectrum of radiation of a rested muscle may be considered as an expression of the processes regulating the steric parameters of nonequilibrium constellations. In this respect the problem of its origin arises.

It was assumed that the existence of only one radiation maximum is the result of the successive changes of the substrate during the early stages of postnatal development. Therefore the radiation intensities under different temperatures during the first weeks of the postnatal development of rabbit muscles have been compared and at the same stages the spectral analysis of the radiation under normal temperature have been performed.

As seen in 22.1, the radiation at normal temperature decreases as the animal grows and after two weeks takes the constant intensity which is specific for the adult. The changes of degradational radiation caused by cooling are of reverse manner: as the animal grows, the radiation intensity increases and after two weeks becomes stable.

These data still do not permit evalution of the state of a substrate but they become important when being compared with further observations. It should be mentioned

Table 22.1: Spectral analysis of the radiation of m. gastrocnemius[a]

Length of waves in nm	Frog 15-16°C	Rabbit 35-38°C	Frog 2-3°C	Rabbit 10-12°C
190-200	−3	−7	3	−3
200-210	1	−1	2	−7
210-220	5	1	12	0
220-230	50	53	42	55
230-240	−7	2	5	−3
240-250	2	−3	−2	—
250-260	−1	−6	11	4
260-270	5	−3	1	−3
270-280	−2	2	4	6
280-290	1	5	8	3
290-300	4	−6	−2	9
300-310	5	1	13	2
310-320	2	8	1	11

[a]The figures give the increase of dividing yeast cells in percent compared to the control ones. Significant average values from several experiments are presented.

that the stable radiation level is established at the same time as the animal's stable pose is acquired, the movements become coordinated and the animals become completely luminescent[11].

In this respect the evolution of the radiation spectra should be considered in more detail[11]. On the first day the spectrum consists of a great number of narrow fringes distributed throughout the whole UV range. At the same time the fringes are concentrated in several groups. It seems probable that the arrangement of the groups reflects further development of the processes. During the second day the spectrum is drastically changed. Its general range is narrowed whereas the separate fringes slowly become (within several days) enlarged.

This is accompanied by slight shifts of the fringes in a limited area of the spectrum. Later on the number of fringes continues to decrease, the remaining fringes enlarge and after two weeks one wide radiation maximum remains which is typical for m. gastrocnemius of an adult rabbit. Therefore, the spectral changes are correlated to developmental processes and for their estimation it is necessary to consider that the spectral evolution reflects the evolution of NMC parameters.

In this context, let us trace the behaviour of the nuclei of the muscle fibres in the prenatal developmental stages as observed by R. Couteaux in his research on the morphogenesis of neuro-muscular junctions in the limb muscles of guinea pig and mouse [13].

The author divides the phenomena observed into two phases:

1. The phase of reinforced growth of neural terminations penetrating into the muscle of simultaneous growth of muscle fibres and accumulation of nuclear divisions in the central fibre zone.

2. The phase when the translocations and reorientations of the musular and neural elements are dominating, specifically for the creation of neuro-muscular connections.

The author stresses that after migrating to the superficial fibre zones the nuclei move away from each other until the same average distance has been established between them. At the same time the neural fibres move close to small zones of the muscle fibre surfaces and the membranes of both fibres merge together.

Pointing out the equidistant nuclei arrangement in sarcoplasmic syncytium, Couteaux defines it as a state of dynamic equilibrium, stressing the appearance of an entire system of nuclei rather then their simple collection. Such a dynamic system spreads to almost the complete fibre surface which is much larger than the area of the motile plates. Therefore one may suggest that the regulations of vast spatially organized nuclei system are required for the arising and establishment of the specialized synaptical structure.

Associating the Couteaux data with our general concept we consider the nonequilibrium dynamic organization of nuclei as zones of their vectorial interactions, that is, as a system of actual fields responsible for the regulation of molecular events.

Let us assume that the mobile dynamics of the muscle nuclei is taking place, most probably as minor oscillations, during their final quasi-equilibrium spatial arrangement as well. In other words, we assume that the actual fields of muscle fibres are also active in adult animals. In order to substantiate this suggestion we may point to the general directiveness in the changes of spectral ranges and radiation intensity, stressing at the same time the tentativity of our proposals. The low level of radiation of rabbit muscle after one day of cooling does not necessarily contradict the existence of nonequilibrium states.

It seems more probable that the latter are actually taking place at this developmental stage as well, but the dynamics of NMC are very high and the energetic inflow during cooling is not sufficient for the restoration of the frequently disturbed NMC. Most probably at this developmental period the muscle NMC are small associations of excited molecules with restricted capacities for migration and accumulation of energy[§]. In the first day of postnatal development the spectrum consists, as was already mentioned, of several groups of narrow fringes.

But later on the tendencies in the changes of both the intensity and the spectral characteristics of radiation clearly emerge. As was already mentioned, the radiation intensity is reduced at the normal temperature and increased after cooling. At the same time the number of fringes is decreased and they become wider. These processes point to the incrased restrictions of freedom degrees which determine the energetic and steric parameters of NMC.

The uniformity of relations between the intensity and the spectra of radiation is demonstrated by the increase of the degradational muscle radiation. as the fringes are narrowed and widened. That means that the regulatory trend of NMC transformations enlarge their possibilities for energy circulation. In other words, their joint energetic levels become more prolonged. The excited states of NMC obtain some degree of metastability and show the tendency to increase. Under these conditions the energy cumulation in some of the NMC links is possible.

By generalizing these facts we come to the following conclusion: The maximum regulation of the spatial NMC parameters, that is, the approaching of their configurations to a certain average sterical state, considerably increases the degree of their nonequilibrium. This leads to the following important conclusion - NMC are formed inside the areas of action of the actual fields which are supported by the spatial interactions of muscle nuclei; the precision of the regulating influence increases as animals develop.

Such a concept is based both upon the processes related to radiation and the biological observations made by Couteaux. After describing these molecular and cellular level events we return to the above, formulating it in a somewhat different way. The necessary and important component of the appearance of any functional state is

[§]In this respect it should be remembered that while we are working with biological detectors we are restricted to studying the UV range only. One cannot exclude the possibility of undetected energy accumulation up to the visible and longer wave ranges in small dynamic NMC.

the dynamics of spatial interactions between the different level elements, embracing both micro- and macrostructural states and having different scale dimensions. These interactions are considered from the point of view of the cellular fields theory.

The above data and concepts seem to us to stand close to the investigations and theoretical constructions proposed by H. Fröhlich, F. Popp, W. Nagl, K. H. Li and other investigators, extending principally the physical approach to the general problems of biology[15-20].

22.3 References

1. A. G. Gurwitsch, *Die histologischen Grundlagen der Biologie* (Fischer Verlag, Jena, 1930) p. 1-310.

2. A. G. Gurvich, *The Principles of Analytical Biology and of the Theory of Cell Fields (russ.)*, (Nauka, Moscow, 1991) p. 1-350.

3. E. S. Bauer, *Theoretical Biology (russ.)*, Ed. Wiem (1935) p. 1-206; Reprint (Russian/English) (Akademiai Kiadó, Budapest, 1982).

4. L. A. Blumenfeld, *The Problems of Biological Physics (russ.)*, (Ed. Nauka Moscow, 1974) p. 1-335.

5. A. G. Gurvich, *The Theory of the Biological Field (russ.)*, (Ed. Sovetskaya Nauka, Moscow, 1944) p. 1-155.

6. A. G. Gurvich, Une Theorie du Champ Biologique Cellulaire, ed. E.J.Brill, *Bibliotheca biotheoretica* **Vol. II S.D.** (Leiden, 1947) p. 1-149.

7. A. G. Gurvich, *Selected Works*, (russ.) Ed. Meditsina (Moscow, 1977) p. 1-348.

8. M. A. Lipkind, Gurwitschs Theorie vom biologischen Feld, in *Fusion* (German edition, 1987), *1. Part:* **No. 4** p. 30-50; *2. Part:* **No. 5/6** p. 53-66.

9. A. G. Gurwitsch and L. D. Gurwitsch, *Die mitogenetische Strahlung* (Fischer Verlag, Jena, 1959) p. 1-308.

10. A. G. Gurvich and L.D. Gurvich, *The Introduction to the Teaching of Mitogenetic Radiation (russ.)*, (Ed. AMN, 1948) p. 1-115.

11. A. A. Gurvich, *The Problem of Mitogenetic Radiation as an Aspect of Molecular Biology (russ.)*, (Ed. Meditsina, Leningrad, 1968) p. 1-238.

12. R. Couteaux, *"Les Jonctions Intertissulaires" en "Les problèmes de structures, d'ultrastructures et de fonctions cellulaires"* (Ed. Masson, Paris, 1955).

13. A. A. Gurvich, S.B. Stefanov, N.N. Lasurkina, The Biological Detector of Mitogenetic Radiation (russ.). *Succ. of Modern Biol.* **99 No.2** (1985) p. 41.

14. N.N. Lasurkina, The Molecular Organization of Sarcoplasm of Skeletal Muscles by Different Functional State (russ.). *Bull. Exp. Biol. and Med.* **94 NII** (1982) p. 43.

15. H. Fröhlich, The Biological Effects of Microwaves and related Questions. *Adv. Electronics and Electron Physics* **53** (1980) p. 85-152.

16. F.A. Popp, *Biologie des Lichts* (Parey Verlag, Berlin, 1984) p. 1-156.

17. F.A. Popp, On the Coherence of ultraweak Photon Emission from living Tissues, in *Disequilibrium and Self-Organization*, ed. Kilmister (1986) p. 207-230

18. F.A. Popp, K.H. Li, W.P. Mei, M. Galle, R. Neurohr, Physical Aspects of Biophotons. *Experientia* **44** (1988) p. 576.

19. W. Nagl, Chromatin Organization and the Control of Gene Activity. *Intern. Rev. of Cytology* **94** (1985) p. 21-56.

20. K.H. Li, F.A. Popp, Nonexponential Decay Law of Radiation Systems with Coherent Rescattering. *Phys. Letters* **93** (1983) p. 262-266.

Chapter 23

Can the Vitalistic Entelechia Principle be a Working Instrument?

(The Theory of the Biological Field of Alexander G. Gurvich)

Michael Lipkind

Vitalism has become a bad and repugnant word for most biologists. Unfortunately, many are already biased when they hear or read the word and, thus, derive themselves of a most positive evaluation.

R. Sattler (1986)[1]

Abstract

The present paper suggests an alternative approach, which is the theory of the vectorial biological field developed by Alexander G. Gurvich (1874-1954), to the biological thought of modern molecular biology. The power of this field theory may cover all levels of biological organizations – morphological, cellular and molecular one. This theory has been used as a tool to explain a wide range of biological phenomena from metabolic processes to the psychic sphere.

23.1 Introduction

For more than a third of a century contemporary biology has been under the powerful influence of the global Crick-and -Watson-double-helix revolution followed by the triplet-molecular-genetic boom. This paradigm has penetrated almost all the branches of the life sciences. There is no need here to describe the great achievements of the triumphant modern molecular biology in gaining fundamental and applied scientific knowledge. However, this dominating trend in biological thought has left no chance for any alternative theoretical approach, so that some attempts to explore alternative directions sounded like "voices in the labyrinth " (Chargaff). This is especially evident in the life sciences related to heredity and ontogenesis which always foreordained an intrinsic biological philosophy.

The present article suggests an alternative approach characterized, however, by some peculiarities. This is a theory which was developed from 1912 to 1954 by a single man who in the 1920's was a candidate for the Nobel prize and in the eighties was included in the ranks of the "false prophets"; this is the only theory based on a purely vitalistic conception (as "Weltanschauung") taken, however, not in the agnostic but in the "working" sense of this notion.

This is a theory whose power of explanation covers all the levels of biological organization – morphological, cellular and molecular – and which was tried by its author as a tool to explain a wide range of biological phenomena from metabolic processes to the psychic sphere; this is the theory, the last version of which was elaborated when its author was living in the Soviet Union in conditions of isolation from western colleagues and in political disgrace, thanks to T. D. Lysenko's medieval obscurantism; this is the theory which, being published mainly in Russian, did not reach the potential western reader and remained practically unknown in the west.

This is the theory of the vectorial biological field by Alexander Gavrilovich Gurvich (1874-1954).

23.1.1 Historical background and conceptual development of the field notion

A. G. Gurvich was one of the first to introduce into biology the notion of "field" belonging to the vocabulary of physics. In his paper "Die Vererbung als Verwirklichungsvorgang"[2] the field notion was associated with the conception of **co-subjecting elements to a common single morphogenetic factor** as opposed to the alternative conception considering the whole morphogenesis as a result of **interactions between the elements**. Since then there has been much successive development of the field conception from the first abstract models describing single morphogenetic phenomena to the general theory of the vectorial biological field.

The profound analysis by A. G. Gurvich of the basic biological concepts was the basis for the development of his theory.

Preformism and epigenesis

Any theory of developmental or hereditary biology, no matter how highly sophisticated, should be inevitably derived from one of two alternative basic theoretical concepts – preformism or epigenesis, which served as "supreme" principles since the beginning of scientific biology (preformistic roots can be retraced to Hippocrates while the epigenetic idea originates from Aristotle). The development of A. G. Gurvich's theory started from the comprehensive analysis of these two concepts.

According to the logical structure of preformism, the zygote contains all the potential prerequisites for the development from itself of a future organism with all its species and individual specific patterns, features and properties. As to the classical (embryological) preformism, the zygote presents a puzzle-like (mosaic) spatial distribution of the entities inside it, each entity representing (being responsible for) a certain morpho-anatomical part of the future organism.

Thus, the development of the embryo is just an evolution of these preformed and predetermined entities. This original, "embryological" definition of the preformistic principle was transformed into a "genetic" definition according to which **all** the observed specific complexity of the organism can be reduced to separate independent features which can then be projected onto strictly determined separate entities contained in the zygote and defined as genes. Accordingly, classic Mendelian genetics can be considered as a pure form of this kind of preformism.

According to the logical structure of epigenesis, any momentary stage of the development of an embryo can be derived from the previous stage but not from a stage several steps earlier. Hence, any stage of the whole chain of development contains a prerequisite only for the immediate next one, such epigenetic **actual** prerequisites being opposed to the preformistic **potential** prerequisites.

Preformism and epigenesis were involved in the logical structure of the two main doctrines of developmental biology – developmental mechanics (Entwicklungsmechanik) founded by Wilhelm Roux in 1895[3] and vitalism founded by Hans Driesch in 1908[4]. The former doctrine considered the embryo as a complicated mechanical gadget in which one part acts on another during the development. The aim of the research is to reveal the network of all the influences of each part of the embryo on any others by changing the local environment - to isolate a certain part, to combine it with another one or to change the external environment.

The crucial experiments by H. Driesch[5] showed the development of **the WHOLE from a PART** (" embryonal regulation " by Driesch) and even formation of the **whole** in normal development in spite of drastic interventions (perturbations in certain parts or mixing together different parts of the embryo, "harmonic regulations"). On the basis of these results, Driesch suggested two principal concepts: equipotentiality and equifinality. The former means that all the elements (cells) of a whole (embryo) are equal in their potential morphogenic capability[37]. The latter means that the final result of the development - formation of a highly specific mature organism - may be reached by several quite different ways.

These considerations led Driesch to his classical general conclusion: **the fate of the element is a function of its location within the whole.** According to Driesch, such a "harmonic whole" is a teleological factor of realization of the purpose of the development. The factor, called by Driesch **"entelechia"** (after Aristotle), which means "having its goal in itself", was defined as a non-material, non-spatial metaphysical factor which leads and organizes the development of the embryo even if its parts are drastically perturbed, mixed or deleted. Thus, designating the entelechia as **vis vitalis** (living force), Driesch gave the name vitalism to his doctrine[4,6] and confronted it with Roux's causal-analytical method.

Apart from the factual contribution of both schools, the confrontation was based on different philosophical backgrounds, namely, different definitions of the notion of cause. The vitalistic conception developed by Driesch was based on the notion of a final ("purpose") cause (Aristotle's "causa finalis ") whereas the causal-analytical method of Roux was based on the notion of an immediate ("acting") cause (Aristotle's "causa efficiens").

Considering the entelechia as a final non-cognizable instance unaccessiable to rational analysis, Driesch himself left experimental science having discouraged any potential followers to explore this direction, while the causal-analytical school founded by Roux and further developed by Spemann[7] subsequently has determined the face of contemporary embryology.

With this background, the field conception developed by A. G. Gurvich became an alternative challenge to the preformistic principle dominating in embryology and, at the same time, gave a positive ("working") meaning to the epigenetic vitalistic principle of **the whole** as a factor organizing both "normal" embryogenesis and harmonic regulations. However, while Driesch himself refrained from further research on the whole which he associated with the agnostic entelechia, for A. G. Gurvich this was just a starting point for the development of his own original method of research.

23.1.2 Elaboration of some quantitative (statistical) criteria confirming the influence of the whole on elements

The first step in the attempt to make the concept of the whole to "work" was the examination of its applicability to the analysis of **a normal development cycle** (without drastic experimental interference). Accordingly, A. G. Gurvich made a unique analysis of the behaviour of elements (cells) during the development of geometrically exact and symmetric forms like sea urchin gastrula, the crystalline lens and cornea of chick embryo, and onion root[8].

The question was whether the exact geometrical definitive "macroform" of the embryo (or of particular organs) is associated with (and due to) the correspondingly strict and coordinated behaviour of the elements. The experiment consisted of the calculation of the numbers of mitoses in both halves of the macroforms and the confrontation of the "empiric curves" with the theoretical Gaussian curve of normal distribution .

The results have shown different kinds of distribution of the empirical curves:

1. "subnormal" , i.e. the empirical deviation (difference between the numbers of mitoses in either half) was less than the standard deviation of the normal distribution;

2. normal (empirical and standard deviations were equal);

3. "supranormal" (the empirical deviation was higher than the standard one).

However, there was no case of mirror symmetry between single mitoses synchronously occurring in both halves. The case of the subnormal distribution of mitoses was designated by A. G. Gurvich as "determination" which means that there is an individual **causal** interaction between the symmetric mitoses. As opposed to that, both normal and supranormal distributions of the microprocesses labelled by A. G. Gurvich as "normating " (an imperfect English translation of the notion designated by A. G. Gurvich in German as "Normierung") show the absence of the immediate **causal** connections between the microprocesses. Instead, the microprocesses are "subordinated" to a "single normating factor" which is common to all the micro-processes.

Thus, the principle of the subordination of "equipotential" elements to a common invariant "normating" factor as an alternative to the principle of immediate (causal) interaction between specifically "predestined" elements was laid down by A. G. Gurvich. As far as the actual influence of the "normating" factor on the elements is associated with their localization within the developing macroform, this notion became a prototype of the field principle. A. G. Gurvich[2] suggested the following criteria for the equipotential elements being under the influence of the whole which is expressed as a field factor:

1. The dependence of the elements on their location within the whole (their "fate", according to Driesch, which can be expressed either by their movements and/or by the type of their differentiation) is fairly certain and mathematically simple;

2. Successive ordering of the mutual arrangement of the elements during development of the embryo (**the whole**) from apparent disorderly dispersion to a more and more regular spatial distribution;

3. "The covering " – the whole – (a geometrically regular contour of the macroform shape) is more precise than "the covered" (the distribution of the elements constituting this contour at the microlevel).

This was the first attempt to make the concept of whole to "work" and, hence, converted it from the "entelechia" which was declared unattainable into a "working principle" available to strict scientific analysis.

23.1.3 Construction of embryonic morphogenetic fields

A. G. Gurvich's idea of the whole as an invariant field factor, subordinating the elements was used for the construction of concrete field models of different ontogenetic developments.

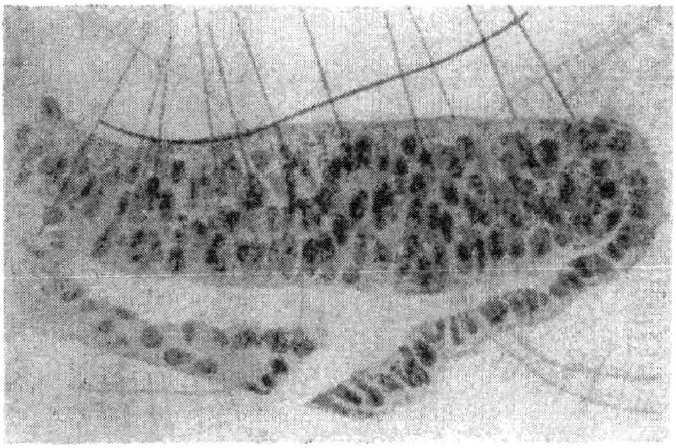

Figure 23.1: *Prognostic singnificance of the orientation of nuclei axes in the mitotic zone of the shark cerebral vesicles.*

In his work "The mechanism of form inheritance"[9] A. G. Gurvich suggested an invariant law describing ("explaining") the movements of embryonic cells and acting over a long period of embryonic development. This was a histoembryological study on the movements of the epithelial cells of the developing shark brain.

The crucial fact found there was that at each given moment of the development of the embryo's cerebral vesicle the long axes of cells in the internal layer of the neural epithelium were not oriented perpendicularly to the layer surface but at an angle of 15–20°. Such orientation turned out to have a prognostic significance: if a curve was constructed normal to the cell axes at a certain stage of development it would coincide with the real shape of the cerebral vesicle at a later stage (Fig. 23.1, 23.2). A. G. Gurvich compares such a prognostic orientation with the turning of a ship's helm which precedes the turning of the ship itself: the latter occurs with a certain delay and settles according to the preceding helm movement.

On the basis of these data, A. G. Gurvich formulated the following abstract invariant regulation.

During the morphogenic process the moving cells are oriented in such a way as if they were attracted by a certain "surface of force" which was designated by A. G. Gurvich as "dynamically preformed morpha " (DPM). According to this, at each given moment the cells are oriented by their long axes along the bisector between a perpendicular to the actual (at the considered moment of time) surface of the devel-

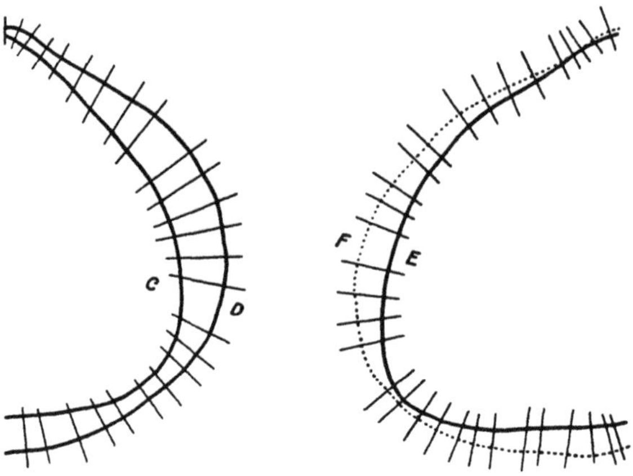

Figure 23.2: *Comparison of the theoretically reconstructed and actually observed forms of the brain vesicle. Coincidence of the predicted (E → F) and actual (C → D.*

oping layer and that of the DPM (Fig. 23.3). If, in the limits of this construction, one analyzed a set of the successive shapes of the developing layer, one could see that some points of the layer are at a minimum distance from the DPM surface , while the others are at a maximum one (Fig. 23.4).

According to the DPM conception, the cells in the former area are under a stronger attracting influence of DPM compared to the latter.Correspondingly, the actual microscopic picture shows strong polarization of the cell nuclei in the areas close to the DPM surface and their disorderly arrangement in areas remote from the DPM surface.This polarization is evident in histological pictures (Fig. 23.5). Thus, the "invariant law" describing the behaviour (movements) of the cells during a relatively long period of development was demonstrated.

The further elaboration of the field principle was done by A. G. Gurvich and his collaborators using various systems: some plant objects[10], differentiation of amphibian retina[11], cytodifferentiation during mammalian spermatogenesis[12], development of the triton phalanx cartilage[13].

These models of A. G. Gurvich based on his primary conception of an embryonic morphogenic field present a rather mixed collection of various abstract constructions. Indeed, the model field can be either attractive ("dynamically preformed morpha" of cerebral vesicles and camomile floscule), or repulsive (triton phalanx), or its describing capacity can "work" independently regardless to whether attraction or repulsion

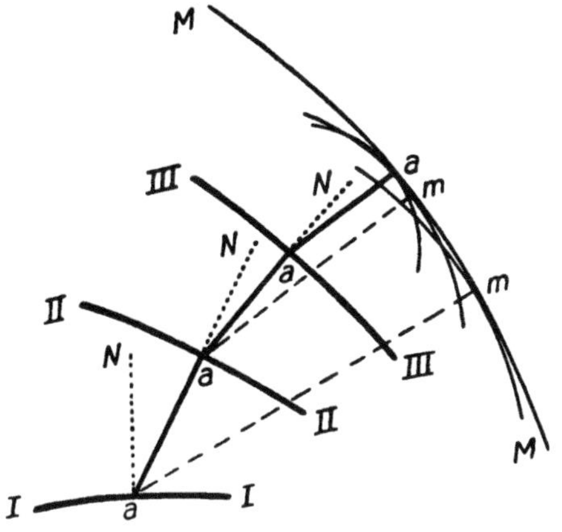

Figure 23.3: *Representation of the movement trajectory of an epithelial cell (nucleus) as a result of DPM influence.* **I, II, III**: *contours of successive stages of the epithelial layer development;* **N**: *normal (perpendicular) to a momentary surface of the layer;* **am**: *the shortest distance from the point of the base of the normal* **N** *to the DPM surface; the actual cell (nucleus) trajectory is "***aaaa***": bisector between* **N** *and* **am**.

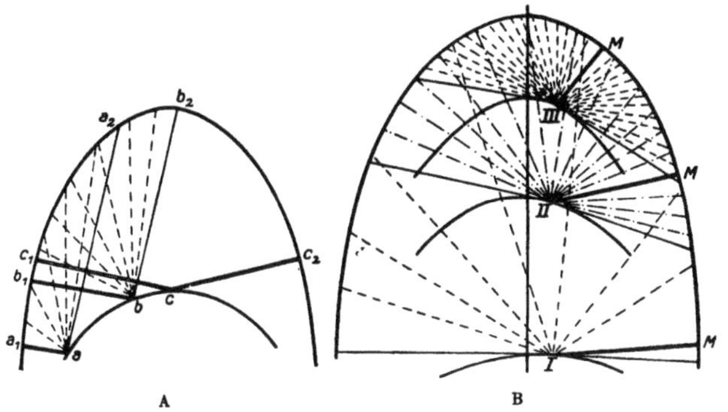

Figure 23.4: *The scheme of the influence of the DPM on different points of the surface of the layer.* **A**: *initial stage;* **B**: *following successive stages:* $I \rightarrow II \rightarrow III$; a, b, c: *different points of the layer surface;* $a_1, b_1, c_1, a_2, b_2, c_2$: *different pionts of the DPM surface;* **M**: *direction of the maximum influence of the DPM.*

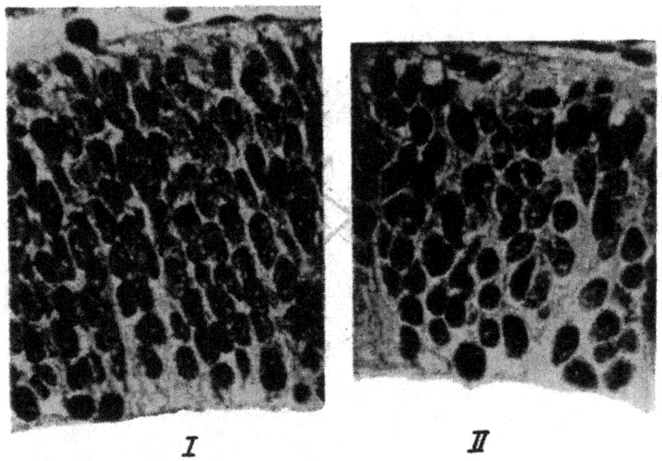

Figure 23.5: *Polarization and dispersion of the cell (nucleus) axes depending on the DPM influence. Two areas of the epithelial layer of the brain vesicle are located at different distances from the DPM surface and correspond to the different coordinate designed in the Fig. 2.6. I-area correponding to the point* **a** *(Fig. 2.6(A)): clearly expressed polarization and slanting arragement of the nuclei axes or oriented in the direction of* **M** *(maximal influence of the DPM). II-area corresponding to the point* **c** *(Fig. 2.6(A)): maximal dispersion of the nuclei axes.*

are ascribed to it (mushroom hats); the field surface can be of parabolic or other regular form (camomile floscule and mushroom hats); it can originate from dot sources (triton phalanx, amphibian retina, spermatogenesis) or it can be a vectorial integral one (mushroom hats).

However, in spite of the mixed character, the above models are united by the same fundamental principle - a single **invariant law** describing the development of a specific embryo part as a complete macroform during a relatively long period.

However, in spite of the elegance and descriptive capacity of the suggested field models, A. G. Gurvich himself realized the shortcomings of his primary field conception. The main one was that although the considered field models were effective in their describing (explaining) ability during a sufficiently long developmental period none of them can serve as an invariant principle for the whole ontogenetic cycle.

23.1.4 "Molecular" prerequisites of the field theory: use of the basic principles of the field theory (equipotentiality, equifinality and normating) for the processes occurring at molecular level

A. G. Gurvich's first work in this direction concerned regulational phenomena in the protoplasm of centrifuged eggs[14–16]. The main evidence was that although after centrifugation all the internal contents of the eggs were reshuffled and all the visible structures in the cytoplasm were destroyed, the first stages of the development like cleavage and blastula formation proceeded. Though the destroyed visible structures were usually restored, the development often proceeded without such restoration.

A. G. Gurvich's conclusion was that when the destruction of the cytoplasmic structure is followed by its restoration it is re-created from certain "amorphous" (invisible) materials which are carriers of living expressions. It was concluded that in the case of reversible perturbations of intracellular structures certain "**dynamic conditions**" in the cytoplasm, responsible for the organization specification, remain undisturbed and restore the damaged molecular organization deliberately involving different material (molecules) .

Application of the regulational principle of Driesch ("harmonic regulations") to molecular level meant that a **single cell** should be considered as that "**whole**". Accordingly, the notion of "equipotentiality" if applied to the molecular level was designated by A. G. Gurvich as "polyreactivity"[17]. That means essentially that there is no pre-determination of intracellular compartments: the subcellular entities behave according to their coordinates in the whole (cell) which, hence, presents a **dynamic non-destructible network** which determines the fate of the locally involved molecules ("**structural process**", by A. G. Gurvich[17]).

These views were supported by the experiments in which application of cooling, starvation and narcosis to a fertilized egg caused chaotic movements of some intracellular organelles: centrosomes, mitotic spindles, centrioles etc.[17].

The conclusion was that the harmonic movements observed in a normal cell are due to a certain factor related to the cell as a whole and this factor is not destroyed or inactivated by the centrifugal destruction of the visible intracellular structures. Hence, space-time connections between separate intracellular structures or processes are not due to any properties of the structures themselves.

A further conclusion was that together with stable **structures** in which the molecules are bound by means of different chemical bonds (covalent, electrostatic, hydrogen and van-der-Waals) there are unstable molecular constellations in which the molecules are not connected with each other by any of the above bonds but where their association is maintained by a continuous influx of energy. This means that if by any reason the energy influx ceases the constellations immediately dissociate.

Such labile energy-dependent molecular associations were designated by A. G. Gurvich as "**unbalanced molecular constellations**". The word "unbalanced" is an inadequate English translation of a Russian equivalent which was first introduced and widely used by A. G. Gurvich who himself had no opportunity to publish the corresponding material in English; later on, either another English variant "**non-equilibrated**"[18] or this variant "**unbalanced**"[19] were used.

As the unbalanced constellations need a continuous influx of metabolic energy for their existence at any moment, they possess the respective amount of accumulated potential energy (position energy). Moreover, the involvement of a high number of protein molecules in a constellation and the belonging of local constellations to the entire continuum create conditions for the formation of **common energetic levels** permitting the migration of common energy within the constellations with its possible **summation** at certain parts of the continuum. Cessation of the energy influx must lead to the destruction of the unbalanced constellations and the liberation of the accumulated energy. Experimentally, this was proved by the phenomenon of **degradational mitogenetic radiation**[20,21].

However, the current influx of metabolic energy is an absolute condition but not the only one for the existence of the unbalanced molecular constellations. Their existence is "thrown on" by a certain **dynamic** factor whose action, although somehow connected to a continuous utilization of metabolic energy, is quite independent. As this factor precisely determines the specific orientation of the protein molecules within the constellation it can be defined as a field factor.

Essentially, the action of such an "orienting" field factor should be expressed by some limitations it imposes on the **free chaotic movements** of the molecules. Hence, the action of the field factor is expressed in a certain **vectorization** of the chaotic movement of the protein molecules. This was the basis for the entirely new general theory based on the epigenetic principle which could now combine,by the same "working principles", the events occurring at all the three levels of biological organization: organismic, cellular and molecular.The first drafts of the new conception were outlined by A. G. Gurvich in autumn 1941[22] in the besieged Leningrad and the completed theory was published in 1944 (in Moscow)[17].

23.2 The Theory of the Vectorial Biological Field

23.2.1 Basic postulates

The postulates of the vectorial biological field are as follows:

1. Each cell is the source of a field generated in the nucleus.

2. The field is of vectorial nature and the vectors are directed centrifugally from the source.

3. The generation of the field is associated with certain processes in the nucleus, namely those somehow related to transformations of "chromatin". The choice of chromatin as a source of field generation, although being arbitrary, is dictated by a "biological common sense" understandable to any biologist: the chromatin's strict succession throughout the life cycle and heredity transfer as well as its incredible in vivo stability as a substance as compared to any other material components of the living systems.

4. There are **elementary "flashes" of the generated field** which are connected with certain elementary acts in chromatin metabolism; it is suggested (without strong obligation) that these acts concern associations of the chromatin with certain kinds of proteins. A total number of such flashes per time unit (its rate) designated as **field intensity**, depends immediately on the intensity of chromatin metabolism and, hence, on general cell metabolism.

5. The elementary flashes of the generated field connected with certain chemical acts of the chromatin metabolism can occur only if these acts proceed within the sphere of influence of the currently existing field. Essentially, this is the expression of the succession of processes in living systems, or, in other words, the proclamation of the same principles as declared by W. Harvey (**"omne vivum ex ovo"**) and L. Pasteur (**denial of a spontaneous generation of life**).

6. The field vectors directed from the nucleus are the result of the total statistical number of the elementary flashes of the field at any moment. Therefore, the field intensity is a completely dynamic fluctuating parameter subtly reacting to metabolic changes.

7. The elementary field is **spatially anisotropic** and this is the main postulate. Its meaning is that the isodynamic surface at which all the vectors are equal **is not spherical** but **ellipsoidal**. The anisotropy of the ellipsoid can be expressed as a particular ratio between its three axes and such a ratio, being species-specific, is considered as an **invariant species constant**. An infinite number of possible axe ratios covers all the potential number of possible species.

8. The field vector has a certain decrement with the distance from the field source. It is reasonable to accept that the value (length) of the vector depends on the square of its distance from the source but a real function of the dependence is a matter for empirical examination. In spite of the decrement, the influence of the field is not limited by the cell boundaries but spreads beyond its borders of the field source cell.

9. Field vectors exert an influence on the **excited** protein molecules (those which have just received a portion of metabolic energy and are in the excitation state) transforming a portion of the molecule's general excitation energy **into directed kinetic energy** and the direction of the movement is determined by the field vector. This is expressed either in the directed movement (flow) of the excited protein molecules along the vector or in specifically directed (by the field vector) deformations of the protein molecules, especially in the case when they are in the state of stable polymerization. This means that in living conditions the field "works" against the chaotic movement (agitation) of protein molecules.

10. The intensity of the field at a certain cell point (the length of the vector at this point) determines what share of the whole molecular excitation energy is transformed into directed kinetic energy. The E_d / E_t ratio, where E_d is the directed kinetic energy and E_t is the total excitation energy of the molecule, displays this share. The intensity of the field does not depend on the amount of chromatin, it depends on its turnover.

11. The vectors from separate field sources can be composed geometrically and the resulting vector at the point of composition will determine the direction of the kinetic constituent of the full molecular excitation energy at the point of the vector application.

    Therefore, in different parts of the embryo consisting of a number of cells and, hence, of the correspoding number of field sources there is an **integral actual**

field resulting from the total geometrical composition of all the vectors issued from all the sources (nuclei). Evidently, in such a composition both the field intensity (being a function of metabolic activity as well as of the distance of the point of the composition from the field source) and the field anisotropy (relation of the point to the nuclei's axes) make contributions to the value (length) of the resulting vector.

Although the question of the biological field is not especially touched upon in the above postulates, two principal comments should be added.

1. A. G. Gurvich's biological field cannot be reduced to any known physical field: it is an immanent property only of living objects. According to the postulate 5, the elementary flash of the biological field is induced **only by the existing field** so that the field is successive and **cannot originate de novo**. This is the full expression of the vitalistic principle.

2. A. G. Gurvich's biological field is **not energetic** which means that no special energy is focused in the field source. The field vector just transforms a portion of the metabolic energy accumulated in the excited protein molecules into directed kinetic energy moving or deforming the molecules. The energy is not supplied by the field to the spot of its action but to the field vector as it **employs the local energy accumulated at this spot.**

23.2.2 The concept and analysis of the actual integral field

The above postulates express the basic logical bases of the conception of the vectorial biological field which is presented as an universally, fundamentally biological (vitalistic) invariant principle. In this form the conception is perfectly logically balanced, i.e. it has no internal contradictions and is free of tautology. However, A. G. Gurvich developed his theory further by introducing the notion of an actual integral field which virtually serves as a working instrument. Some schematic illustrations of the formation and action of the actual fields at molecular, cellular and supracellular levels are presented below.

Molecular level

If a single cell is under consideration, the nucleus as the field source will have a repulsive action on the excited protein molecules in the cytoplasm. This results in

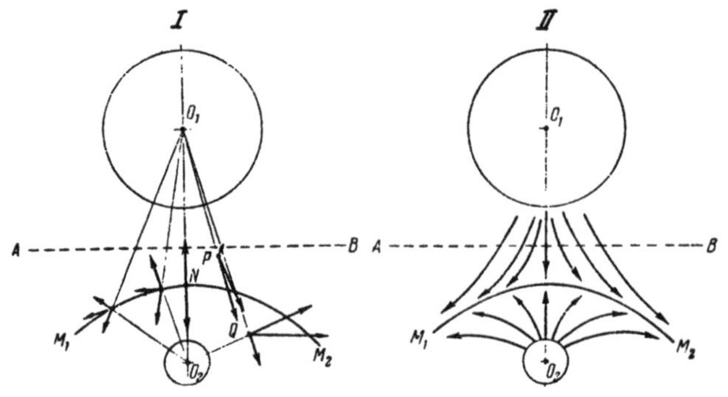

Figure 23.6: *Formation of "zero zone". I-vectorial caculation for some particular points; II-general picture. O_1 and O_2: sources of isotropic fields whose intensities are proportional to the diameters of the circles. $A \rightarrow B$: a border line between the cells with the nuclei O_1 and O_2; $M_1 \rightarrow M_2$: "zero zone"; the resulting vector at the point* **N** *is equal to zero, at the other points the resulting vectors are tangentially directed; at the points* **P** *and* **Q***, which are beyond the "zero zone", the resulting vectors are directed towards the "zero zone".*

increasing the concentration of excited molecules towards the cell periphery. Simultaneously, a certain "counterflow" of non-excited molecules must exist, increasing their concentration in perinuclear zones. In such conditions the most balanced position of the nucleus should be in the geometrical centre of the cell.

Interaction between two field sources (not only between the nuclei of two adjacent cells but also between any extranuclear chromatin-containing entities, such as different chromosomes or between nucleus and centrosome within the same cell) will result in a geometric composition of the opposite vectors. This will lead to the formation of a "**zero zone**" where the values of the resulting vectors will be equal to zero.

Since the excited molecules are driven together into these zones by the opposite vectors from the two interacting sources the concentration of the molecules in these zones will be maximum (Fig. 23.6). Such "**condensation zones**", according to A. G. Gurvich, favour different molecule interactions and facilitate different reactions.

However, the essential result of the field action is the formation of unbalanced molecular constellations. Insofar as the constellations form an entire unbalanced continuum which stretches for a certain space, the actual field within this space can have a complex configuration corresponding to the local geometry of field vectors which, in turn, depends on local interrelations between the field sources. Therefore, the character of the unbalanced molecular constellations (their steric configurations) depends on the actual fields. On the other hand, the constellations are not only the result but also the object of the actual field action.

The facilitation of the molecular reactions occurring within the "zero condensation zones" can promote the formation of some stable, subcellular, submicroscopic structures (called "**vestigia**", by A. G. Gurvich) which may become visible microscopic ones being constituents of intracellular cytological structural formations. Then such structural "vestigia" can be considered as traces of the increased molecular activities in certain **specific** "hot" points of the "condensation zones", the specificity being determined by the particular configurations of the actual field.

Thus, the analysis of the formation of the actual integral fields and of their actions at the molecular level, although being based on the concept of the unbalanced molecular constellations which, in turn, is based on the phenomenon of degradational mitogenetic radiation, seems to be rather speculative. On the other hand, the analysis of the actual field formation at the cellular level is more evidently displayable and can be supported by more immediate evidence.

Cellular and supracellular level

Transference (movements) and deformations of cells (nuclei) under the influence of the actual field vectors should be considered as a result of the transference of the whole molecular content of the cell, this being compatible with some physical (mechanical) properties of the cell as a whole, such as its incompressibility.In accordance with this, A. G. Gurvich performed his analysis of the actual integral field using the formal rules of the geometric composition of the field vectors and found some remarkable regularities.

The simplest case of the formation of the actual field is that presented by the interaction of two single cells (nuclei). It can be seen (Fig. 23.7), that the mutual influence of the opposite vectors results in both the divergence and deformation of the cells (nuclei).These effects decrease with the increasing distance between the interacting cells.

The formation and actions of the integral actual field in epithelial layers and large three-dimensional cell complexes are of the main interest as these are the "developing" parts recognizable in morphogenesis.

The remarkable peculiarity of the actual synthetic field of an epithelial layer of

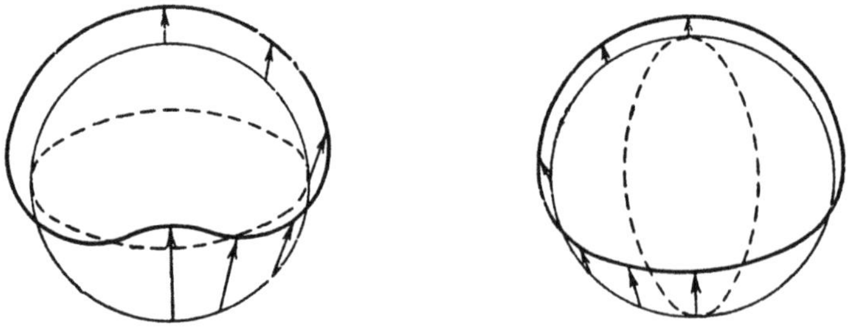

Figure 23.7: *Interaction of fields of two single cells. The cells inially are of spherical form. Anisotropy of their fields is designated by ellipses inscribed in the cells and repiprocally perpendicilar directions of the long axes of the ellipses are considered.*

sufficient extent is that it is in each point (cell) much stronger (the resulting vector is longer) than the field (vector) of the local single cell would be. At the same time, the actual field **outside** the layer (the **external** field, according to A. G. Gurvich) depends on both the extent and configuration of the layer (Fig. 23.8). It can be seen (Fig. 23.9), that the larger the extent of the layer the stronger the external actual field (the longer the vector of the actual field at the same distance from the layer). This means that the actual field in this case is practically without decrement and, hence the notion of **the long-distance effect (absence of the decrement)** of the actual field should be accepted (the model of the "effective cone", according to A. G. Gurvich).

As to the actual fields of large cell complexes, there is a difference between a "flat" and a spherical cell complex. In the former case almost all the cells (except the marginal ones) have similar "field conditions": the actual field in each does not even depend on the cell coordinates within the cell complex. Contrary to it, in a three-dimensional (spheric) complex the field vectors at a certain internal point would differ considerably depending on the coordinate changing (Fig. 23.10).

The above considered models describing some regularities of the actual field formation are static ones. Together with these, A. G. Gurvich considers some dynamic situations. In particular, he analyzes the mechanism of the curving of epithelial layers - one of the main morphogenic manifestations. Application of the vectorial composition shows that insignificant initial displacement of one of the nuclei relative to the adjacent ones is sufficient for starting the formation of a continuously growing concavity (Fig. 23.11). A similar consideration can be applied to the opposite case - the protruding of the epithelial layer [for example, development of the sea urchin

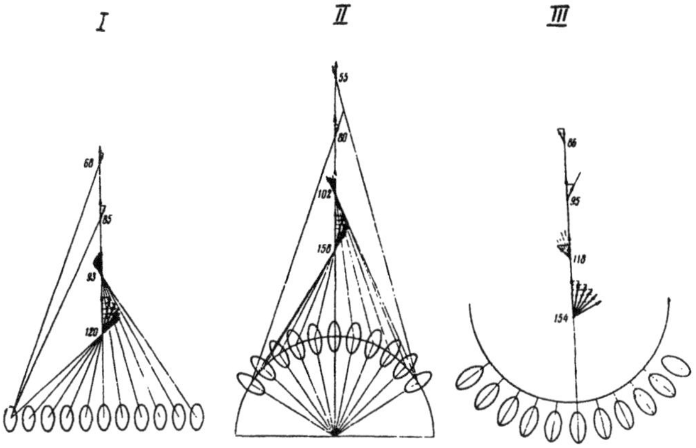

Figure 23.8: *Actual integral field from the layers of different configurations. I, I.I, III-right, convex and concave configurtions of the layers; the figures are the values of the resulting vectors desingnated in relative units at the four equidistant points.*

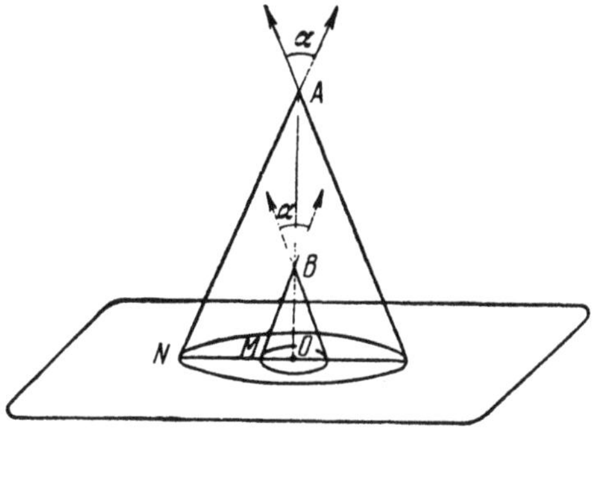

Figure 23.9: *Scheme demonstrating a long-range effect of the actual integral field from a layer of unlimited extent (practical absence of the field decrement). Within the angle α the field vectors do not deviate much from the normal line (perpendicular) to the layer; moving off the layer in the direction $B \rightarrow A$ is compensated by contribution of the field vectors from the cells within the ring* **MN** *area.*

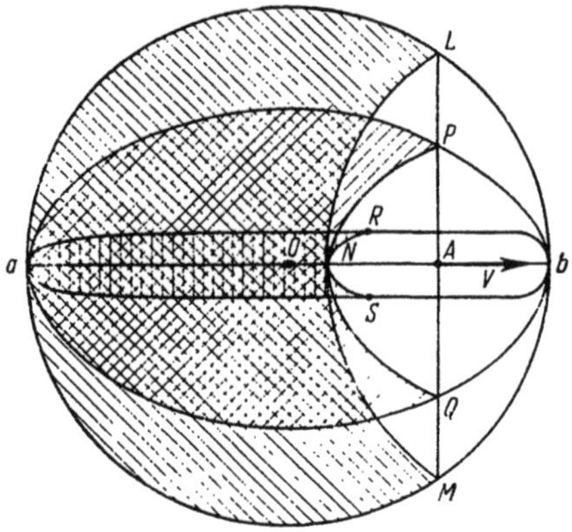

Figure 23.10: *Actual integral field from spherical and "flat" three-dimensional cell complexes.*

gastrula (Fig. 23.12)].

The next model demonstrates the significance of field anisotropy for the configuration of the actual field. It is evident that even small differences in the ratio of the ellipsoid axes designated as an anisotropy index can have significant influence on the actual field configuration (Fig. 23.13).

23.3 Conclusion

The theory considered here was employed by A. G. Gurvich and his collaborators and followers as a working tool for the analysis of different biological phenomena, namely, morphogenesis, differentiation and histogenesis, mitosis, reversible physiological processes (metabolic and neuromuscular), and of the psychical sphere[17–19,23–34]. This principal aspect is beyond the present consideration and one can be referred to the recently published review[35].

One point, however, should be emphasized.

The concept of "field" introduced into biology by A. G. Gurvich was used in the 1920's and 30's without any strict definition of the notion and without pragmatic common sense (see review[36]). In certain cases of observed equipotentiality, the involvement of the field was declared in a vague general form which was not far from mere semantic tautology.

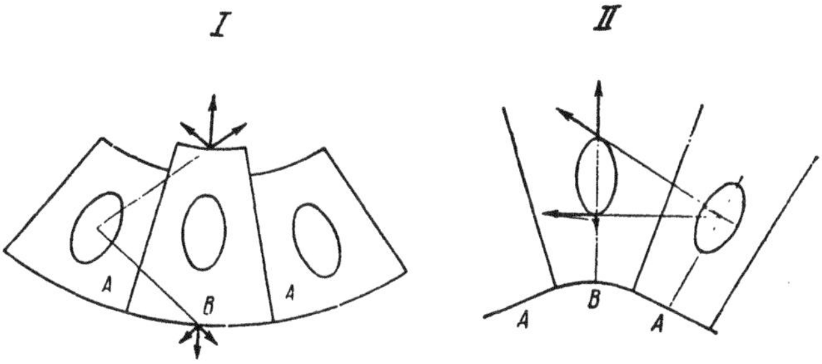

Figure 23.11: *Initiation and progress of curving of epithelial layer as a result of the field vector composition. I-influence of the field vectors from the cells* **A** *on a larger cell* **B***: the resulting vector directed to the concave side of the layer is prevailing over the opposite one. II-further progress of the initiated process.*

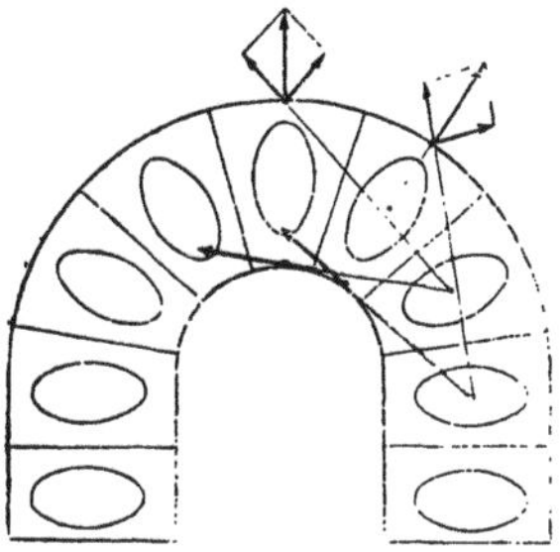

Figure 23.12: *The actual field at the top curve of the convex layer: the vectors show the probability of the cell protruding.*

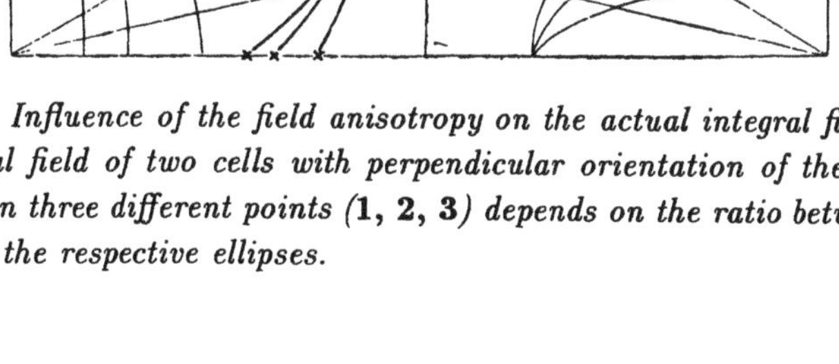

Figure 23.13: *Influence of the field anisotropy on the actual integral field. Influence of the integral field of two cells with perpendicular orientation of the long axes of their nuclei on three different points (***1, 2, 3***) depends on the ratio between long and short axes of the respective ellipses.*

The field conception, as developed by A. G. Gurvich, based first on the elaboration of particular models of morphogenesis using specific invariant laws (dynamically preformed morpha) and then on the development of the unitary vectorial field theory, using specific strictly defined postulates, marked substantial progress towards the creation of an epigenetic conception of development. The vitalistic conception of H. Driesch which was considered by A. G. Gurvich as logically irreproachable and based on remarkable experiments was led by its author to a deadlock due to a metaphysical definition of entelechia which was beyond rational scientific analysis.

Therefore, the great achievement of A. G. Gurvich was that he transferred the vitalistic theory from the heavens of an immaculate logical building to the grounds of a "working" instrument helping to explain the biological phenomenology displayed at all the levels of biological organization - molecular, cellular and organismic. The concept of the **actual integral fields** can serve as such a working tool.

The relation of Gurvich's field to physical and chemical laws and the principles of field action were comprehensively discussed in Gurvich's last work "Principles of Analytical Biology"[24]. The following quotation from it is worth being noted.

"The dominating conviction is that different chains of events bringing living manifestations to an observable state will in some distant future boil down to the level of those data and notions which are in use in physics and chemistry.As opposed to this conviction, our suggestion is that the confluence will occur at a higher level, **specific to living manifestations**, which means that biology can possess its own **specific fundamental notion** not contradicting the fundamental notions of physics and chemistry but also not reduced to them. The field theory is an attempt to formulate such a fundamental notion...".

In the limits of A. G. Gurvich's theory of the biological field, the vitalistic entelechia principle can be a powerful tool for generating working hypotheses in various branches of life sciences and medicine.

23.4 References

1. R. Sattler, *Biophilosophy: Analytic and Holistic Perspectives* (Springer-Verlag, Berlin, 1986).

2. A. G. Gurvich, Die Vererbung als Verwirklichungsvorgang. *Biologisches Zentralblatt* **32** (1912) p. 458-486.

3. W. Roux, *Gesammelte Abhandlungen ueber Entwicklungsmechanik der Organismen.* **1, 2** (Leipzig, 1895).

4. H. Driesch, Studien ueber das Regulationsvermoegen der Organismen. *W. Roux' Archiv fuer Entwichlungsmechanik* **26** (1908) (Festschrift fuer H. Driesch).

5. H. Driesch, Entwicklungsmechanische Studien. 1. Der Werth der beiden ersten Furchungszellen in der Echinodermenentwicklung. Experimentelle Erzeugung von Theil- und Doppelbildungen. *Zeitschrift fuer Zoologie* **53** p. 160-178.

6. H. Driesch, Vitalism, *Its History and System.* (In Russian, the authorized translation from German by A. G. Gurvich) (Moscow, 1915).

7. H. Spemann, Experimentelle Beitraege zu einer Theorie der Entwicklung. (Berlin, 1936), and H. Spemann & H. Mangold: Ueber Induction von Embryonal Anlage durch Implantation artfremder Organisatoren. *Archiv fuer Entwicklungsmechanik der Organismen* **100** (1924) p. 599-638.

8. A. G. Gurvich, Ueber Determination, Normierung und Zufall in der Ontogenese. *W. Roux' Archiv fuer Entwicklungsmechanik* **30** (1910) p. 133-193.

9. A. G. Gurvich, Der Vererbungsmechanismus der Form. *W. Roux' Archiv fuer Entwicklungsmechanik* **39** (1914) p. 516-577.

10. A. G. Gurvich, Ueber den Begriff des embryonalen Feldes. *W. Roux' Archiv fuer Entwicklungsmechanik* **51** (1922) p. 383-415.

11. L. D. Felicine-Gurvich, Die Verwertung des Feldbegriffes zur Analyse Embryonaler Differenzierungsvorgaenge. *W. Roux' Archiv fuer Entwicklungsmechnik* **101** (1924) p. 40-52.

12. A. G. Gurvich, Weiterbildung und Verallgemeinerung des Feldbegriffes. *W. Roux' Archiv fuer Entwicklungsmechanik* **112** (1927) p. 433-454.

13. A. W. Anikin, Das morphogene Feld der Knorpelbildung. *W. Roux' Archiv fuer Entwichlungsmechanik* **114** (1929) p. 549-577.

14. A. G. Gurvich, Ueber die Zerstoerbarkeit und Restitutionsfaehigkeit des Protoplasmas in Echinodermeneiern und Amphibieneiern. *Verhandlungen der Anatomischen Gesellschaft* (1904) p. 146-151.

15. A. G. Gurvich, Ueber die Zerstoerbarkeit und Restitutionsfaehigkeit des Protoplasmas des Amphibieneies. *Anatomischer Anzeiger* **27** (1905) p. 481-497.

16. A. G. Gurvich, Regulation phenomena in protoplasm. *Proceedings of the Sanct-Petersburgh Naturalist Society* **37** (1908) p. 140-189 (In Russian).

17. A. G. Gurvich, *The Theory of the Biological Field. "Sovetskaya Nauka" Publishing House*, (Moscow, 1944) (In Russian).

18. Anna Gurvich, Attempt of constructing a model of nonbalanced molecular constellations and degradational radiation, in *Collection of Works on Mitogenesis and the Theory of the Biological Field* (in russian), ed. A. G. Gurvich, USSR Academy of Medical Sciences Publishing House, (Moscow, 1947) p. 92-101.

19. Anna Gurvich, Unbalanced molecular orderliness of the substrate of living systems and its significance for the regulation of certain metabolic processes. *Enzymologia* (1962) p. 237-250.

20. A. G. Gurvich, Degradational radiation from central nervous system. *Arkhiv Biologicheskikh Nauk (Archives of Biological Sciences)* **45** (1937) p. 53-57 (In Russian).

21. A. G. Gurvich and L. D. Gurvich, Degradational mitogenetic radiation. *Byulleten Experimental'noy Biologii Meditsiny (Bulletin of Experimental Biology and Medicine)* **4** (1937) p. 459-460 (In Russian).

22. A. G. Gurvich, *Autobiographic diary* (Unpublished) (1941).

23. A. G. Gurvich, The concept of the "whole" in the light of the cell field theory, in *Collection of Works on Mitogenesis and the Theory of the Biological Field*, ed. A. G. Gurvich, USSR Academy of Medical Sciences Publishing House, (Moscow, 1947) p. 141-147 (In Russian).

24. A. G. Gurvich, *Principles of Analytical Biology and of the Theory of Cellular Fields.* (Nauka Publishers, Moscow, 1991) (In Russian)

25. Anna Gurvich, The problem of mitogenetic radiation as an aspect of molecular biology. *Meditsina*, Leningrad Branch, USSR Academy of Medical Sciences, (Leningrad, 1968) (In Russian).

26. E. Puchalskaya, Morphological alteration of mitotic figures resulting in their interaction (an attempt to an analysis of mitosis in the light of field theory), in *Collection of Works on Mitogenesis and the Theory of Biological Field*, ed. A. G. Gurvich, USSR Academy of Medical Sciences Publishing House, (Moscow, 1947) p. 148-173 (In Russian).

27. L. V. Beloussov, The origin, development and perspectives of the theory of the biological field, in *Physical and Chemical Grounds of Living Phenomena (Historical Essays)*, USSR Academy of Science (Moscow, 1963) (In Russian).

28. L. V. Beloussov, Alteration of growth direction of rudiments as a result of their interactions. *Doklady Akademii Nauk SSSR (Proceedings of the USSR Academy of Sciences)* **160** 475-478 (1965) (In Russian).

29. L. V. Beloussov, Interactions between neural tubes artificially drawn together. *Doklady Akademii Nauk SSSR (Proceedings of the USSR Academy of Sciences)* **176** (1967) (In Russian).

30. L. V. Beloussov, On the mechanism of the curving of epithelial layers. *Zhurnal Obshchey Biologii (Journal of General Biology)* **28** (1976) p. 580-589 (In Russian).

31. L. V. Beloussov, Computations of certain cell movements in the morphogenesis of hydroid polyps. *Biologicheskiye Nauki (Biological Sciences)* **3** (1968) p. 7-16 (In Russian).

32. L. V. Beloussov, Orientated interactions of opposite cell layers. *Doklady Akademii Nauk SSSR (Proceedings of the USSR Academy of Sciences)* **187** (1969) (In Russian).

33. L. V. Beloussov, Some problems of embryonal morphogenesis. *Uspekhy Sovremennoy Biologii (Achievements of Modern Biology)* **67** (1969) p. 127-146 (In Russian).

34. L. V. Beloussov, *The Problem of Embryonal Morphogenesis.* Moscow University Publishing House, (Moscow, 1971) (In Russian).

35. M. Lipkind, Gurwitschs Theorie vom biologischen Feld (A. G. Gurvich's theory of the biological field.) **I.** Major prerequisites, origin, logical structure and conceptual development; **II.** Analysis of different pheomena in living systems by means of the theory of the vectorial biological field. *Fusion* (German edition), **8 (4)** p. 30-49, **8 (5/6)** p. 53-65 (1987) (In German).

36. C. H. Waddington, Fields and Gradients, in *Major Problems in Developmental Biology* (Academic Press, New York-London, 1966).

37. Gurvich, A. G, Der Begriff der Aequipotentialitaet in seiner Anwendung auf physiologische Probleme. *W. Roux' Archiv fuer Entwicklungsmechanik* **116** (1929) p. 20-35.

Index

D

Y

Z

www.ingramcontent.com/pod-product-compliance
Ingram Content Group UK Ltd.
Pitfield, Milton Keynes, MK11 3LW, UK
UKHW020404150926
8624IPUK00004B/343

9 789810 208554